INTRODUCTION TO FINANCIAL MATHEMATICS

Advances in Applied Mathematics

Series Editor: Daniel Zwillinger

Published Titles

Green's Functions with Applications, Second Edition *Dean G. Duffy*

Introduction to Financial Mathematics *Kevin J. Hastings*

Linear and Integer Optimization: Theory and Practice, Third Edition
  Gerard Sierksma and Yori Zwols

Markov Processes *James R. Kirkwood*

Pocket Book of Integrals and Mathematical Formulas, 5th Edition
  Ronald J. Tallarida

Stochastic Partial Differential Equations, Second Edition *Pao-Liu Chow*

Advances in Applied Mathematics

INTRODUCTION TO FINANCIAL MATHEMATICS

Kevin J. Hastings

Knox College
Galesburg, Illinois, USA

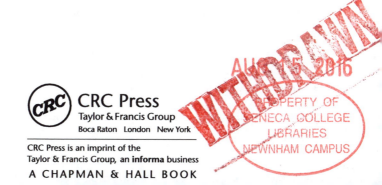

CRC Press
Taylor & Francis Group
Boca Raton London New York

CRC Press is an imprint of the
Taylor & Francis Group, an **informa** business

A CHAPMAN & HALL BOOK

CRC Press
Taylor & Francis Group
6000 Broken Sound Parkway NW, Suite 300
Boca Raton, FL 33487-2742

© 2016 by Taylor & Francis Group, LLC
CRC Press is an imprint of Taylor & Francis Group, an Informa business

No claim to original U.S. Government works

Printed on acid-free paper
Version Date: 20150814

International Standard Book Number-13: 978-1-4987-2390-9 (Hardback)

Visit the Taylor & Francis Web site at
http://www.taylorandfrancis.com

and the CRC Press Web site at
http://www.crcpress.com

To Gay Lynn, with whom no amount of time will ever be enough

Contents

Preface

A number of years ago I started to develop a keen interest in mathematical problems in economics and finance. It was a natural progression for a probabilist with an applied bent, because activity in this area has been accelerating and it provides a rich set of materials to which undergraduates can apply themselves. I was always impressed with the text resources that I found and learned from, many of which are listed in the bibliography. But I found it curious that they differed so much from one another. In addition, it seemed to me that books were mostly pitched at the very advanced undergraduate or introductory graduate level, given their emphasis on continuous models and stochastic calculus. So I began to think that a resource at a more introductory undergraduate level would be a valuable addition to the field, and as well might serve to standardize subject matter in a useful way and prepare students better to embark on their studies in financial mathematics.

The organization of the book is as follows.

Chapter 1 covers fairly standard material on the mathematics of interest, including compound interest and present value, annuities and loans, several versions of the rate of return on an investment, and interest in continuous time. These topics appear in courses even at the level of finite mathematics, although I have attempted to be more rigorous here and to emphasize real problem solving as opposed to template "plug-and-chug" drill. Such problem solving depends on understanding basic concepts well so as to apply them in new situations, and the examples and exercise sets take that approach. I have tried to roughly follow the pattern established in the excellent books by Federer Vaaler [5] and Broverman [2]. The material in this and the following chapter is particularly appropriate for students who want to get started preparing for the FM (Financial Mathematics) Actuarial Exam.

In Chapter 2 we show how to value bonds, at their issue dates, at coupon times, between coupon times, and in the case where the bonds may be terminated early by either the issuer or the holder. There is also a brief subsection on bond duration, a kind of average time of payment of a bond. This leads to a lengthy section on the term structure of interest rates, in which the concepts of spot rates and forward rates of interest play starring roles.

As a probabilist, I am rather surprised at the degree to which problems in the theory of interest are restricted to deterministic situations where interest rates are known perfectly in advance, nobody ever defaults on loans or skips payments, dividends are fixed and known, etc. So one of the distinctive features of this book is Chapter 3 on discrete probability for finance. It is a rapidfire pass through the main ideas and techniques of discrete probability: sample spaces and probability measures, random variables and distributions, expectation, conditional probability, and independence. Throughout this chapter, there is emphasis on probability in the context of financial problems. For example, problems from Chapter 1 involving interest rates that change occasionally are generalized to problems in which these rates are random variables instead of known constants. The chapter also lays the groundwork for the ensuing material on portfolios of risky assets and valuation of simple derivatives in the next two chapters. The binomial branch model for the motion of risky assets over time is introduced here and used as a common theme. There is a bonus section at the end of Chapter 3 on estimation and simulation. In this chapter, we illustrate the use of data to estimate means, variances, and covariances of rates of return on assets, and we develop Monte Carlo simulation algorithms to study the behaviors of financial assets. The latter is important for the increasingly prevalent applications in which closed form answers are difficult to get.

I have always been interested in the problem of optimal portfolio selection, which integrates ideas and techniques from calculus, finance, and probability. It also seems strange to me that this subject matter does not find its way into the financial mathematics curriculum anywhere. In Chapter 4 we introduce the basic terminology of stocks and stock trading, and then go on to derive the rate of return on a portfolio, and use the idea of risk aversion to model the investor tradeoff between risk and return. Students will solve problems with multiple independent assets, with a risk-free asset like a bond included in the mix, and with correlated assets. There is a peek at the deeper idea of the market portfolio, which is a key idea in the famous Capital Market Theory which is covered in more advanced courses. Besides the earlier material and the introduction to probability, this is another important way in which this book serves as a very good bridge to advanced work.

Much of the thrust in modern finance has been in the direction of valuing assets that are derivative of some more basic underlying asset in the market. This is the topic of Chapter 5, the last chapter. The easiest examples are futures and European options, but the results in this chapter apply to more general derivatives. As is normal in the literature, the theme is the use of anti-arbitrage assumptions to enforce initial values of these derivatives. Both single-period and multiple-period derivatives are valued, assuming a binomial branch model for the underlying asset. The technique of chaining is used heavily in the chapter, which my experience indicates is intuitively appealing to many students. Again a glimpse of deeper things is given to the student in the form of martingale characterization of solutions, which is the modern approach to the problem. Another bonus section appears at the end of the book, in which some non-vanilla options are valued (e.g., American options and barrier options), and the use of simulation to value derivatives is practiced.

Problem solving in this book assumes at least the use of a good calculator, but my point of view is that more sophisticated tools are widely available and should not be ignored. Many of the most interesting examples in the book involve numerical solution of complicated non-linear equations; others ask students to produce algorithms which beg to be implemented as programs. For maximum flexibility, I have produced the print version of this text without adhering to any particular computational platform. There are several popular symbolic algebra, numerical and graphical packages, but my particular favorite is *Mathematica*®. This book can be obtained digitially in the form of *Mathematica* notebooks, one per chapter, with enhanced material and computation. Besides its adaptable general purpose functions that are perfect for numerical solution of equations, graphics, and animation, *Mathematica* has some special financial functions, such as TimeValue for time valuations of payments, and FinancialData for importing current and historical financial data in real time. *Mathematica* also is a programming language with sophisticated random number simulation that works well for the kinds of simulations illustrated in this book. The experience of studying and interacting with the text is greatly enhanced by the use of *Mathematica*.

It is my sincere hope that this text will be a valuable resource for students, and moreover that it will stimulate the creation of new coursework in mathematics departments that will help to draw students into the fascinating and growing area of financial mathematics.

I would be remiss not to thank the folks at Taylor & Francis for their help, including Marsha Pronin and Bob Ross. I also have a special fondness for all of the students I have taught, and who have taught me and kept me honest. This project benefited tremendously from their input, and they have helped me to find quite a few errors. Any errors that are left are entirely my responsibility, not theirs. But don't tell them that.

Kevin J. Hastings

Chapter 1

Theory of Interest

We begin our study of financial mathematics in this chapter by introducing the key concepts of **rate of return** and **present value**. With the stage thus set, we will be prepared to investigate a number of important ideas in the theory of interest: compound interest, annuities, loan repayment, and the valuation of bonds. Prerequisite mathematics for the first few sections of this chapter, in which time moves discretely, is simply standard precalculus material with an emphasis on geometric sequences and series (which will be reviewed). At the end of the chapter we will see how the ideas extend to continuous time models, for which both differential and integral calculus will be necessary. By the end of the chapter students should be prepared to make educated decisions in many financial situations in their personal lives. The groundwork for further study in financial mathematics is laid as well, including study of actuarial mathematics and preparation for the FM exam that is administered by the major actuarial societies. For this purpose, you may find the references Federer Vaaler [5] and Broverman [2] to be very useful. Both Chapters 1 and 2 of this book are heavily influenced by these works. Furthermore, the financial services industry is concerned with how to value firms and tradable assets, and students will acquire the background to continue the pursuit of problems related to this important area.

1.1 Rate of Return and Present Value

In this section we explore the two most fundamental and far-reaching ideas in financial mathematics: the rate of return on an investment or a loan, and the present value of a future amount of money. Let us consider an investment situation in which there are two times: "now" and "later," symbolized by times t_0 and t_1. For simplicity, we may sometimes assume that $t_0 = 0$ and $t_1 = 1$. An amount of money $A_0 = A(t_0)$ is invested at time t_0, and its value at time t_1 is $A_1 = A(t_1)$. In general, we will use a function $A(t)$ to represent the total accumulated value of an investment at time t. We can measure the performance of the investment by its gain in value per dollar invested, which gives rise to the following definition.

Definition 1. The **rate of return** or **effective rate of interest** on the investment is:

$$R = \frac{A_1 - A_0}{A_0} = \frac{A(t_1) - A(t_0)}{A(t_0)}. \tag{1.1}$$

Notice that, if we solve for the final, **future value** $A(t_1)$ in Equation (1.1), we obtain:

$$A(t_1) = A(t_0)(1 + R). \tag{1.2}$$

Remark 1. We can consider a loan of A_0 dollars to a borrower as an investment by the lender at time t_0. If the borrower agrees to pay back A_1 to the lender at time t_1, then the lender's rate of return on this investement would be given by Equation (1.1). But, from the borrower's point of view, the right side of this equation is also called the **effective interest rate** charged to the borrower for this loan on the time interval $[t_0, t_1]$.

Example 1. (**Simple interest**) A bank offers interest on a checking account such that the gain in value over a period in which the beginning balance is B and no deposits or withdrawals are made is proportional to both the balance and the time elapsed t (in units of years). The proportionality constant r is called the **simple interest rate**. Then the gain in value under simple interest is:

$$A(t) - A(0) = r \cdot B \cdot t. \tag{1.3}$$

So the final value at time t is

$$A(t) = B + r \cdot B \cdot t = B(1 + r \cdot t). \tag{1.4}$$

For simple interest situations, the value of the investment is therefore a linear function of time. If the balance begins in a particular month at $1000, and the account offers a 1% per dollar per year interest rate, then the account value after a week (i.e., 1/52 of a year) of no account activity is:

$$A(1/52) = \$1000\left(1 + \frac{.01}{52}\right) = \$1000.19. \blacksquare$$

Example 2. Suppose now that there are two times t_1 and t_2 after the initial time when we observe the value of an investment. Then we could talk about the rates of return over time intervals $[t_0, t_1]$, $[t_1, t_2]$, or $[t_0, t_2]$, which would be, respectively,

$$R_{01} = \frac{A(t_1) - A(t_0)}{A(t_0)}, R_{12} = \frac{A(t_2) - A(t_1)}{A(t_1)}, R_{02} = \frac{A(t_2) - A(t_0)}{A(t_0)}.$$

If \$500 is invested initially and the rate of return on $[t_0, t_1]$ is 3% while the rate of return on $[t_1, t_2]$ is 2%, find the rate of return on $[t_0, t_2]$.

Solution. The strategy is to first find $A(t_1)$ using the given rate of return on $[t_0, t_1]$, then use it and the rate of return on $[t_1, t_2]$ to find $A(t_2)$, and finally substitute that into the formula for R_{02}.

$$
\begin{aligned}
A(t_1) &= A(t_0)(1 + R_{01}) \\
&= \$500(1.03) \\
&= \$515
\end{aligned}
$$

$$
\begin{aligned}
A(t_2) &= A(t_1)(1 + R_{12}) \\
&= \$515(1.02) \\
&= \$525.30
\end{aligned}
$$

Therefore,

$$
R_{02} = \frac{A(t_2) - A(t_0)}{A(t_0)} = \frac{\$525.30 - \$500}{\$500} = .0506. \blacksquare
$$

If you look at the details of the computation in Example 2, you can generalize an approach to find the final value and the rate of return on the larger interval $[t_0, t_2]$ in terms of the rates of return in the subintervals. We had both:

$$
A(t_1) = A(t_0)(1 + R_{01}) \text{ and } A(t_2) = A(t_1)(1 + R_{12}),
$$

and thus

$$
A(t_2) = A(t_0)(1 + R_{01})(1 + R_{12}), \tag{1.5}
$$

which gives the overall rate of return as:

$$
\begin{aligned}
R_{02} &= \frac{A(t_2) - A(t_0)}{A(t_0)} \\
&= \cdot\,\frac{A(t_0)(1 + R_{01})(1 + R_{12}) - A(t_0)}{A(t_0)} \\
&= (1 + R_{01})(1 + R_{12}) - 1. \tag{1.6}
\end{aligned}
$$

Checking the numerical result in the example,

$$
R_{02} = (1 + .03)(1 + .02) - 1 = .0506.
$$

Equations (1.5) and (1.6) are easy to extend to multiple time periods.

Example 3. If a loan of \$1000 is made at simple interest with a repayment of \$1100 at the end of 2 years, what is the corresponding simple interest rate per year?

Solution. By Equation (1.4), the interest rate r satisfies the equation

$$A(2) = \$1100 = \$1000(1 + r \cdot 2) \Longrightarrow r = \frac{1}{2}\left(\frac{1100}{1000} - 1\right) = .05. \ \blacksquare$$

Next we introduce the key element in the valuation of many financial quantities. It provides a common yardstick against which we can compare the values of different sums earned at different times.

Definition 2. The ***present value*** of a future amount $A(t_1)$ at a current time t_0 using an effective rate of R in $[t_0, t_1]$ is the amount that would need to be invested at t_0 in order to accumulate to $A(t_1)$, which is

$$A(t_0) = (1 + R)^{-1}A(t_1). \tag{1.7}$$

Equation (1.7) follows directly from Equation (1.2). We sometimes call the expression $(1 + R)^{-1}$ the ***discount factor*** that converts the future value $A(t_1)$ to the present value $A(t_0)$.

Example 4. Find the present value under simple interest at rate 4% per year of an amount of \$10,000 in 10 years.

Solution. Equation (1.4) gives the relationship between the present and future value under simple interest. Since the elapsed time in general is the difference between the ending and starting times $t_1 - t_0$, we have:

$$A(t_1) = A(t_0)(1 + r \cdot (t_1 - t_0)) \Longrightarrow A(t_0) = (1 + r \cdot (t_1 - t_0))^{-1}A(t_1).$$

For our problem parameters, the present value is:

$$A(t_0) = (1 + .04 \cdot 10)^{-1}\$10,000 = \$7142.86. \ \blacksquare$$

Example 5. Suppose as in Example 2 that the rate of return on $[t_0, t_1]$ is 3% and the rate of return on $[t_1, t_2]$ is 2%. Find the time t_0 present value of an amount of \$5000 at time t_2.

Solution. The rate of interest to use in discounting was found to be $R = (1 + R_{01})(1 + R_{12}) - 1$. Then

$$(1 + R)^{-1} = (1 + R_{01})^{-1}(1 + R_{12})^{-1},$$

hence the present value is

$$A(t_0) = (1+R)^{-1}A(t_2) = (1+R_{01})^{-1}(1+R_{12})^{-1}A(t_2)$$
$$= (1.03)^{-1}(1.02)^{-1} \cdot \$5000$$
$$= \$4759.19. \quad \blacksquare$$

Important Terms and Concepts

Rate of return or Effective interest rate - Change in value of an investment divided by initial value: $R = \frac{A(t_1)-A(t_0)}{A(t_0)}$.

Future value - The amount that a current investment is worth at a later time: $A(t_1) = (1+R)A(t_0)$.

Simple interest - Interest earned is proportional to both the initial amount and the time elapsed: $A(t_1) - A(t_0) = r \cdot A(t_0) \cdot (t_1 - t_0)$.

Present value - The amount that must be invested to achieve a given value in the future at a given time with a given effective interest rate: $A(t_0) = (1+R)^{-1}A(t_1)$.

Exercises 1.1

1. If the accumulated value of an investment at time t years is of the form $A(t) = 100(1.05)^t$, then find the effective rate of interest over the first year, first two years, and first three years.

2. An investment of \$1000 earns simple interest at rate 4% for the first year, decreasing by half a percent each year until the interest rate is zero. (a) What is the final value of the investment? (b) What is the effective rate of interest on the investment?

3. In order for an investment of \$500 to reach a value of at least \$800 in 10 years at simple interest, at least what must the interest rate per year be?

4. Find the rate of return on each of the time intervals $[t_0, t_1]$, $[t_0, t_2]$, $[t_0, t_3]$, $[t_1, t_2]$, $[t_2, t_3]$ on an investment that begins at value \$800 and takes on intermediate values \$825, \$850, and \$875 at times t_1, t_2, and t_3. Make note of the pattern of the rates, in particular whether the rates of return are constant or varying with time.

5. Tom loans Tim an amount of \$3000 at 7% simple interest per year for a period of 3 years, with repayment of the full balance occurring at the end of the 3 years.

Find the amount that Tim owes Tom at the end of each year, and the effective rate of interest in each individual year. Does simple interest seem to favor the lender or the borrower?

6. For an investment earning a constant effective rate of $R = .04$ in each of four consecutive periods, find the initial value of a final value of $5000.

7. At least how much must the yearly simple interest rate be so that the present value of an amount of $1200 earned at the end of 6 months is no more than $1100?

1.2   Compound Interest

To understand basic financial objects, such as savings accounts, loans, annuities, and bonds, which investors encounter most frequently, it is necessary to have a thorough grounding in the idea of compound interest. So, you will find this section to be one of the most important parts of this book. You should read the examples very carefully, because they illustrate critical ways of thinking financially and very useful computational techniques.

For most of this section we will restrict to the case where an investment earns a constant rate of return r in each time interval of unit length. This will allow us to easily derive very useful and simple formulas for future and present values of an investment. But to accomplish this, first it is appropriate to review some background on geometric sequences and series.

1.2.1   Geometric Sequences and Series

In general, a **sequence** is a (possibly infinite) list $a_0, a_1, a_2, \ldots$. A **geometric sequence** is a list of the form

$$a, a \cdot q, a \cdot q^2, a \cdot q^3 \ldots$$

In other words, each term in a geometric sequence after the initial term a is a common multiple q of the preceding term, as shown below.

$$
\begin{aligned}
a_0 &= a \\
a_1 &= q \cdot a_0 = a \cdot q \\
a_2 &= q \cdot a_1 = a \cdot q^2 \\
a_3 &= q \cdot a_2 = a \cdot q^3 \\
&\vdots
\end{aligned}
$$

The general term of a geometric sequence is therefore $a_n = a \cdot q^n$. The constant q is also called the **common ratio** between consecutive terms.

Example 1. (a) Write out the first few terms of the geometric sequence with initial term 1 and common ratio $1/2$.
(b) Is either of the following sequences geometric? If so, identify a and q.
(i) $3, 5, 7, 9, ...$ (ii) $2, -4, 8, -16, 32, ...$

Solution. (a) The first few terms of this geometric sequence are

$$1, 1/2, 1/4, 1/8, ...$$

It is easy to see that the nth term is of the form $\left(\frac{1}{2}\right)^n$, $n = 0, 1, 2, ...$

(b) In sequence (i) consecutive terms have a common difference, not ratio, of 2. This kind of sequence is called an **arithmetic sequence**. Sequence (ii) is geometric; the initial term is $a = 2$ and the common multiple is -2. A closed form expression for the nth term of the sequence is therefore $a_n = 2 \cdot (-2)^n$. ∎

Certain financial problems involve adding the terms of a geometric sequence:

$$a + aq + aq^2 + \cdots + aq^m \equiv \sum_{j=0}^{m} a \cdot q^j. \qquad (1.8)$$

(Remember that in general the sigma notation $\sum_{j=m}^{n} a_j$ is shorthand for the sum of terms $a_m + \cdots + a_n$ of a sequence.) The quantity in (1.8) is called a **finite geometric series**. How can we evaluate a sum like this? Fortunately, there is an easy formula, as Theorem 1 shows.

Theorem 1. The sum of the geometric series in Equation (1.8) is

$$\sum_{j=0}^{m} a \cdot q^j = a \cdot \frac{q^{m+1} - 1}{q - 1} = \frac{1 - q^{m+1}}{1 - q}. \qquad (1.9)$$

Proof. It suffices to factor out the initial term a and try to sum the series $s = \sum_{j=0}^{m} q^j$. Note that:

$$
\begin{aligned}
s - q \cdot s &= \left(1 + q + q^2 + \cdots + q^m\right) - q\left(1 + q + q^2 + \cdots + q^m\right) \\
&= \left(1 + q + q^2 + \cdots + q^m\right) - \left(q + q^2 + q^3 + \cdots + q^{m+1}\right) \\
&= 1 - q^{m+1}.
\end{aligned}
$$

Therefore we have the following equation for the unknown sum s.

$$s(1 - q) = 1 - q^{m+1} \implies s = \frac{1 - q^{m+1}}{1 - q}.$$

This proves (1.9), in light of the fact that the middle expression in the formula is simply obtained from the right-hand term by multiplying by -1 in both numerator and denominator. ∎

Example 2. (a) Find the sum of the first eight terms of the geometric sequence from Example 1(a) with initial term 1 and common ratio $1/2$.
(b) Find the sum of the first six terms of the geometric sequence from Example 1(b): $2, -4, 8, -16, 32, -64, \ldots$

Solution. (a) The first eight terms begin with the 0th term and end with the 7th, so Equation (1.9) gives

$$\sum_{j=0}^{7} 1 \cdot \left(\frac{1}{2}\right)^j = \frac{1 - (1/2)^8}{1 - (1/2)} = 2 \left(1 - (1/2)^8\right) = 1.99219.$$

(b) The first term is 2 and the common ratio is -2, hence Equation (1.9) yields that the sum is

$$\sum_{j=0}^{5} 2 \cdot (-2)^j = 2 \cdot \frac{(-2)^6 - 1}{-2 - 1} = -\frac{2}{3} \left((-2)^6 - 1\right) = -42. \ \blacksquare$$

Parts (a) and (b) of the previous example illustrate the two possibilities that can happen as more and more terms are added to a geometric series. As the middle formula $a \cdot \frac{q^{m+1}-1}{q-1}$ in (1.9) shows, when the common ratio q is at least 1 in magnitude as in part (b), the series cannot converge to a finite number as $m \longrightarrow \infty$. But when $|q| < 1$ as in (a), the right-hand formula $a \cdot \frac{1-q^{m+1}}{1-q}$ of (1.9) gives us the following result for the infinite geometric series:

$$\sum_{j=0}^{\infty} a \cdot q^j = \lim_{m \to \infty} \sum_{j=0}^{m} a \cdot q^j = a \cdot \frac{1}{1-q} \text{ if } |q| < 1. \tag{1.10}$$

1.2.2 Compound Interest

Suppose that the rate of return on an investment is a constant r on any time interval of length 1. Then, by Equation (1.2) of Section 1, the values $A(n)$ at integer times n satisfy:

$$
\begin{aligned}
A(1) &= (1+r)A(0) \\
A(2) &= (1+r)A(1) \\
A(3) &= (1+r)A(2)
\end{aligned}
$$

etc., so that the terms $A(n)$ form a geometric sequence with common ratio $1 + r$ and initial term $a_0 = A(0)$. Notice that the increase in value between consecutive integer times is $A(n + 1) - A(n) = r \cdot A(n)$, so that interest is being added on the entire current balance, including the part of it which is interest that has been accumulated earlier. Value is being gained by so-called (discrete) **compound interest**. Because of what we know about geometric sequences, the future value of the investment at an integer time n is given by:

$$A(n) = a_0 \cdot (1+r)^n, n = 0, 1, 2, \ldots \tag{1.11}$$

Figure 1.1 shows the exponential increase in the value of the investment for an initial investment of \$2000 and a per period interest rate of $r = 6\%$. So far we have only referred to integer values n for time periods, but the function in Equation (1.11) can be extended to real values of time t as well, and we have superimposed a continuous graph on top of the graph of the points $(n, A(n))$ that shows this idea.

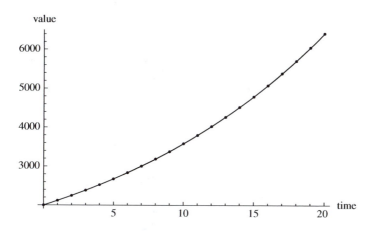

Figure 1.1: Growth in value under compound interest, $a_0 = 2000$, $r = .06$.

Remark. The number r above is the interest rate per compounding period. If there is a stated or **nominal** yearly interest rate y, but the compounding method compounds m times per year, then $r = y/m$. The value of the investment at time t years, which includes $m \cdot t$ compoundings, would be:

$$A(t) = a_0 \left(1 + \frac{y}{m}\right)^{m \cdot t}. \tag{1.12}$$

You may hear the word **convertible** applied to compound interest problems, as in "convertible quarterly" for compounding that is done four times per year, or "convertible semiannually" for compounding twice per year.

Note that the total interest earned on an investment with interest rate r per compounding period over n time periods is:

$$
\begin{aligned}
\text{total interest earned} \quad &= \quad \text{future value} - \text{initial value} \\
&= \quad a_0(1 + r)^n - a_0, \tag{1.13}
\end{aligned}
$$

which can be rewritten as $a_0\left((1 + r)^n - 1\right)$. Then the rate of return for the n-period investment is:

$$\text{overall rate of return} = (1 + r)^n - 1. \tag{1.14}$$

The **effective rate of interest** per year is the actual rate of return on an investment over a one-year period. So, if there are m compoundings per year and the nominal interest rate is y, then Equation (1.12) easily gives:

$$\text{effective rate} = \left(1 + \frac{y}{m}\right)^m - 1. \tag{1.15}$$

Example 3. The future value of an investment of $2000 after 3 years if interest is convertible quarterly at a nominal yearly rate of 6% is

$$\$2000 \left(1 + \frac{.06}{4}\right)^{12} = \$2391.24.$$

The interest earned is $391.24, and so the total rate of return over the 3-year period is $391.24/$2000 $\approx$ 19.6%. How long will it take such an investment to double in value? We want the value to reach $4000 after a certain number of periods t. Thus,

$$4000 = 2000 \left(1 + \frac{.06}{4}\right)^t \implies 2 = \left(1 + \frac{.06}{4}\right)^t \implies t = \frac{\ln(2)}{\ln(1.015)} \approx 46.5.$$

Since the time period is in units of quarter-years, it will take about $46.5/4 = 11$ 5/8 years to double. ■

Remark. If you look carefully at the derivation of the doubling time in the last example, you will note that it is not necessary to know the initial amount invested in order to find the doubling time t, since the doubling time would satisfy the equation:

$$2A(0) = A(0) \left(1 + \frac{y}{m}\right)^t \implies 2 = \left(1 + \frac{y}{m}\right)^t.$$

Given the yearly nominal rate y and the number of compoundings per year m we can solve exactly for the number of periods t. As in the example, taking the natural logarithm of both sides would produce the formula:

$$\text{doubling time (in periods) } t_d = \frac{\ln(2)}{\ln\left(1 + \frac{y}{m}\right)}. \tag{1.16}$$

Example 4. Some interest-bearing accounts only give simple interest on portions of a time period. Suppose that Doug deposits $6000 in an account that gives 5% interest compounded semiannually at the end of the period, but for any partial period simple interest at 5% per year is added on. Find the value in the account at the end of $2\frac{1}{4}$ years, and compare it to the value that would be computed if interest is compounded for partial years.

Solution. The interest rate per full period is 2.5%. In $2\frac{1}{4}$ years there are four full compoudings, and then a half period goes by during which simple interest is earned. The ending value would then be

$$\$6000 \cdot (1.025)^4 (1 + .025(.5)) = \$6705.66.$$

If interest had been fully compounded, the total balance at time t half-years would be $A(t) = \$6000 \cdot (1.025)^t$. Time t would equal 4.5, hence Doug would have had

$$\$6000(1.025)^{4.5} = \$6705.15.$$

Does this answer surprise you? (See Exercise 11.) ■

Example 5. An investment of $1000 is held in an account for 6 years; for 2 years the yearly interest rate was 3%, and after the 2nd year the yearly interest rate changed. If the amount at the end of the 6 years was $1070, then what must have been the interest rate for the final 4 years?

Solution. In the first 2 years the value would accumulate to $A(2) = \$1000(1.03)^2$. Then the interest rate becomes an unknown value r for the next 4 years. The amount at the end of 6 years would be:

$$A(6) = \$1070 = \$1000(1.03)^2(1 + r)^4.$$

Thus,

$$(1+r)^4 = \frac{\$1070}{\$1000(1.03)^2} \Longrightarrow r = \sqrt[4]{\frac{\$1070}{\$1000(1.03)^2}} - 1 = .00214. \ ■$$

Example 6. What is the effective rate of an investment compounding semi-annually with a nominal yearly rate of 5%? To what nominal yearly rate does an effective yearly rate of 6% correspond, if interest is compounded monthly?

Solution. By Equation (1.15), the effective rate corresponding to nominal 5% yearly is:

$$\left(1 + \frac{.05}{2}\right)^2 - 1 = .050625.$$

To answer the second question, if y is the nominal yearly rate, then $y/12$ is the effective rate per month. The effective rate for a year is determined in two ways: from the problem statement it is 6%, but it is also the rate of return on an investment compounding 12 times per year at monthly rate $y/12$ from which the equation below follows.

$$.06 = \left(1 + \frac{y}{12}\right)^{12} - 1 \Longrightarrow y = 12\left(\sqrt[12]{1.06} - 1\right) = .0584106. \ ■$$

Using the effective rate, we can combine investments or compare investments to each other, as the next examples show.

Example 7. If an investor has a choice of investing money at 6% convertible daily or 6 1/8% convertible quarterly, which is the better choice?

Solution. We can compare the effective yearly rates for the two investments to make the decision.

Investment 1:

$$\left(1 + \frac{.06}{365}\right)^{365} - 1 = .0618313;$$

Investment 2:

$$\left(1 + \frac{.06125}{4}\right)^{4} - 1 = .0626713.$$

Hence, investment 2 with the higher interest rate is better. ■

Example 8. An investor invests half of an original amount of $10,000 in municipal bonds earning 6% per year compounded yearly, and the other half in certificates of deposit, earning 3% per year compounded every 2 months. How much is the investment worth in 10 years? What interest rate, compounded yearly, was earned on the whole investment?

Solution. A total of $5000 is invested in each asset. The bonds grow to

$$\$5000(1.06)^{10} = \$8954.24.$$

The CD grows to

$$\$5000\left(1 + \frac{.03}{6}\right)^{60} = \$6744.25.$$

The rate of return, computed on a yearly basis, must be the interest rate r such that

$$\$10,000(1 + r)^{10} = \text{bond future value} + \text{CD future value} = \$15,698.49$$

Therefore the effective yearly rate on the whole investment satisfies

$$r = \sqrt[10]{\frac{15698.49}{10000}} - 1 = .0461303. \quad ■$$

1.2.3 Discounting

One framework under which a loan may be carried out is for the lender to view the loan as an investment made at compound interest at a given rate for a given number of periods. If the amount borrowed is L, the rate is r, and the number of periods is n, then the value of the loan at the end of its term would be:

$$L(1+r)^n. \tag{1.17}$$

If no other payments are made, the borrower would have to repay the lender this amount after n periods. We will have much more to say about loans later in the chapter. There is another way in which two parties may agree to a loan, called **lending at a discount**. A borrower seeking $\$K$ and a lender willing to loan it may sometimes make an agreement in which interest is deducted in advance at a given rate (called the **discount rate**), and the borrower actually receives $\$K$ minus the interest and pays back $\$K$ when the loan is due. Consider the following example.

Example 9. Suppose that Phil borrows $\$1000$ for 1 year from Maureen at a discount rate of $d = 5\%$. This means that Phil receives $\$1000 - .05(\$1000) = \$950$ at the start of the year and pays back $\$1000$ at the end. Considered as an investment for Maureen, her effective annual interest rate r satisfies:

$$950(1+r) = 1000 \implies r = (1000 - 950)/950 = 50/950 \approx 5.26\%. \ \blacksquare$$

In general, for a loan made at a discount at rate d over time period $[n-1, n]$, the present and future values $A(n-1)$ and $A(n)$ satisfy the relationship:

$$A(n-1) = (1-d)A(n), \tag{1.18}$$

hence the discount rate satisfies:

$$d = \frac{A(n) - A(n-1)}{A(n)}. \tag{1.19}$$

Notice that the rate of return over this interval differs in the sense that it would have $A(n-1)$ in the denominator instead.

Working as in Example 9, suppose that an amount $\$L$ is loaned at a discount rate of d. What is the **equivalent interest rate** r under which the lender could invest the amount actually given to the borrower at the starting time and receive $\$L$ at the end? An amount $(1-d)L$ must increase in value to L at the end of the time period, which requires:

$$(1-d)L(1+r) = L \implies (1+r)(1-d) = 1. \tag{1.20}$$

You can solve for the equivalent interest rate r by solving Equation (1.20).

Example 10. Which of the following is more beneficial to the borrower: a loan made at a discount rate of 8% or a loan made at an interest rate of 8.5%?

Solution. The equivalent interest rate to the 8% discount rate is the solution of:

$$(1+r)(.92) = 1 \Longrightarrow r = \frac{1}{.92} - 1 = .08696.$$

So a borrower pays a higher rate of interest under the discounting plan than under the compound interest plan. ■

From (1.20) we see that the present value single period discount factor $(1 + r)^{-1}$ is the same as $1 - d$. So if loans are made at a discount of 5%, for example, then the factor converting a future value in one period to the present value would be 95%.

1.2.4   Present Value and Net Present Value

Recall that the idea of present value is the amount invested now at a known interest rate and compounding policy that is needed to achieve a target value at a future time. In a case of discrete compounding at a rate of r per period, the relationship between the present value P at time 0 and the target future value F at time t is:

$$F = (1+r)^t P, \tag{1.21}$$

hence we have that the present value of an amount F at a time t is:

$$P = F(1+r)^{-t}. \tag{1.22}$$

Example 11. An educational trust fund is being set up by a single payment. The goal is to have the fund be worth $50,000 in 15 years. Assuming 7% monthly interest, how much money should be put into the fund?

Solution. This is just the present value of $50,000 future dollars. In 15 years there will be $15 \cdot 12$ compounding periods, and the rate per period is $.07/12$. So the present value is:

$$\$50,000 \left(1 + \frac{.07}{12}\right)^{-15 \cdot 12} = \$17,550.30. \blacksquare$$

Example 12. A *zero-coupon bond* is an investment instrument in which the buyer pays a fixed amount at the start to receive an amount called the *face value of the bond* at a future time called the *maturity date* of the bond. Suppose that you purchase a bond whose face value is $500, which has a maturity of 10 years and an interest rate of 5% compounded semiannually. How much do you pay for it?

Solution. You should pay the present value of the bond's face value. Let us find the present value of a zero-coupon bond in general first, then plug in the problem parameters. Denote

$$
\begin{aligned}
y &= \quad \text{the nominal rate of interest per year;} \\
m &= \quad \text{number of compoundings per year;} \\
t &= \quad \text{number of years to maturity;} \\
F &= \quad \text{face value of the bond.}
\end{aligned}
$$

We set up a simple **equation of value**, based on our earlier work on future value under discrete compounding:

$$
\text{future value} = F = \text{present value} \left(1 + \frac{y}{m}\right)^{mt}
$$

$$
\Longrightarrow \text{present value} = F \left(1 + \frac{y}{m}\right)^{-mt}.
$$

For our question,

$$
\text{price of bond} = \$500 \left(1 + \frac{.05}{2}\right)^{-20} = \$305.14. \quad \blacksquare
$$

We will be dealing with bonds again later, in the case that the bond delivers regular payments to its holder. As in this example, our concern will be to value the bond at the present time.

The concept of present value extends to streams of payments at different times. When comparing two different streams, a consistent basis upon which the comparison can be made is the total present value of all of payments. We are led to the following definition.

Definition 1. The **net present value** at time 0 of a sequence of future payments $P(t_1), P(t_2), P(t_3), \ldots, P(t_n)$ using an effective rate of r per period is:

$$
A = \sum_{i=1}^{n} (1 + r)^{-t_i} P(t_i). \tag{1.23}
$$

Example 13. Stock A will pay dividends of \$100, \$150, \$100, and \$50 each quarter this year, and stock B will pay semiannual dividends of \$200 each. Which one is preferable, assuming 6% per year compounding monthly? What would the (equal) stock B dividends have to be so that you are indifferent between the two?

Solution. Compute the net present value of each stream of dividends. Note that the monthly rate is .5% = .005. For stock A the payments are at months 3, 6, 9, and 12 and the present value is:

$$
\$100(1.005)^{-3} + \$150(1.005)^{-6} + \$100(1.005)^{-9} + \$50(1.005)^{-12} = \$386.80.
$$

For stock B the payments are made at months 6 and 12. So the present value is:

$$\$200(1.005)^{-6} + \$200(1.005)^{-12} = \$382.49.$$

So stock A is a slightly better investment. This seems to have occurred because more of the dividend value occurred earlier for stock A than for stock B. For the second question, set up an equation of net present values. Let x be the common unknown dividend payment for stock B occurring at months 6 and 12. Then, to make the present value equal to that of stock A, we have:

$$x(1.005)^{-6} + \$x(1.005)^{-12} = \$386.80$$

$$\Longrightarrow x = \frac{\$386.80}{(1.005)^{-6} + (1.005)^{-12}} = \$202.26. \ \blacksquare$$

Important Terms and Concepts

Geometric sequence - A list in which each member is a common multiple of its predecessor: $a_n = a \cdot q^n$.

Finite geometric series - The sum of finitely many terms of a geometric sequence: $\sum_{j=0}^{m} a \cdot q^j = a \cdot \frac{1-q^{m+1}}{1-q}$.

Infinite geometric series - The sum of infinitely many terms of a geometric sequence: $\sum_{j=0}^{\infty} a \cdot q^j = a \cdot \frac{1}{1-q}$ if $|q| < 1$.

Compound interest - Interest earned on interest as well as original investment: $A(n) = a_0 \cdot (1+r)^n$, $A(t) = a_0 \left(1 + \frac{y}{m}\right)^{m \cdot t}$.

Effective (yearly) rate of return - The rate of return over 1 year for an investment compounding m times per year at nominal rate y: $\left(1 + \frac{y}{m}\right)^m - 1$.

Lending at a discount - A loan of face value L in which interest is charged in advance, so that the borrower receives only $(1 - d)L$ at the start.

Equivalent interest rate to a discount rate - Interest rate r for which an ordinary loan at rate r is equivalent to a loan at a discount rate d: $(1 + r)(1 - d) = 1$.

Present value - The current value of a future payment at time t assuming per period rate r: $P = F(1 + r)^{-t}$.

Net present value - The total present value of a future stream of payments: $A = \sum_{i=1}^{n} (1 + r)^{-t_i} P(t_i)$.

Zero coupon bond - An investment object in which an amount of money is paid now in order to receive the face value F of the bond at its maturity:

$$F\left(1+\tfrac{y}{m}\right)^{-mt}.$$

Exercises 1.2

1. Which of the following are geometric sequences? For those that are, identify the common ratio and give a formula for the general term.
   (a) $6, -2, 2/3, -2/9, \ldots$
   (b) $2,\ 4,\ 16,\ 256,\ \ldots$
   (c) $-1, 3, 7, 11, \ldots$

2. Compute the sum of the terms from the 5th to the 60th of the geometric series

$$\frac{1}{2}+\left(\frac{1}{2}\right)^{2}+\left(\frac{1}{2}\right)^{3}+\cdots$$

3. Consider the geometric series

$$4+\frac{4}{3}+\frac{4}{9}+\frac{4}{27}+\cdots$$

How many terms are necessary to add in order for the sum to exceed 5.8? Can the sum ever exceed 7?

4. Derive a general formula for $\sum_{j=1}^{m} a\cdot q^{j}$.

5. How many time periods will it take an investment to grow to 2.5 times its initial value if the interest rate per period is 1%?

6. An investment earns interest convertible quarterly at a nominal rate of 6% per year. How many years will it take for the investment to triple in value? What is the yearly effective rate of interest?

7. Find the future value of an investment of $2000 earning 6% nominal yearly interest after 10 years if interest is compounded (a) yearly, (b) monthly, (c) daily. Try to find the limit as the number of compoundings approaches infinity.

8. Insurance companies recommend buying policies for an amount at least three times the policyholder's annual income (and will for a fee adjust for inflation). How much insurance should a person earning $50,000 now buy to protect against death in 30 years if inflation occurs at an effective yearly rate of 4%?

9. An investment convertible quarterly earns an overall rate of return of 12.3% over a period of 3 years. What is the rate of return per quarter? The yearly effective rate of return?

10. Jim observes that the comprehensive fee to attend his alma mater Trey Zexpensive College is currently $40,000, and in the previous 2 years was, respectively, $38,300 and $36,200. If he assumes that the rate of growth will be similar to the rates in the past 2 years, how much does he expect the fee to be in 14 years when he hopes his child will go to college?

11. From Example 4, show that for $x, r \in (0, 1)$ $(1+r)^x \leq 1+rx$. Conclude that it is better to receive simple interest for time intervals less than a full period than to receive compound interest.

12. What nominal interest rate convertible monthly would be equivalent to a nominal rate of 5% convertible weekly?

13. For compound interest with interest rate r in every period and initial investment a, we know that the accumulated value of the investment after k periods is $A(k) = a \cdot (1+r)^k$, $k = 0, 1, 2, ...$ Conversely, show that if an investment has this accumulated value function, then its rate of return is constantly r in every period.

14. A sports star has a contract that pays a base salary now of $4,000,000 per year and is structured to increase by a certain percentage r this year, then by $2r$ the year after that, and then by $3r$. If his final salary is $10,000,000 per year, find r, and find his salary in the second and third years.

15. An investor invests $20,000 in stock A, which in 4 years has increased in value to $25,000; $10,000 in stock B, which in the same time has increased to $11,000, and another $10,000 in stock C, which has actually declined to $8,000 during that time. Find the effective yearly rate of return on the total investment.

16. Kay lends her husband George $1000 for new fishing equipment, at a discount rate of 5% per year. The proceeds of the loan turn out to be enough to cover the equipment. They agree that George will pay Kay back at the end of a year. George finds his equipment more cheaply on line for $800, invests the leftover money in an account earning 5% compounded monthly, and uses the proceeds plus cash to repay the money that he owes Kay at the end of the year. Had he not found the cheaper equipment, what would have been the effective yearly rate of interest on the loan? How much cash will George have to come up with on his own, given his on line triumph and investing wisdom?

17. What amount must be invested now at a nominal rate of 6% per year in order to have $20,000 in 10 years if compounding occurs (a) monthly; (b) weekly; (c) daily (assuming a 360 day year)?

18. What amount, assuming quarterly compounding at nominal yearly rate 4%, must be invested at the start of the year in order to have the same value at the end of a year as an investment of $1000 compounded yearly at 6%?

19. How would you adjust the analysis in Example 11 if the $50,000 is the goal in today's dollars, but there is an effective annual inflation rate of 4%?

20. If a 1-year loan of $1000 is issued at a discount rate of 10%, what is the equivalent effective yearly rate of interest earned by the lender?

21. Herb owns a share of stock valued at $150 and expects to receive a dividend of $5 at the end of the first quarter, increasing by 10% each successive quarter. Assuming monthly compounding at a nominal yearly rate of 4%, what is the present value of the dividend stream over the course of a year?

22. A company offers to buy out the retirement contract of its outgoing president for a one-time lump sum payment of $15,000,000. The contract would have paid the president $2.4 million per year at the end of each of the next 10 years. Assuming a discount rate of 5%, should the president accept the offer?

23. The sports star in Exercise 14 uses a discount rate of 5% per year to value his contract. He has a competitive offer of $6,500,000 per year each year for these 4 years. Which contract should he pick?

24. You have loaned an amount of money and are indifferent between two repayment plans. In one, you receive $3000 at the end of 3 years, and in the other, you receive an equal payment at the end of each of the first 3 years. Assuming an effective yearly rate of 5%, what is the equal payment amount?

25. To fund a construction project, a college issues a bond in the amount of $5,000,000 in which the principal is to be paid in full in 10 years. The college must pay the bondholders $50,000 per year at the beginning of the year, and in addition it wants to set aside enough money at the beginning of each year to pay off the bond principal when it is due. Assuming the college can earn a fixed rate of return of 8% compounded yearly, how much will it have to pay in total at the beginning of each year?

26. (Inflation and the Real Rate of Return) Most of the time the economy is in a state of inflation, where there is some (effective) rate i applied that increases purchase prices of goods by a factor of $1 + i$ in a year's time. If an investment is earning an effective rate of r per year, the simplest way of thinking about the **inflation-adjusted rate of return** is:

$$\text{inflation-adjusted rate of return (simple-minded version)} = r - i.$$

But we need to look closer at the purchasing power of money earned at the end of a year. Consider an investment of $1000 which earns an effective rate of 10% in an economy in which inflation runs at 5%. The simple-minded adjusted rate is therefore $10\% - 5\% = 5\%$.

(a) Suppose that a product costs \$1 at the beginning of the year. Then our investor has the purchasing power of 1000 such products at the beginning. Under 5% inflation, how many such products can the investor purchase at the end of the year? What is the relative change in the purchasing power?

(b) A better way of looking at inflation-adjusted rate of return on an initial investment of K is this:

$$\text{inflation-adjusted rate} = \frac{\text{value of real return in year-end dollars}}{\text{value of invested amount in year-end dollars}}.$$

Show that this is the same as:

$$\text{inflation-adjusted rate of return} = \frac{r - i}{1 + i}.$$

Check for the numbers in part (a) that you get the same result with this formula as the relative change in purchasing power.

(c) Show in general that the relative change in purchasing power equals the expression in part (b) for the inflation-adjusted rate of return.

1.3   Annuities

A repeated investment or payment of a fixed amount occurring at equal intervals of time is called an **annuity**. The annuity may be paid out to you, as in the collection of retirement funds, or paid out by you, as in a loan or savings program. For simplicity, the payment occurs in synch with the compounding period. Our main objectives in this section are to derive the present and future values of annuities, and to use these as problem-solving tools in investment decision making.

If the payment is made at the end of the time period, then the annuity is called an **ordinary annuity** or **annuity immediate**. If it is made at the start, it is an **annuity due**. Each of these cases will have its own subsection below, and we will also consider several variations of annuities in which payments or interest rates are not constant, or payments are not always made at every period. We will also study **perpetuities**, in which the payments continue indefinitely. In all cases, we will see the power of the geometric series and sound financial reasoning.

1.3.1   Ordinary Annuities

As indicated above, an **ordinary annuity** is a sequence of payments $P_1, P_2, P_3, \ldots$ that are made at time instants 1, 2, 3, $\ldots$, each earning compound interest at rate r per period. For the time being, we will assume a constant payment value

$P_i = P$. Notice that we view the ith payment P_i as being made at the end of the time interval $[i-1, i]$. The ***future value*** or ***amount of an annuity*** is the value to which the total of all payments has accumulated at a given future time. The ***present value of an annuity*** is the net present value of all of the payments that occur from the current time onward.

FIgure 1.2, which is intentionally set up to mimic the way in which a spreadsheet could compute annuity values, shows how an ordinary annuity accumulates in time. (In a spreadsheet, you would set up formulas that make cell values depend on other cell values; for example, the payment 2-time 4 cell would be the product of $(1+r)$ and the payment 2-time 3 cell.) Think of each payment as a separate investment made at compound interest rate r and accumulating until time n. Reading across the row to the last column, the value of the ith payment is $P(1+r)^{n-i}$, since it is able to compound for $n-i$ periods. Totaling each column gives the total future value of the annuity at each intermediate time until, at the final time n, we get the future value of the annuity as:

$$\text{FV annuity immediate} = \sum_{i=0}^{n-1} P(1+r)^i. \tag{1.24}$$

time	0	1	2	3	$\cdots$	n
payment1		P	$P(1+r)$	$P(1+r)^2$	$\cdots$	$P(1+r)^{n-1}$
payment2			P	$P(1+r)$	$\cdots$	$P(1+r)^{n-2}$
payment3				P	$\cdots$	$P(1+r)^{n-3}$
$\vdots$					$\ddots$	$\vdots$
payment$n$						P
total		P	$\sum_{i=0}^{1} P(1+r)^i$	$\sum_{i=0}^{2} P(1+r)^i$	$\cdots$	$\sum_{i=0}^{n-1} P(1+r)^i$

Figure 1.2: Valuing an ordinary annuity.

But this is a finite geometric series that we know how to sum.

$$\sum_{i=0}^{n-1} P(1+r)^i = P \cdot \frac{(1+r)^n - 1}{(1+r) - 1} = P \cdot \frac{(1+r)^n - 1}{r}. \tag{1.25}$$

We have now proved the following.

Theorem 1. The future value of an ordinary annuity that makes constant payments P at the end of each unit time interval is:

$$\text{FV ordinary annuity} = P \cdot \frac{(1+r)^n - 1}{r}. \tag{1.26}$$

where r is the constant interest rate per period. Consequently, if there are m compoundings per year with nominal yearly interest rate y, the future value of the ordinary annuity at time t years is:

$$P \cdot \frac{(1 + y/m)^{mt} - 1}{y/m}. \tag{1.27}$$

Example 1. Suppose that a young couple with a new baby decides to start a college fund with a monthly deposit (at the end of each month) of \$100 in an account compounding monthly at a nominal yearly rate of 5%. How much money will be in the account in 18 years?

Solution. It may be valuable to go back over the main ideas instead of simply substituting into the formula in Theorem 1. In this case, the time period is months, and since the annuity runs for 18 years, the final time is 18·12 months. The initial \$100 compounds for $18(12) - 1$ months at monthly rate $r = .05/12$, yielding a value of:

$$\$100 \cdot \left(1 + \frac{.05}{12}\right)^{18 \cdot 12 - 1}$$

at the end. The next \$100 compounds for $18(12) - 2$ months yielding a value of:

$$\$100 \cdot \left(1 + \frac{.05}{12}\right)^{18 \cdot 12 - 2}$$

at time $n = 18 \cdot 12$. The next \$100 compounds for $18(12) - 3$ months, etc., until finally the last \$100 deposit is made at the end of the time horizon and gathers no interest.

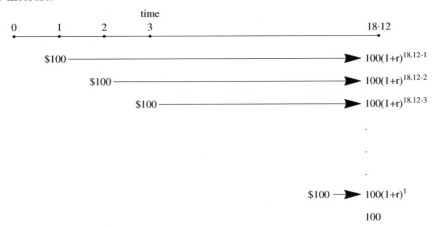

Figure 1.3: Future value stream for \$100 ordinary annuity with 18·12 periods, $r = .05/12$.

Another way of displaying the process is to use a time line as in Figure 1.3. The interpretation of this line is similar to the table in Figure 1.2. A time axis at the top shows the periods from time 0 to time 18·12 months. Beneath the time line are lines for each separate deposit indicating when that deposit was made and what its future value is at time 18·12. The diagram shows that the total future value of the annuity is:

$$\$100(1+.05/12)^{18(12)-1} + \$100(1+.05/12)^{18(12)-2} + \cdots + 100(1+.05/12)^0$$

$$= \$100 \cdot \sum_{j=0}^{18(12)-1} (1+.05/12)^j.$$

In light of the geometric series theorem, the future value is:

$$\$100 \cdot \frac{(1+.05/12)^{18(12)} - 1}{(1+.05/12) - 1} = \$100 \cdot \frac{(1+.05/12)^{18(12)} - 1}{.05/12} = \$34,920.20.$$

The couple is wise to allow compound interest to work for them for a long time. Without the interest, if they would have stashed $\$100 \cdot 12 = \$1200$ per year in a shoebox instead of investing it, their savings would have been

$$\$1200 \cdot 18 = \$21,600$$

at the end of 18 years, so by investing they made over $13,000 in interest. ∎

Notation. The factor in (1.25) that multiplies the common payment P in order to get the future annuity value comes up so often that a notation $s_{n\rceil r}$ has been introduced as a shorthand for it:

$$s_{n\rceil r} \equiv \frac{(1+r)^n - 1}{r} \implies \text{FV ordinary annuity} = P \cdot s_{n\rceil r}. \tag{1.28}$$

Example 2. For the young couple in Example 1, how many months will it take until their annuity is worth at least $25,000?

Solution. We want to solve the inequality

$$100 \cdot s_{n\rceil r} = 100 \cdot \frac{(1+.05/12)^n - 1}{.05/12} \geq 25,000$$

$$\implies (1+.05/12)^n - 1 \geq \frac{25,000 \cdot (.05)}{1200}$$

$$\implies n \cdot \log(1+.05/12) \geq \log(\frac{25,000 \cdot (.05)}{1200} + 1)$$

$$\implies n \geq \frac{\log(\frac{25,000 \cdot (.05)}{1200} + 1)}{\log(1+.05/12)} = 171.661 \text{ months} \approx 14.3 \text{ years.} \blacksquare$$

Our next problem is to develop a formula for the present value of an ordinary annuity. This is easy; it is just the total present value of all of the payments.

Again suppose that constant payments of P are made at the end of each period, with interest rate r per period. The present value of the first payment, which occurs at the end of the first period, is $P(1+r)^{-1}$. The present value of the second is $P(1+r)^{-2}$, etc. (see Figure 1.4). The payment made at time n discounts back for a full n time periods. Then,

$$\text{PV ordinary annuity} = P(1+r)^{-1} + P(1+r)^{-2} + \cdots + P(1+r)^{-n}. \quad (1.29)$$

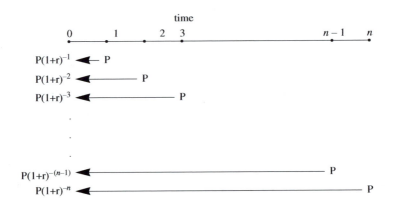

Figure 1.4: Present value stream for $\$P$ ordinary annuity with n periods, rate r per period.

Summing the geometric series with initial term $P(1+r)^{-1}$ and common ratio $(1+r)^{-1}$, we can obtain the following result.

Theorem 2. The present value of an ordinary annuity that makes constant payments P at the end of each unit time interval is:

$$\text{PV ordinary annuity} = P \cdot \frac{1-(1+r)^{-n}}{r}, \quad (1.30)$$

where r is the constant interest rate per period. Consequently, if there are m compoundings per year over t years with nominal yearly interest rate y, the present value of the ordinary annuity is:

$$P \cdot \frac{1-(1+y/m)^{-mt}}{y/m}. \quad (1.31)$$

Proof. From formula (1.29) we get

$$\begin{aligned}
\text{PV} &= P(1+r)^{-1} \cdot \sum_{i=0}^{n-1} \left(\frac{1}{1+r}\right)^i \\
&= P(1+r)^{-1} \cdot \frac{1 - \left(\frac{1}{1+r}\right)^n}{1 - \left(\frac{1}{1+r}\right)} \\
&= P \cdot \frac{1 - \left(\frac{1}{1+r}\right)^n}{(1+r) - 1} \\
&= P \cdot \frac{1 - (1+r)^{-n}}{r}. \ \blacksquare
\end{aligned}$$

Notation. As with the future value, there is a standard notation $a_{\overline{n}|r}$ for the quantity in (1.30) that multiplies P in order to get the present value of an ordinary annuity.

$$\text{PV ordinary annuity} = P \cdot \frac{1 - (1+r)^{-n}}{r} \equiv P \cdot a_{\overline{n}|r}. \qquad (1.32)$$

Example 3. Alesha wishes to take out a personal loan to cover some expenses to be incurred in moving into her new apartment. She can afford to make monthly payments on this loan of $100 at the end of each month, and she has also saved $1000 that she is willing to use up front on her expenses. She qualifies for a 36-month loan with a nominal interest rate of 6.9% convertible monthly. How much money will she have to use for her move?

Solution. Alesha's initial amount would be the $1000 savings plus the present value of her monthly payment stream:

$$\$1000 + \$100 \cdot a_{\overline{36}|.069/12} = \$1000 + \$100 \cdot \frac{1 - \left(1 + \frac{.069}{12}\right)^{-36}}{\frac{.069}{12}} = \$4243.45. \ \blacksquare$$

Example 4. Myron just got his first real job and wants to accumulate $10,000 in 4 years for a vacation in Las Vegas. Suppose that he starts a savings plan in which he makes monthly payments at the end of each month at a nominal annual rate of 4% into an account. At least how much should those payments be to achieve the goal?

Solution. There are two ways to solve this problem. If the payments are each equal to P, then we have the equation of value:

$$\$10,000 = P \cdot s_{\overline{n}|r},$$

where the number of payments is $n = 4 \cdot 12 = 48$ and the monthly rate is $r = .04/12$. Then we would have the following equation for P:

$$P = \frac{\$10,000}{s_{\overline{48}|.04/12}} = \frac{\$10,000}{\frac{(1+.04/12)^{48}-1}{.04/12}} = \$192.46.$$

If Myron had not saved money into this interest bearing account, he would have only had $\$192.46 \cdot 48 = \$9238.08$ at the end of 48 months.

The other way to solve the problem is to cast it in terms of present value. The present value of an annuity of payments of $\$P$ should equal the present value of the desired future value of \$10,000 in order for the payments to be adequate to fund the vacation. This sets up the equation of value:

$$
\begin{aligned}
\$10,000 \left(1 + \frac{.04}{12}\right)^{-48} &= \text{PV annuity} \\
&= P \cdot a_{\overline{48}|.04/12} \\
&= P \cdot \frac{1 - \left(1 + \frac{.04}{12}\right)^{-48}}{\frac{.04}{12}}.
\end{aligned}
$$

You should check that multiplying both sides of this equation by $\left(1 + \frac{.04}{12}\right)^{48}$ gives the same equation as we solved using the future value approach. (The connection between present and future value is explored further in Exercise 6.) ∎

Example 5. How long will it take to pay off a \$4000 credit card debt with effective yearly rate 12.9% using monthly payments of \$100? If the credit card company specifies a \$50 minimum monthly payment and the debtor simply pays that, how long will the person be in debt? What monthly payment will yield an unending debt of \$4000?

Solution. Consider the loan payments as an annuity to the lender; its present value is the current balance on the credit card of \$4000. The number of months n that will be required to pay off the debt satisfies the equation of value:

$$\text{PV} = \$4000 = \$100 \cdot a_{\overline{n}|r} = \$100 \cdot \frac{1 - (1+r)^{-n}}{r},$$

where r denotes the monthly interest rate. But, since the effective yearly rate is given to be .129, the monthly effective rate r would satisfy the equation $(1 + r)^{12} = 1.129$, whose solution is $r = .0101623$.

Now that we have a value for r, using the equation above we may write:

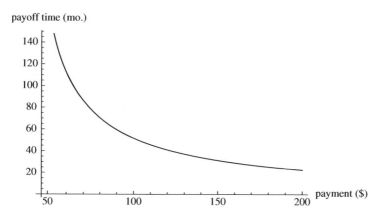

Figure 1.5: Payoff time for a $4000 debt as a function of monthly payment.

$$\$4000\frac{r}{\$100} = 1 - (1+r)^{-n} \implies (1+r)^{-n} = 1 - 40r$$
$$\implies -n\log(1+r) = \log(1-40r)$$
$$\implies n = \frac{-\log(1-40r)}{\log(1+r)} \approx 51.6 \text{ months.}$$

The loan will be paid off in 52 months, with a partial payment occurring in the last month.

If the minimum payments of $50 are used, the payoff time satisfies:

$$\$4000\frac{r}{\$50} = 1 - (1+r)^{-n} \implies (1+r)^{-n} = 1 - 80r$$
$$\implies -n\log(1+r) = \log(1-80r)$$
$$\implies n = \frac{-\log(1-80r)}{\log(1+r)} \approx 165.8 \text{ months.}$$

Cutting the monthly payment in half more than tripled the time required to pay off the debt. In general, the same computations show that with a monthly payment of P the time until the debt is eliminated is

$$n = \frac{-\log(1 - (4000/P)r)}{\log(1+r)}$$

The graph in Figure 1.5 shows the dependence of the payoff time on R. Note the rapid descent of the time as the payment level increases.

Finally, if the monthly payments are made so that they only pay the monthly interest $P = r \cdot \$4000 = \$40.65$ then the principal of the loan will never be reduced. Notice in the formula above that as P decreases to $4000r$, the negative of the log in the numerator converges to ∞. ∎

1.3.2   Annuities Due

Now let us derive a formula for the future value of an annuity due in which a payment of $P is made at the beginning of each compounding period for a total of n periods, with effective yearly interest rate r. As with ordinary annuities, a table can facilitate your understanding. In the table in Figure 1.6, the payment is made at the "left edge" of the box, i.e. payment i occurs at time $i - 1$, so for instance payment 1 at time 0 has n full periods to compound, payment 2 at time 1 has $n - 1$ full periods, etc., and the last payment occurs at the beginning of time interval $[n - 1, n]$, and hence it has one period to compound. The time line in Figure 1.7 depicting the final values of each payment is similar to the one in Figure 1.3.

time	0	1	2	$\cdots$	n
payment 1	P	$P(1+r)$	$P(1+r)^2$	$\cdots$	$P(1+r)^n$
payment 2		P	$P(1+r)$	$\cdots$	$P(1+r)^{n-1}$
payment 3			P	$\cdots$	$P(1+r)^{n-2}$
$\vdots$				$\ddots$	$\vdots$
payment $n-1$					$P(1+r)^2$
payment n					$P(1+r)$
total before payment	$-$	$P(1+r)$	$\sum_{i=1}^{2} P(1+r)^i$	$\cdots$	$\sum_{i=1}^{n} P(1+r)^i$

Figure 1.6: Valuing an annuity due.

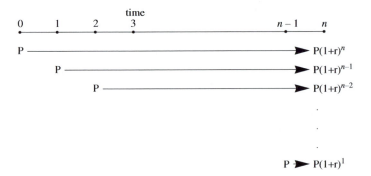

Figure 1.7: Future value stream for annuity due; payments P, interest rate r.

It is easy to see that the future annuity value will be:

$$\text{FV annuity due} = \sum_{i=1}^{n} P(1 + r)^i. \tag{1.33}$$

This can be simplified to yield the following formulas. You are asked to complete the proof in Exercise 9.

Theorem 3. The future value of an annuity due that makes constant payments P at the beginning of each unit time interval is:

$$\text{FV annuity due} = P(1+r) \cdot \frac{(1+r)^n - 1}{r} = P(1+r)s_{n\rceil r}, \qquad (1.34)$$

where r is the constant interest rate per period. Consequently, if there are m compoundings per year with nominal yearly interest rate y, the future value of the annuity due at time t years is:

$$P \cdot (1 + y/m) \cdot \frac{(1 + y/m)^{mt} - 1}{y/m}. \qquad (1.35)$$

Example 6. Find and interpret the relationship between the two future value formulas for ordinary annuities and annuities due.

Solution. Since $\text{FV} = P\frac{(1+r)^n - 1}{r}$ for ordinary annuities, it is obvious that

$$\text{FV annuity due} = (1 + r) \cdot \text{FV ordinary annuity}.$$

This is natural, because the annuity due has the same payments of $\$P$ as the ordinary annuity, shifted left by one time unit, hence each payment gains in value by a factor of $1 + r$ as compared to the corresponding ordinary annuity payment. This causes the total future value for the annuity due to be a factor of $1 + r$ times the total future value of the ordinary annuity. ■

In light of the relationship between future values for the two types of annuites, you should be willing to believe that the present value of an annuity due is the factor $(1 + r)$ times the present value of an ordinary annuity. The time line of payments in Figure 1.8 helps you to see this.

Theorem 4. The present value of an annuity due that makes constant payments P at the start of each unit time interval is:

$$\text{PV annuity due} = P(1 + r) \cdot \frac{1 - (1 + r)^{-n}}{r}, \qquad (1.36)$$

where r is the constant interest rate per period. Consequently, if there are m compoundings per year over t years with nominal yearly interest rate y, the present value of the annuity due is:

$$P\left(1 + \frac{y}{m}\right) \cdot \frac{1 - (1 + y/m)^{-mt}}{y/m}. \qquad (1.37)$$

Proof. Figure 1.8 makes it clear that the present value is

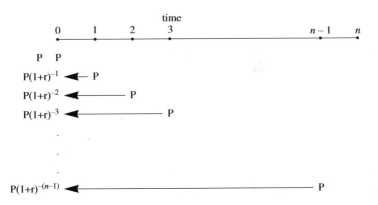

Figure 1.8: Present value stream for P annuity due with n periods, rate r per period.

$$
\begin{aligned}
\mathrm{PV} &= P \cdot \sum_{i=0}^{n-1} \left(\frac{1}{1+r} \right)^{i} \\
&= P \cdot \frac{1 - \left(\frac{1}{1+r} \right)^{n}}{1 - \left(\frac{1}{1+r} \right)} \\
&= P(1+r) \cdot \frac{1 - \left(\frac{1}{1+r} \right)^{n}}{(1+r)\left(1 - \left(\frac{1}{1+r} \right) \right)} \\
&= P(1+r) \cdot \frac{1 - (1+r)^{-n}}{r} . \quad \blacksquare
\end{aligned}
$$

An alternative approach to finding the present value of an annuity is possible, by setting up an equation of value. Consider the case of an annuity due with nominal annual rate y convertible m times per year and time horizon t years. We should be indifferent between receiving a lump sum of PV dollars now (which can be deposited earning compound interest under the same terms as the annuity) to consume at the end of t years, and receiving the future value of the annuity in t years. (Donors to organizations sometimes have this decision to make, or businesses which purchase loans from other businesses.) This means that:

future value of lump sum deposit $=$ future value of annuity

$$\Longrightarrow \text{PV}\left(1 + \frac{y}{m}\right)^{mt} = P\left(1 + \frac{y}{m}\right)\frac{(1 + \frac{y}{m})^{mt} - 1}{y/m}$$

$$\Longrightarrow \text{PV} = P\left(1 + \frac{y}{m}\right)\frac{1 - (1 + \frac{y}{m})^{-mt}}{y/m}.$$

Example 7. Two brothers, Moe and Curly, want to donate to their alma mater. Moe has just won the lottery. He will give \$20,000 immediately. Not wanting to be outdone, Curly decides to give a smaller donation at the beginning of the year each year for the next 20 years in such a way that the value of his donations at the end of the 20 years equals that of Moe. What does Curly have to pay at the beginning of each year, if the effective yearly interest rate is 6%?

Solution. This time we are setting the present value of an annuity equal to \$20,000, and asking what regular deposit P will give equal future values. The future value of the lump sum is \$20,000$(1.06)^{20}$, and we can just equate that to the future annuity value from Equation (1.34)

$$\$20,000(1.06)^{20} = P(1.06)\frac{1.06^{20} - 1}{.06}$$

$$\Longrightarrow P = \frac{\$20,000(1.06)^{20} \cdot .06}{1.06\,(1.06^{20} - 1)} = \$1644.99. \ \blacksquare$$

Example 8. Glenn is a very good planner, who starts on his 25th birthday to contribute \$200 at the beginning of each month to an annuity for his retirement. He will stop doing this when he reaches his 60th birthday in order to set aside some cash for daily use. He will begin to consume his annuity on his 65th birthday, and will receive a constant payment at the beginning of each month. Assuming for simplicity a constant nominal yearly interest rate of 4% during both the investment period and the consumption period, what would Glenn's monthly payouts be during his retirement if he wants the annuity to last for 20 years?

Solution. In this problem time periods are months, and the effective rate per period is $r = .04/12 = .00333$. He will contribute for 35 full years, which is $n_1 = 420$ months. On his 60th birthday the annuity will be worth

$$\$200(1 + r)s_{\overline{420}|r} = \$200(1 + r) \cdot \frac{(1 + r)^{420} - 1}{r} = \$183,355.$$

But this is also able to earn compond interest for 5 more years, or 60 months, so that by the time that Glenn is ready to start drawing on his annuity he has:

$$\$183,355 \cdot \left(1 + \frac{.04}{12}\right)^{60} = \$223,876.$$

The $20 \cdot 12 = 240$ payments that he takes on retiring form an annuity due whose present value must equal \$223,876. Let P be the constant monthly payment; then by Formula (1.37), P must satisfy:

$$\$223,876 = P \left(1 + \frac{.04}{12}\right) \cdot \frac{1 - (1 + .04/12)^{-240}}{.04/12},$$

which implies that Glenn's monthly payout is:

$$P = \frac{\$223,876 \cdot (.04/12)}{\left(1 + \frac{.04}{12}\right) \cdot (1 - (1 + .04/12)^{-240})} = \$1352.14. \ \blacksquare$$

In this example, Glenn chose to defer the consumption of the annuity that he created. We will have more to say about deferred annuities later in this section.

1.3.3   Variations on Annuities

Perpetuities

In a ***perpetuity*** there is no upper limit on the number of periods over which the annuity is paid. If P is again the common payment made at the end of each period, we call the infinite sequence of payments a ***perpetuity immediate***. As usual, let r be the constant per period effective interest rate. Then recall that the present value of an n period annuity is:

$$\text{PV annuity immediate} = P \cdot \frac{1 - (1 + r)^{-n}}{r}.$$

In the limit as $n \to \infty$, the present value of the perpetuity immediate converges to P/r. There is a similar notion of a ***perpetuity due***, in which payments are made indefinitely at the beginning of each period. Since the present value of an n period annuity due is

$$\text{PV annuity due} = P(1 + r) \cdot \frac{1 - (1 + r)^{-n}}{r},$$

the present value of a perpetuity due would be the limit of this, namely, $P(1 + r)/r$. Thus, we have the following result.

Theorem 5. The present value of a perpetuity immediate in which payments of P are made at the end of each time interval, and the per period interest rate is r is:

$$\text{PV perpetuity immediate} = \frac{P}{r}. \tag{1.38}$$

The present value of a perpetuity due in which the payments are made at the beginning of each period is:

$$\text{PV perpetuity due} = \frac{P(1 + r)}{r}. \tag{1.39}$$

Notice that, for a perpetuity immediate in which there is an initial balance of A, the amount that can be withdrawn at the end of each period forever without changing the principal in the account is Ar, because this is the interest that accrues in each period, which can be taken out without affecting the balance for the next period.

Example 9. Melati would like to set up a perpetuity for her retirement that allows her to withdraw $50,000 per year at the beginning of each year. Assuming constant effective interest rates of 4% per year, and assuming that she has 30 years to save, how much should she deposit at the beginning of each year to achieve the retirement goal?

Solution. In order for her to consume $50,000 at the beginning of each year, this must be the interest earned per year on her principal at the beginning of her retirement. This means that to support her goal, she needs to set up a perpetuity due with present value

$$\text{PV} = \frac{P(1+r)}{r} = \frac{\$50,000(1.04)}{.04} = \$1,300,000.$$

So now we must determine what amount P is necessary to deposit each year during the savings period to form an annuity due with the future value $1,300,000 after 30 years. By Theorem 3, the payment P satisfies

$$P(1+.04) \cdot s_{\overline{30}|.04} = \$1,300,000 \implies P = \frac{\$1,300,000}{1.04 \cdot s_{\overline{30}|.04}} = \$22,287.60. \ \blacksquare$$

Non-constant payments

Returning to annuities with a finite time horizon, let us first generalize the assumption of constant, regular payments. Suppose that an annuity consists of non-constant payments in amounts $P_1, P_2, ..., P_n$ for n periods and assume that the effective interest rate per period is still a constant denoted by r. Some of the P_i's may be zero in cases where payments are skipped. Then the future value of the ordinary form of the annuity is:

$$\text{FV ordinary annuity} = P_1(1+r)^{n-1} + P_2(1+r)^{n-2} + \cdots + P_n. \quad (1.40)$$

(See, for instance, Figure 1.2, amended so that the payments down the diagonal of the table are not constant.) The present value of such an annuity would be:

$$\text{PV ordinary annuity} = P_1(1+r)^{-1} + P_2(1+r)^{-2} + \cdots + P_n(1+r)^{-n}. \quad (1.41)$$

In Exercise 16 you are asked to derive similar formulas for the case of an annuity due.

It is simple enough to compute these values for small problems, as in the next example, but to reduce (1.40) and (1.41) to closed form for larger problems requires the payments to have some structure, as we will see.

Example 10. This year Joe expects to receive dividends on a stock of $20 at the end of the first quarter, increasing by $5 each successive quarter. Assuming monthly compounding at a nominal yearly rate of 6%, what is the present value of the dividend stream? What is the future value at the end of 1 year?

Solution. Let the time units be months. The dividends are $20 at the end of month 3, $25 at the end of month 6, $30 at the end of month 9, $35 at the end of month 12, and zero otherwise. Thus, the present value at the beginning of the year is:

$$\$20 \left(1 + \frac{.06}{12}\right)^{-3} + \$25 \left(1 + \frac{.06}{12}\right)^{-6} + \$30 \left(1 + \frac{.06}{12}\right)^{-9}$$

$$+\$35 \left(1 + \frac{.06}{12}\right)^{-12} = \$105.62.$$

Since the $20 dividend earns interest for 9 months, the $25 dividend for 6 months, etc., the future value at the end of the year is:

$$\$20 \left(1 + \frac{.06}{12}\right)^{9} + \$25 \left(1 + \frac{.06}{12}\right)^{6} + \$30 \left(1 + \frac{.06}{12}\right)^{3}$$

$$+\$35 \left(1 + \frac{.06}{12}\right)^{0} = \$112.13. \ \blacksquare$$

Example 10 suggests the case of payments made in **arithmetic progression**, that is, changing by a fixed amount in each period. (This example didn't quite fit the mold, since the period of compounding was not in synch with the payments.) This is one case in which closed formulas for future and present values are possible to derive. We will not attempt a comprehensive encyclopedia of such situations here, being content to leave to the reader's developing financial intuition the task of deriving results as needed.

To illustrate the reasoning, for the case of the present value of an annuity due, suppose that an initial payment of P is made at the start of the investment period, and it is increased by ΔP each period. Let r be the effective interest rate per period. The ith payment is $P + i \cdot \Delta P$, made at time i, for $i = 0, 1, 2, ..., n-1$. Therefore the present value of this annuity is:

$$P + (P + \Delta P)(1+r)^{-1} + (P + 2\Delta P)(1+r)^{-2} + (P + 3\Delta P)(1+r)^{-3}$$

$$+ \cdots + (P + (n-1)\Delta P)(1+r)^{-(n-1)}.$$

Separating the P factors from the Q factors, we get:

$$\text{PV annuity due} \quad = \quad P\sum_{k=0}^{n-1}\frac{1}{(1+r)^k} + \Delta P\sum_{k=1}^{n-1}\frac{k}{(1+r)^k}$$

$$= \quad P\cdot\frac{1-(1+r)^{-n}}{1-(1+r)^{-1}} + \Delta P\cdot\frac{1}{1+r}\cdot\sum_{k=1}^{n-1}k\cdot\left(\frac{1}{1+r}\right)^{k-1}.$$

The first term is $P(1+r)\cdot a_{\overline{n}|r}$. To simplify the second, consider the series $\sum_{k=1}^{n-1}kx^{k-1}$. This is the derivative of the series

$$\sum_{k=0}^{n-1}x^k = \frac{1-x^n}{1-x}.$$

Therefore,

$$\sum_{k=1}^{n-1}kx^{k-1} \quad = \quad \frac{d}{dx}\left(\sum_{k=0}^{n-1}x^k\right)$$

$$= \quad \frac{d}{dx}\left(\frac{1-x^n}{1-x}\right)$$

$$= \quad \frac{(1-x)\left(-nx^{n-1}\right) - (1-x^n)(-1)}{(1-x)^2}$$

$$= \quad \frac{1-nx^{n-1}+(n-1)x^n}{(1-x)^2}.$$

Substituting $x = (1+r)^{-1}$ and plugging the result into the present value formula above, we obtain the following formula for the present value of an annuity due with initial payment P and subsequent payments increasing by ΔP each period:

$$\text{PV annuity due} = P(1+r)\cdot a_{\overline{n}|r} + \frac{\Delta P}{1+r}\cdot\frac{1-n(1+r)^{1-n}+(n-1)(1+r)^{-n}}{\left(1-(1+r)^{-1}\right)^2}.$$

$$(1.42)$$

Formulas for the future value of an annuity due, and the present and future values of ordinary annuities with payments in arithmetic progression can be derived in like manner (see Exercise 21).

Payments in Geometric Progression

The closed formula in the last section for the present value of an annuity due with payments in arithmetic progression took some skill to derive and was rather cumbersome in the end. Another special case of non-constant payments occurs when the payments themselves form a **geometric progression**, and both the derivation and the final result are simpler. This generalization might happen in an investment plan in which the investor wants to compensate for inflationary devaluation of money by contributing an additional percentage. Suppose that the interest rate per period is r, the initial payment is P, and payments increase by a percentage $q \times 100\%$ per period. Thus, the sequence of payments is:

$$P_1 = P, P_2 = P(1+q), P_3 = P(1+q)^2, \ldots, P_n = P(1+q)^{n-1},$$

made at times $1, 2, 3, \ldots, n$ for an annuity immediate. The present value of the payments is therefore:

$$\text{PV} = P(1+r)^{-1} + P(1+q)(1+r)^{-2} + \cdots + P(1+q)^{n-1}(1+r)^{-n}.$$

This formula can be used as is, but a closed form is possible. This is a geometric series of the form:

$$P \cdot \sum_{j=0}^{n-1} (1+q)^j (1+r)^{-(j+1)} = \frac{P}{1+r} \cdot \sum_{j=0}^{n-1} \left(\frac{1+q}{1+r} \right)^j.$$

The case where $q = r$ is studied in Exercise 19. But in the case $q \neq r$, the present value simplifies to:

$$\text{PV} = \frac{P}{1+r} \frac{((1+q)/(1+r))^n - 1}{(1+q)/(1+r) - 1}. \tag{1.43}$$

Rewriting this formula slightly, we have derived the present value formula in the next theorem. You are asked to prove the future value formula in Exercise 20.

Theorem 6. The present value of an annuity immediate in which payments form a geometric progression $P_i = P(1+q)^{i-1}$, $i = 1, 2, 3, \ldots n$ and the per period interest rate is $r \neq q$ is:

$$\text{PV} = P \frac{((1+q)/(1+r))^n - 1}{q - r}. \tag{1.44}$$

The future value of this annuity is:

$$\text{FV} = P \cdot \frac{(1+q)^n - (1+r)^n}{q - r}. \tag{1.45}$$

In Exercise 22 you will prove the corresponding results for annuities due, stated below as Theorem 7.

Theorem 7. The present value of an annuity due in which payments form a geometric progression $P_i = P(1+q)^i$, $i = 0, 1, 2, \ldots, n-1$ and the per period interest rate is $r \neq q$ is:

$$\text{PV} = P(1+r) \cdot \frac{((1+q)/(1+r))^n - 1}{q - r}. \tag{1.46}$$

The future value of this annuity is:

$$\text{FV} = P(1+r) \cdot \frac{(1+q)^n - (1+r)^n}{q-r}. \tag{1.47}$$

Example 11. Andy sets up a 20-year annuity which he begins by putting \$10,000 at the end of year 1 into an account earning 8% compounded annually. He predicts inflation to occur at a rate of 5% per year, and so he will increase his deposits by that percentage each year to protect his investment against inflation. What is the present value of his total contributions?

Solution. The initial contribution in Andy's ordinary annuity is $P = \$10,000$ and the problem gives us that $n = 20$, $q = .05$, and $r = .08$. By Formula (1.44), the present value is:

$$\text{PV} = P\frac{((1+q)/(1+r))^n - 1}{q-r} = \$10,000 \cdot \frac{(1.05/1.08)^{20} - 1}{.05 - .08} = \$143,580. \blacksquare$$

Example 12. An avaricious software conglomerate (AVSOFT) plans to acquire a small one-product software firm so that it doesn't have to spend its own money on research and development of a similar product. The small firm has net yearly profits of 2.5 hundred thousand dollars currently, which AVSOFT believes will grow at a rate of 10% per year for the next 5 years, after which the product will probably become obsolete. Valuing money at an effective yearly discount rate of $d = 5\%$, if AVSOFT wants to offer the small firm half of what it is worth, what should their offer be?

Solution. We compute half of the present value of the profit stream. Recall that if $d = .05$, then the present value factor that values next year's value in terms of this year's value is $1 - d = .95$. We model the profits as receivable at the end of a year, which allows us to find the present value of the firm by valuing an ordinary annuity of profits in which the payments are in geometric progression with $q = .1$. Let us write out an appropriate sum longhand this time since there are only five periods involved. The present value of the small firm (in \$hundred thousand) is:

$$
\begin{aligned}
(.95)2.5 + (.95)^2(2.5)(1.1) + \cdots & \\
+ (.95)^5(2.5)(1.1)^4 &= 2.5(.95)\sum_{i=0}^{4}((.95)(1.1))^i \\
&= 2.5(.95) \cdot \frac{1 - ((.95)(1.1))^5}{1 - (.95)(1.1)} \\
&= \$12.99.
\end{aligned}
$$

So the offer would be half that, or about \$650,000. $\blacksquare$

A problem related to Example 12 that would require numerical solution techniques is the following. Suppose that AVSOFT's offer is again half of what the small software company is worth in present value terms, the offer is $6.2 hundred thousand, and the company's current profit level is again $2.5 hundred thousand. What is the implied rate of growth of the company's profits?

In view of the computation in the example, the rate of growth q would satisfy the equation:

$$\text{present value of firm} = 2 \cdot 6.2 = 2.5(.95)\frac{1 - (.95(1+q))^5}{1 - .95(1+q)}.$$

The solution turns out to be $q \approx .0754$.

Non-Level Interest Rates

Thus far we have supposed that the interest rate per compounding period stays constant. The interest rate may well change while an annuity is progressing. To find the future value of the annuity, the strategy is to find the future value of each payment at the time when the interest rate changes, then let that balance earn compound interest at the new rate until the rate changes again. Continue to earn compound interest using the new rate, etc., until the end of time is reached.

Figure 1.9 illustrates a case where the payment is $100, the time horizon is 8, the rate is $r_1 = .04$ for the first two periods, then $r_2 = .03$ for the next two periods, and then $r_3 = .035$ for the last four periods. The last five payments are straightforward, since from time 4 through 8 the interest rate stays the same at 3.5%. The first three payments show the thinking process. Payment 1 made at time 1 accrues interest at 4% for one period to reach a value of $100(1.04)$, then that balance earns interest at 3% for two periods to bring the value up to $100(1.04)(1.03)^2$, and finally that amount earns interest for the remaining four periods at 3.5%. The total value at time 8 for that line is $100(1.04)(1.03)^2(1.035)^4$. You should have no trouble verifying similarly that the $100 deposit at time 2 when the interest rate has already changed to 3% will have a final value of $100(1.03)^2(1.035)^4$ at time 8, and that the $100 deposit at time 3 reaches a value of $100(1.03)(1.035)^4$.

Actually, to generalize the strategy, it is useful to lump the payments into blocks determined by the times when the interest rate changes. Payments 1 and 2 form an ordinary annuity with final time $n_1 = 2$ and rate $r_1 = .04$, so these two payments together are worth $\$100s_{\overline{n_1}|r_1}$ at time 2, and they subsequently earn compound interest with one change in rate for a final value of $\$100s_{\overline{n_1}|r_1}(1.03)^2(1.035)^4$ at time 8. Payments 3 and 4 are another ordinary annuity with $n_2 = 2$ periods and rate $r_2 = .03$, and they reach the value $\$100s_{\overline{n_2}|r_2}$ at time 4, after which they earn interest at rate 3.5% for four periods, to achieve a final value of $\$100s_{\overline{n_2}|r_2}(1.035)^4$. The final four payments are their own ordinary annuity starting at time 4 with $n_3 = 4$ periods and rate $r_3 = .035$, so they reach a value of $\$100s_{\overline{n_3}|r_3}$ at time 8. The total value of this annuity is therefore:

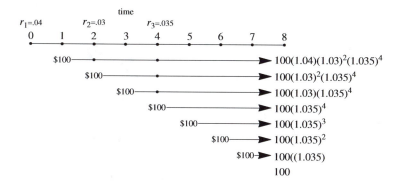

Figure 1.9: Future value stream for $100 ordinary annuity with changing interest rates.

$$\$100s_{\,n_1]r_1}(1.03)^2(1.035)^4 + \$100s_{\,n_2]r_2}(1.035)^4 + \$100s_{\,n_3]r_3}. \tag{1.48}$$

The following theorem should now be apparent.

Theorem 8. Consider an ordinary annuity with constant payments P and non-level interest rates such that, for the first n_1 periods the rate is r_1, for the next n_2 periods it is r_2, etc., and the rate for the last n_k periods is r_k. Then the future value of the annuity is:

$$\begin{aligned}
\text{FV} = {}& P\, s_{\,n_1]r_1}\left(1+r_2\right)^{n_2}\left(1+r_3\right)^{n_3}\cdots\left(1+r_k\right)^{n_k}\\
& + P\, s_{\,n_2]r_2}\left(1+r_3\right)^{n_3}\cdots\left(1+r_k\right)^{n_k}\\
& + \cdots\\
& + P\, s_{\,n_k]r_k}. \tag{1.49}
\end{aligned}$$

Example 13. Gary buys a $1000 certificate of deposit at the end of every year for 10 years. Initially the interest rate is 3.5% per year. At time 3 years the rate rises to 4.0%, and at time 6 years it falls to 3.2%. Find the total value at the end of the tenth year. (Assume that the CDs roll over at their maturity to whatever interest rate is in effect at the time.)

Solution. This is a direct application of the ideas discussed above. The CDs purchased at times 1, 2, and 3 form an ordinary annuity with interest rate $r_1 = .035$ and horizon $n_1 = 3$. They roll over to earn interest at rate $r_2 = .04$ for 3 years, after which they earn interest at rate $r_3 = .032$ for the last 4 years. The second block of CDs is purchased at times 4, 5, and 6; that block has time horizon $n_2 = 3$, and its future value at time 6 continues to earn interest for 4 years at rate r_3. The payments made at times 7, 8, 9, and 10 make up an ordinary annuity with rate r_3 and time horizon $n_3 = 4$. Hence the total future value of all certificates at time 10 is:

$$\begin{aligned}
\text{FV} \;&=\; \$1000s_{3\rceil.035}(1.04)^3(1.032)^4 + \$1000s_{3\rceil.04}(1.032)^4 + \$1000s_{4\rceil.032} \\[4pt]
&=\; \$1000\,\frac{(1.035)^3 - 1}{.035}(1.04)^3(1.032)^4 + \$1000\frac{(1.04)^3 - 1}{.04}(1.032)^4 \\[4pt]
&\quad +\$1000\frac{(1.032)^4 - 1}{.032} \\[4pt]
&=\; \$11,700.10. \;\blacksquare
\end{aligned}$$

How would we go about calculating the present value of an ordinary annuity with non-level interest rates? The idea is to look backward on a time line diagram such as the one in Figure 1.9. Find the present value of each block of payments at the time when the block started, then discount back to time 0, noting changes in interest rate. For example, in that problem, the last four payments formed an ordinary annuity with time horizon $n_3 = 4$ starting at time 4, with interest rate $r_3 = .035$. Its present value at time 4 would therefore be $\$100a_{n_3\rceil r_3}$. To find the present value of this block at time 0, discount back for two time units with discount factor $(1.03)^{-1}$ and then for two more time units with discount factor $(1.04)^{-1}$ to yield a time 0 present value of:

$$\$100a_{n_3\rceil r_3}(1.03)^{-2}(1.04)^{-2}.$$

Do the same for the other two blocks and add to find the total present value.

In general, the effect of these computations is to divide every term in Equation (1.49) by $(1 + r_1)^{n_1}(1 + r_2)^{n_2}(1 + r_3)^{n_3}\cdots(1 + r_k)^{n_k}$ to convert future value to present value. Noting that $s_{n_i\rceil r_i}/(1 + r_i)^{n_i} = a_{n_i\rceil r_i}$ for each i, we get the present value formula:

$$\begin{aligned}
\text{PV} =\; &Pa_{n_1\rceil r_1} + Pa_{n_2\rceil r_2}(1 + r_1)^{-n_1} \\[4pt]
&+ Pa_{n_3\rceil r_3}(1 + r_1)^{-n_1}(1 + r_2)^{-n_2} \\[4pt]
&+\cdots \\[4pt]
&+ Pa_{n_k\rceil r_k}(1 + r_1)^{-n_1}(1 + r_2)^{-n_2}\cdots(1 + r_{k-1})^{-n_{k-1}}. \quad (1.50)
\end{aligned}$$

Example 14. A company treasurer is trying to account properly for all of the current liabilities of the company. Among them is a payoff settlement for a lawsuit, in which the terms of the settlement call for the company to pay the plaintiff $\$2,000$/month for 20 years. The current nominal yearly interest rate is 3% per year, but the treasurer expects that to grow by .1% per year. What is the present value of this liability?

Solution. We assume that the $\$2000$ payments are made at the end of the month so that it is necessary to compute the present value of an ordinary annuity with $\$2000$ payments and non-level interest rates $r_1 = .03/12$, $r_2 = .031/12$, $r_3 = .032/12$, etc. up to $r_{20} = (.03 + 19(.001))/12 = .049/12$, changing every 12 months. Each block forms a 12-month annuity, for which we can find the present value using its particular effective monthly interest rate, and then discount back to time 0 by the product of the previous discount factors. The present value is the total of the individual present values as in formula (1.50), which is:

$$\$2000\ a_{\overline{12}|.03/12} + \$2000\ a_{\overline{12}|.031/12}\left(1 + \frac{.03}{12}\right)^{-12}$$

$$+ \$2000\ a_{\overline{12}|.032/12}\left(1 + \frac{.03}{12}\right)^{-12}\left(1 + \frac{.031}{12}\right)^{-12}$$

$$+ \cdots + \$2000\ a_{\overline{12}|.049/12}\left(1 + \frac{.03}{12}\right)^{-12}\left(1 + \frac{.031}{12}\right)^{-12}\cdots\left(1 + \frac{.048}{12}\right)^{-12}.$$

You will probably want to use a form of technology (such as a spreadsheet or *Mathematica*) to compute this; the result comes out to $342,597. Even though the cash value is $2000 \cdot 12 \cdot 20 = \$480,000$, in present value terms the liability is much less. ■

Deferred Annuities and Perpetuities

We have one more annuity variation to examine. It is possible that annuity payments begin at a point more than one time period after the annuity is purchased. This kind of cash flow is called a **deferred annuity**, and a **deferred perpetuity** is similar, except that it has an infinite time horizon. There are two fairly easy ways to give the present value of such a financial object. The first strategy is to make a standard computation of the time 0 present value of a stream of payments beginning at a later time t. You can formulate and sum an appropriate geometric series to do this. The other method is to compute the value of a full annuity starting from time 0, then subtract from it the value of the omitted payments from time 0 to time t.

Example 15. Alex is about to retire from his full-time job, so he decides to use his life savings to purchase from an insurance company a perpetuity to help support him with payments of $2000 per month at the beginning of every month for as long as he lives. But after receiving a quote for the price of the perpetuity, he decides to take a part-time job for the next 3 years, deferring payments until 36 months after the purchase of the perpetuity. Assume an annual nominal yearly interest rate of 4%. Also assume that the insurance company sets its prices as 102% of the present value of payments in order to have a margin of profit. How much would the original perpetuity have cost him had he not taken the part-time job, and how much does the deferred annuity cost?

Solution. Let time 0 be the time at which the deferred perpetuity is purchased. The time unit is months, and the effective monthly rate to use in discounting is $r = .04/12$. The following chart shows the payment stream, and note that payments are made at the beginning of the month.

$$
\begin{array}{lcccccccccc}
\text{time} & 0 & 1 & 2 & \cdots & 35 & 36 & 37 & 38 & \cdots \\
\text{payment} & 0 & 0 & 0 & \cdots & 0 & 2000 & 2000 & 2000 & \cdots
\end{array}
$$

Working from first principles, the net present value in dollars will be:

$$
\sum_{i=36}^{\infty} 2000 \left(1 + \frac{.04}{12}\right)^{-i} = 2000 \left(1 + \frac{.04}{12}\right)^{-36} \sum_{i=0}^{\infty} \left(1 + \frac{.04}{12}\right)^{-i}
$$

$$
= 2000 \left(1 + \frac{.04}{12}\right)^{-36} \cdot \frac{1}{1 - \left(1 + \frac{.04}{12}\right)^{-1}}
$$

$$
= 534,033.
$$

The cost to Alex will be $1.02(\$534,033) = \$544,714$. You could think of this as a full perpetuity due with payments of \$2000 each month, with the annuity due consisting of the first 36 payments subtracted out. So the present value is also:

$$
\sum_{i=36}^{\infty} 2000 \left(1 + \frac{.04}{12}\right)^{-i} = 2000 \left(\sum_{i=0}^{\infty} \left(1 + \frac{.04}{12}\right)^{-i} - \sum_{i=0}^{35} \left(1 + \frac{.04}{12}\right)^{-i} \right)
$$

$$
= 2000 \left(\cdot \frac{1}{1 - \left(1 + \frac{.04}{12}\right)^{-1}} - \left(1 + \frac{.04}{12}\right) a_{\overline{36}|.04/12} \right)
$$

$$
= 534,033.
$$

The first quantity in the difference is the present value of the non-deferred perpetuity, and so the cost to Alex for that retirement plan is:

$$
1.02 \left(\$2000 \cdot \frac{1}{1 - \left(1 + \frac{.04}{12}\right)^{-1}} \right) = \$614,040.
$$

Deferring his annuity saved Alex about \$80,000. ∎

Important Terms and Concepts

Ordinary annuity (annuity immediate) - A regular payment of a constant amount at the end of each time period up to a finite time horizon.

Annuity due - A regular payment of a constant amount at the beginning of each time period up to a finite time horizon.

Future value of an annuity - The value to which the total of all payments has accumulated at a given future time.

Present value of an annuity - The net present value of all of the payments that occur from the current time onward.

Perpetuity - A regular payment of a constant amount with infinite time horizon. It can be either of the "immediate" or "due" variety.

Annuities with non-constant payments - The value of the ith payment is P_i, not necessarily equal to the value of all other payments.

Payments in arithmetic progression - The ith annuity payment is $P + i \cdot \Delta P$.

Payments in geometric progression - The ith annuity payment is $P(1+q)^{i-1}$.

Annuities with non-level interest rates - An annuity in which the per period interest rate changes occasionally.

Deferred annuities - Annuity payments begin at a time after the time the annuity is purchased.

Exercises 1.3

1. What payments to an ordinary annuity are necessary to give it the same present value as a payment of $10,000 made 10 years in the future? Assume a constant yearly effective interest rate of 5%.

2. Draw a time line and find an expression for the present value of an ordinary annuity of $50 per month for 10 years, assuming a nominal 6% yearly rate compounded monthly. You should write a correct expression in summation notation and evaluate the sum in closed form.

3. At age 21, when he began his first real job, Edgar started an annuity with an insurance company to which he contributed $100 per month at the end of each month. Assume there is a nominal yearly interest rate of 5% convertible monthly. He has now reached age 65, and will stop contributing to the annuity and retire. If he (or his heirs) consume an equal amount of money from this annuity at the end of each month for 20 years until it is exhausted, how much can be paid out each month?

4. Lois is in the market for a used car, and she wants to have payments of no more than $300 per month occurring at the end of the month for a 60-month loan. She will not make a down payment. How expensive a car can she buy at a nominal yearly interest rate of 8%? Find a general formula for the affordable

price as a function of the nominal yearly rate of interest that she can negotiate.

5. Tim and Tom are two highly competitive brothers. Tim, the elder by 2 years, wants to set up a retirement annuity of value $500,000 for his retirement in 35 years. He finds an insurance company willing to give a 4% nominal rate compounded monthly on his investment. Tom simply wants to have $1 more than his brother upon his retirement, but that is 37 years away. The best deal that Tom can find gives him a 3.5% nominal rate compounded monthly. Find the monthly payment that each will have, assuming the annuities are ordinary. What larger payment must Tom make if he wishes to retire in the same year as his brother Tim with $500,001?

6. What is the algebraic relationship between $a_{n\rceil r}$ and $s_{n\rceil r}$? Between the present and future values of an ordinary annuity? Give the economic intuition behind the relationship.

7. Find the future value of a 30-year ordinary annuity convertible monthly in which the nominal yearly rates are graduated: 3% in the first 10 years, 4% in the next 10 years, and 5% in the final 10 years. Suppose that the monthly payments remain at $200 throughout the annuity. Suppose also that past contributions continue to earn interest at the rate that was active when the contributions were made.

8. Sam wants to supplement his expected social security payment by withdrawals from his retirement account. To compensate for inflation, Sam wants to increase his consumption level by .5% each month when he retires, starting with $1000 (assume that consumptions occur at the end of each month). Suppose that on retirement he will have saved $1,000,000, and the retirement account continues to earn 2% nominal yearly interest, convertible monthly, during his consumption period. How long will his savings last? What will be his final consumption amount?

9. Prove Theorem 3.

10. Scott is a collector of vintage pinball machines, and he wants to take out a loan to buy another one. In light of the fact that he is still paying for some of his other toys, he can only afford monthly payments of $80, to be paid at the beginning of each month. At one bank he can get a nominal yearly interest rate of 4.5% on a 4-year loan, and at another he can get a rate of 5.8% on a 5-year loan. Which bank should he deal with to be able to afford the most expensive machine possible?

11. A company is attempting to cut its future obligations by buying out the retirement annuities of some of its top executives using a one-time lump sum payment. One particular executive has a contract that stipulates that the company will pay $100,000 per year on retirement at the beginning of each year for

a period of 15 years. The company offers the person $1,200,000 immediately to retire and void the contract. If the interest rate per year is 2%, how much is the company saving with this action?

12. Sidney is anticipating saving $500,000 for her retirement, and is choosing between two insurance companies to which to give this money to manage and distribute monthly perpetuity payments to her throughout her post-retirement life. The first company reports an effective yearly discount rate of $d = .04$, and the second reports an effective yearly discount rate of $d = .035$. Assume that the payments are at the end of each month. Compute the monthly payments for each company. Which gives her the higher monthly payments?

13. A college wants to build a new dormitory, and plans to raise enough money to fund in perpetuity the utilities and maintenance on the building, which it expects to be $5000 per month (to be spent at the start of each month). If the school can earn 6% per year compounded monthly, at least how much money will it have to raise?

14. Emily has been given a gift by her uncle Gary for use in college in which he promises to send her $200 at the end of her first month, then $180 at the end of her second month, $160 at the end of her third month, etc., until the amount decreases to zero. Assume a nominal yearly interest rate of 2.5%. If Emily wants to tell her uncle that she prefers the money up front, what amount should she ask for?

15. Barney and Betty strike a deal with their local bank for a so-called **reverse mortgage**. Their house, currently valued at $200,000, is paid free and clear. The bank will assume ownership of the house in 25 years and in return will pay the couple a monthly stipend at the beginning of each month, using an interest rate of 4% nominal per year. But Barney expects to need more money earlier than later, and so he and Betty agree to let their monthly stipend be decreased by .2% per month. Find their stipends during the first 12 months.

16. Write and justify formulas analogous to (1.40) and (1.41) for the future and present values of an annuity due with non-constant payments.

17. Lars buys a municipal bond of face value $1000 and maturity 5 years, for which he earns payments called *coupons* (at the beginning of the year) at a rate of 1% of the face value for the first year, 1.5% for the second, 2% for the third, 2.5% for the fourth, and 3% for the fifth year. He invests these coupon payments in an account earning a nominal yearly rate of 4% compounded monthly. He also cashes the bond in for its face value at the end of the 5 years. What is the total future value at the time of maturity of the bond and all of these payments?

18. Harvey begins an annuity immediate on his 30th birthday with regular yearly payments of $1000 at the end of each year (meaning that the first of them will

occur on the eve of his 31st birthday). The interest rate begins at 4% effective per year. After the first 4 years, the rate changes to 5%, and then after the seventh year the rate changes to 6%. What is the annuity worth after the tenth year (i.e., on the eve of his 40th birthday)?

19. How is the formula (1.43) for the present value of an ordinary annuity with payments in geometric progression different when $q = r$? Find a similar result for an annuity due with payments in geometric progression.

20. Derive formula (1.45) for the future value of an annuity immediate with payments in geometric progression.

21. Derive a formula for the future value of an ordinary annuity with n periods, rate r per period, with payments in arithmetic progression. Let P be the initial payment, and let ΔP be the increment in the payment in each period.

22. Prove Theorem 7, the results for present and future value of an annuity due with payments in geometric progression.

23. Herb owns a share of stock valued at $150, and expects to receive a dividend of $5 at the end of first quarter, increasing by 10% each successive quarter. Assuming monthly compounding at a nominal yearly rate of 4%, what is the present value of the dividend stream over the course of a year?

24. Find a formula analogous to (1.48) for the future value of an annuity due with $100 payments, time horizon 8, and the same changes in interest rate as in that example.

25. Find the present value of Gary's certificates of deposit in Example 13.

26. Explain how Theorem 8 might be generalized if changes in payment value are allowed at the times when interest rate changes occur.

27. Tom has a retirement account in which he begins with a monthly deposit of $200 on January 1, and then each subsequent January 1 he increases his monthly payment by 2%. The account begins at an interest rate of 4% nominal per year, convertible monthly, but changes three times, on July 1 of the second year, July 1 of the twelfth year, and July 1 of the sixteenth year, to 3.5%, 3.2%, and 3.7% respectively. Write an expression for the value the account on December 31 of the twentieth year. If you have access to technology, compute the final value of the account.

28. Find both the present and future values of an annuity due of $5000 with effective yearly rate 5% and maturity 10 years that begins 4 years after the present time.

29. A college has raised \$2,000,000 to fund a half-time position in sustainability, but they are unable to successfully hire a person for 4 years. The principal was invested at 4% effective interest per year. Assuming that they plan for the position to have a base salary of S with 3% increases per year indefinitely, what is S?

1.4 Loans

Several examples in the last section indicated that you can think of a loan as an annuity of payments to the lender, and thereby solve for such important quantities as the per period payment required to pay off the loan in equal installments, or the effect of changing interest rates on the term of the loan. In this section we want to shine the spotlight directly on loan problems, utlizing much of the same financial reasoning as we applied to annuities. An additional quantity of interest with loans is the **outstanding balance** remaining to pay off at any time, and we will derive two equivalent formulas for this. We will also look at a means called a **sinking fund** for paying off a debt by paying interest only until the end, and setting up an annuity in order to pay off the initial principal at the loan's due date.

1.4.1 Loan Payments

Our main interest is the **amortization** of a loan, that is, the scheduling of a regular series of payments which will offset the principal plus interest of a loan in a given length of time. The first question is: What payment is necessary?

Let L be the amount of the loan, let n be its duration (in periods), let r be the per period interest rate, and let P be the regular payment, which can be made at the end or the beginning of the period depending on the loan agreement. View the repayment of the loan as an annuity of payments to the lender whose present value is L. Since

$$
L = \begin{cases} P \cdot a_{n\rceil r} = P \cdot \dfrac{1-(1+r)^{-n}}{r} & \text{end of period} \\[2mm] (1+r)P \cdot a_{n\rceil r} = (1+r)P \cdot \dfrac{1-(1+r)^{-n}}{r} & \text{beginning of period} \end{cases}
$$

the regular payment would be:

$$
P = \begin{cases} \dfrac{L}{a_{n\rceil r}} = L \cdot \dfrac{r}{1-(1+r)^{-n}} & \text{end of period} \\[2mm] \dfrac{L}{(1+r) \cdot a_{n\rceil r}} = L \cdot \dfrac{r}{(1+r)\left(1-(1+r)^{-n}\right)} & \text{beginning of period} \end{cases} \tag{1.51}
$$

Remark. If the loan is taken on for t years at a nominal yearly rate of y convertible m times per year (typically monthly), then $r = y/m$ and $n = mt$ in the formulas above.

Another way of arriving at the same formula is by setting up an equation of future values. The institution granting the loan should be indifferent between just keeping L dollars and earning compound interest on that money with rate r for n periods, or issuing the loan to the individual, and consequently receiving payments of P every period from the borrower. Then in the annuity immediate case, equating future values gives the equation:

$$L(1+r)^n = P\frac{(1+r)^n - 1}{r}$$

$$\implies P = L \cdot \frac{r(1+r)^n}{(1+r)^n - 1}. \tag{1.52}$$

This is equivalent to the form above in (1.51), as you can see by dividing the numerator and denominator by $(1+r)^n$.

For example, an \$80,000 30-year mortgage on a house at nominal yearly interest rate 7.5% would have monthly payments:

$$P = \$80,000 \cdot \frac{\frac{.075}{12}\left(1 + \frac{.075}{12}\right)^{30 \cdot 12}}{\left(1 + \frac{.075}{12}\right)^{30 \cdot 12} - 1} = \$559.37.$$

Example 1. A college borrows \$2 million on a bond issue, which is due to be paid back with interest in 20 years. How much should they pay at the end of each year to do this, if they can get a 7% effective annual interest rate? How much total interest is paid? If they are able to repay \$.5 million each year, in how many years can the bond be retired?

Solution. To answer the first question, note that $L = \$2$ million, $r = .07, n = 20$, hence the yearly payment is:

$$P = \frac{L}{a_{\overline{n}|r}} = \$2 \text{ million} \cdot \frac{.07}{1 - (1.07)^{-20}} = \$188,786.$$

Then the total amount paid is $20 \cdot \$188,786 = \$3,775720$. The difference below is the interest paid by the college:

$$20 \cdot \$188,786 - \$2,000,000 = \$1,775,720.$$

The final question asks for the number of periods such that an ordinary annuity with payments of \$.5 million retires a debt of \$2 million. The period must be long enough so that the present value of the payments is at least \$2 million. So we want the smallest value of n that satisfies $2 \leq .5 \cdot \frac{1 - (1.07)^{-n}}{.07}$. Solving for n we obtain:

$$\frac{2(.07)}{.5} \leq 1 - (1.07)^{-n} \implies (1.07)^{-n} \leq 1 - \frac{2(.07)}{.5}$$

$$\implies n \geq -\frac{\log\left(1 - \frac{2(.07)}{.5}\right)}{\log(1.07)} = 4.85532.$$

So five periods are required, and probably the last payment is less than $.5 million. When we see how to compute the outstanding balance at any time, we will be able to find out what the last payment is. ∎

Remark. As Example 1 indicated, the total interest paid on a loan is always the difference between the total amount of the payments and the initial value of the loan:

$$\text{Total interest paid} = nP - L. \tag{1.53}$$

Example 2. A family currently has \$50,000 left on a mortgage that has 12 years remaining. Their interest rate was 8.5% nominal per year, and now they can get a new mortgage for 7.75% nominal, but there would be \$1000 of closing costs to add to the loan. Should they take out the new mortgage? If not, suppose they were willing to go to a 15-year mortgage. Would their monthly payments be reduced significantly?

Solution. Their current monthly payment must be:

$$P = \frac{\$50,000}{a_{\overline{12 \cdot 12}|.085/12}} = \$555.03.$$

If they renegotiate their loan keeping the same term, then, because the closing costs are added on, their monthly payment would change to:

$$P = \frac{\$51,000}{a_{\overline{12 \cdot 12}|.0775/12}} = \$545.08.$$

Since their monthly payments go down by about \$10, there is no reason not to change, but they may be reluctant to go to the trouble to renegotiate for such little cash flow benefit. If they do choose to lengthen the term to 15 years, their monthly payment goes down to:

$$P = \frac{\$51,000}{a_{\overline{15 \cdot 12}|.0775/12}} = \$480.05.$$

This is a much more substantial monthly saving of about \$75. They should weigh the short-term benefit of reducing their current expenses against the longer term cost of additional interest when considering the 12-year term vs. the 15-year term. Assuming they do renegotiate, the total prices paid for the two plans are:

$$12 - \text{year} : 12 \cdot 12 \cdot 545.08 = \$78,491.50$$

$$15 - \text{year} : 15 \cdot 12 \cdot 480.05 = \$86,409.$$

Nearly $8000 of extra interest is paid over the course of the 15-year loan, which could cause them to decide not to take this route. ∎

1.4.2   Loan Amortization

The loan amortization process is often summarized on an **amortization table**, which tabulates the current balance of the loan, balance plus new interest, payment, and final balance for each time period. The new interest is r times the current balance, and the final balance is the current balance plus interest minus the payment. Then the amount of principal repaid in a period is the monthly payment P minus the interest in that period, which increases as the periods go by, since the balance and interest decrease. The table in Figure 1.10 shows the first 20 payment periods for a loan of initial amount $80,000 at a nominal yearly rate of 7.5% convertible monthly, and a term of 30 years or 360 months, which was the data used in the earlier example on a home mortgage.

Balance	Interest	New Balance	Payment	Final Balance
80000.	500.	80500.	559.37	79940.6
79940.6	499.63	80440.3	559.37	79880.9
79880.9	499.26	80380.1	559.37	79820.8
79820.8	498.88	80319.7	559.37	79760.3
79760.3	498.5	80258.8	559.37	79699.4
79699.4	498.12	80197.5	559.37	79638.2
79638.2	497.74	80135.9	559.37	79576.5
79576.5	497.35	80073.9	559.37	79514.5
79514.5	496.97	80011.5	559.37	79452.1
79452.1	496.58	79948.7	559.37	79389.3
79389.3	496.18	79885.5	559.37	79326.1
79326.1	495.79	79821.9	559.37	79262.5
79262.5	495.39	79757.9	559.37	79198.6
79198.6	494.99	79693.5	559.37	79134.2
79134.2	494.59	79628.8	559.37	79069.4
79069.4	494.18	79563.6	559.37	79004.2
79004.2	493.78	79498.	559.37	78938.6
78938.6	493.37	79432.	559.37	78872.6
78872.6	492.95	79365.6	559.37	78806.2
78806.2	492.54	79298.7	559.37	78739.3

Figure 1.10: Loan amortization table for $80,000, 30-year loan at 7.5%.

To see the acceleration in the paydown of a loan's principal as time progresses, consider the shorter 10-period loan in Figure 1.11, in which $3000 is borrowed for a period of 10 years with yearly payments of $388.51 at the end of the year at effective rate 5% per year. During the first few periods, the amount of interest paid goes down by about $12, then by about $12.50, then by about $13.00, etc., so the decline in interest paid is faster than linear. Since payments are constant and the residual after interest is paid is applied to the principal, the part of the yearly payment applied to the principal increases at a rate faster than linear. Figure 1.12 contains a point graph of principal and interest paid as a function of time, in which the phenomena that we noted are apparent.

Balance	Interest	New Balance	Payment	Final Balance
3000.	150.	3150.	388.51	2761.49
2761.49	138.07	2899.56	388.51	2511.05
2511.05	125.55	2636.6	388.51	2248.09
2248.09	112.4	2360.49	388.51	1971.98
1971.98	98.6	2070.57	388.51	1682.06
1682.06	84.1	1766.16	388.51	1377.65
1377.65	68.88	1446.53	388.51	1058.02
1058.02	52.9	1110.92	388.51	722.41
722.41	36.12	758.53	388.51	370.01
370.01	18.5	388.51	388.51	0.

Figure 1.11: Interest declines super-linearly with time, accelerating payoff.

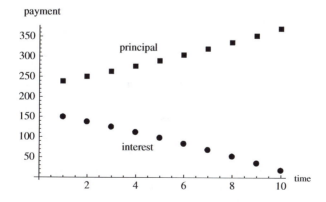

Figure 1.12: Interest paid decreases as principal paid increases.

In Figure 1.13 we show how an amortization table can be implemented in a spreadsheet. (Details of formula syntax may vary by software; these formulas are in standard Excel syntax.) In this display we have the problem parameters in an area in the upper left of the sheet, so that the user enters the initial loan balance, the nominal yearly rate, the number of compoundings per year, and the term of the loan in cells B1, B2, B3, and B5, respectively. Cell B4 is a computed cell that computes the effective interest rate per period. Cell C5 implements the formula for $a_{\overline{n}|r}$, and then cell B6 computes the monthly payment using formula (1.51). The amortization table itself begins in row 8 and extends from columns A through E. After a header line in row 8, the initial balance is copied from cell B1 to cell B9. The interest for the first period is the current balance in cell B9 times the rate in cell B4. The payment is copied from cell B6 to cell D9, and the resulting balance after the payment is computed in cell E9. Then in row 10, the period is incremented in cell A10 and the previous final balance E9 is used as the initial balance in cell B10. But now the use of the $ operator for direct addressing should allow you to copy the formulas from cells C9 through E9 and paste down to cells C10 through E10, and cell references should be adjusted automatically. In fact, line 10 can now be copied all at once to lines 11, 12, etc., to form the rest of the amortization table.

	A	B	C	D	E
1	initial balance	B			
2	yearly rate	y			
3	$\frac{\text{comps}}{\text{year}}$	m			
4	$\frac{\text{rate}}{\text{period}}$	$= \frac{\$B\$2}{\$B\$3}$			
5	term	n	$= \frac{(1-(1+\$B\$4)^{(-\$B\$5)})}{\$B\$4}$		
6	payment	$= \frac{\$B\$1}{\$C\$5}$			
7					
8	period	initial balance	interest	payment	final balance
9	1	$= \$B\$1$	$= B9 * \$B\$4$	$= \$B\$6$	$= B9 + C9 - D9$
10	$= A9 + 1$	$= E9$	$= B10 * \$B\$4$	$= \$B\$6$	$= B10 + C10 - D10$
11	$\vdots$	$\vdots$	$\vdots$	$\vdots$	$\vdots$

Figure 1.13: Template loan amortization table spreadsheet.

In the past we have been viewing annuities and loans globally over their whole lifetime, but understanding amortization tables encourages us to think in a useful way about the repayment of a loan one period at a time. There is a current balance or principal of the loan, and interest is added to that, after which a payment is subtracted to produce the balance for the next period. We see such thinking frequently in daily life, as the next example shows.

Example 3. A credit card company produces a monthly bill for one of its card-holders with the following information:

Previous Balance:	$15.07
Payments:	$15.07
Other Credits:	$0.00
Purchases, Balance Transfers, & Other Charges:	$57.00
Interest Charged:	$0.65
New Balance:	$57.65
Minimum Payment:	$3.25

Infer the company's nominal yearly rate of interest on this account, and its policy with regard to the determination of the minimum payment. If the minimum payment was met and there were no further charges the next month, what would the next bill say?

Solution. The information about the previous bill is not relevant for these questions; apparently there had been a balance of just over $15 which was paid off in full. The rest is similar to a line of an amortization table. But the fact that there was a debt of $57.00 incurred this month and the interest charged was $0.65 allows us to conclude that the monthly effective rate was $.65/57 = .0114$, hence the nominal yearly rate must be $.0114 \cdot 12 = .1368 = 13.68\%$.

We would suppose that the minimum payment would be a percentage of the outstanding debt; if so, the company's policy must be that the minimum payment per month is $3.25/57.65 = .0564 = 5.64\%$ of the debt.

If only the $3.25 minimum payment was applied, and no further charges were incurred, then before interest the new balance would be $57.65 - $3.25 = 54.40, and interest at the rate of 1.14% would be charged on that balance, which is $54.40 \cdot .0114 = 0.62. That would bring the balance to $54.40 + $0.62 = 55.02. The minimum payment on that would equal $55.02 \cdot .0564 = 3.10, hence the next bill would look like:

Previous Balance:	$57.65
Payments:	$3.25
Other Credits :	$0.00
Purchases, Balance Transfers, & Other Charges:	$0.00
Interest Charged:	$0.62
New Balance:	$55.02
Minimum Payment:	$3.10 ∎

1.4.3 Retrospective and Prospective Forms for Outstanding Balance

It would be helpful not to be forced to produce a complete amortization table in order to answer such questions as (1) What is the outstanding loan balance at a given time? and (2) How much interest and principal are paid in one or more particular periods? It turns out that the second question can be answered if we can answer the first. There are two ways of thinking that enable us to find the amount of the balance still outstanding at the end of period k, which we will denote by OB_k. We use the notation above for the other problem parameters: L for the initial amount of the loan, r for the per period rate, n for the term of the loan, and P for the payment at the end of each period.

In the **retrospective form** for outstanding balance, we think back to time 0 and find the future value at k of the initial balance $OB_0 = L$ less the total value at k of all the payments that have been made so far. To see the reasoning, suppose that the loan is taken out at time 0, and, as the payments come in, the lender puts them in an account with the same compound interest specifications as the loan. At the end of period k, the lender compares the amount that the loan of value L has accrued to, namely, $L(1 + r)^k$, to the future value at time k of the annuity that he has received from the borrower, namely, $P \cdot s_{k\rceil r}$. The difference is the outstanding balance.

So we have proved the following.

Theorem 1. Consider a loan of L for n periods with rate r per period and payments of P made at the end of each period. The outstanding balance at the end of period k is:

$$\text{retrospective form of } OB_k = L(1 + r)^k - P \cdot s_{k\rceil r}$$

$$= L(1 + r)^k - P \cdot \frac{(1 + r)^k - 1}{r}. \tag{1.54}$$

The other way of thinking about the outstanding balance is to view the outstanding balance at time k as the present value (at that time) of all future payments. The lender should be indifferent between accepting a lump sum payoff of the outstanding balance, and receiving an annuity of payments of P starting at the end of month $k + 1$ and extending until time n. (Note that, at period k, there are $n - k$ remaining periods.) The corresponding equation of value produces the **prospective form** of the outstanding balance listed in the next theorem.

Theorem 2. Consider a loan of L for n periods with rate r per period and payments of P made at the end of each period. The outstanding balance at the end of period k is:

$$\text{prospective form of } OB_k = P \cdot a_{n-k\rceil r} = P \cdot \frac{1 - (1 + r)^{-(n-k)}}{r}. \tag{1.55}$$

In Exercise 15 you are asked to develop similar formulas for retrospective and prospective forms in the case that payments are made at the beginning of the month. Also in Exercise 16 you will show that the two formulas are algebraically equivalent.

Example 4. Recall the 30-year loan of $80,000 at nominal rate 7.5% convertible monthly in our earlier example. The outstanding balance at the end of the 15th year is:

$$P \cdot a_{n-k]r} = \$559.37 \cdot a_{30 \cdot 12 - 15 \cdot 12].075/12}$$

$$= \$559.37 \cdot \frac{1 - \left(1 + \dfrac{.075}{12}\right)^{-(30 \cdot 12 - 15 \cdot 12)}}{\dfrac{.075}{12}}$$

$$= \$60,341.20.$$

With patience or a computer algebra system, you can repeat the computation for each year. The graph of the outstanding balance at the end of the year as a function of year is in Figure 1.14 and you will observe a rather steep decline in later years.

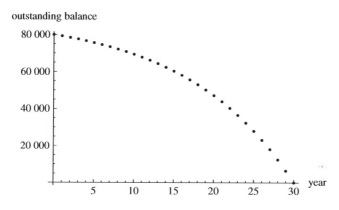

Figure 1.14: End-of-year outstanding balance on $80,000 loan.

Example 5. A 5-year car loan in the amount of $20,000 is taken out at an effective yearly interest rate of 4%, convertible monthly. What is the outstanding balance at the end of the first year? The second year? How much interest is paid in the sixth monthly payment?

Solution. First, the effective monthly (per period) rate r satisfies:

$$(1 + r)^{12} = 1 + .04 \implies r = 1.04^{1/12} - 1 = .00327.$$

The monthly payment is:

$$P = \frac{\$20,000}{a_{60].00327}} = \$367.65.$$

The question asks for the outstanding balance after the 12th payment. Using the retrospective form:

$$OB_{12} = \$20,000(1.00327)^{12} - \$367.65 \cdot s_{12].00327} = \$16,307.10.$$

Similarly, after the 24th payment, the outstanding balance is:

$$OB_{24} = \$20,000(1.00327)^{24} - \$367.65 \cdot s_{24]\,.00327} = \$12,466.$$

The interest in the 6th payment is the difference between the monthly payment $P = \$367.65$ and the reduction in principal of the loan, that is, the outstanding balance change for the time period $[5,6]$:

$$
\begin{aligned}
\text{interest in } 6^{\text{th}} \text{ payment} \;=\;& P - (OB_5 - OB_6) \\
=\;& \$367.65 \\
& -((\$20,000(1.00327)^5 - \$367.65 \cdot s_{5]\,.00327}) \\
& -(\$20,000(1.00327)^6 - \$367.65 \cdot s_{6]\,.00327})) \\
=\;& \$60.43.
\end{aligned}
$$

Although the question did not specifically ask, the principal paid in the 6th payment is:

$$
\begin{aligned}
\text{principal in } 6^{\text{th}} \text{ payment} \;=\;& (OB_5 - OB_6) \\
=\;& ((\$20,000(1.00327)^5 - \$367.65 \cdot s_{5]\,.00327}) \\
& -(\$20,000(1.00327)^6 - \$367.65 \cdot s_{6]\,.00327})) \\
=\;& \$307.22,
\end{aligned}
$$

which of course is also the difference between the monthly payment of $367.65 and the interest payment of $60.43. ■

1.4.4   Sinking Fund Loan Repayment

Some loans are not amortized and paid off in regular installments. In this subsection, we look at a case where a borrower and a lender agree for a loan of amount L to be repaid by paying only the interest (at per period rate r) at the end of each period, and the original principal L is repaid in a lump at the end. A *sinking fund* is an annuity set up by the borrower for himself (at possibly a different per-period rate j) to make the interest payments, and accumulate to the principal by the end of the loan in order to pay off the loan principal.

What would be the total outlay per period for the borrower? It must cover the interest payment Lr, plus the annuity payment $L/s_{n]j}$ required for the annuity to be worth $\$L$ after n periods. Hence, we have:

$$\text{sinking fund payment} = Lr + \frac{L}{s_{n]j}}. \tag{1.56}$$

Example 6. A $10,000 loan is to be repaid by five annual payments beginning 1 year after the loan is made. The lender demands interest payments only at effective yearly rate 8%, then repayment of the original $10,000 at the end of the 5 years. The borrower sets up a sinking fund, earning interest at 6% effective per year to cover the debt. (a) What is the total annual outlay, including interest payments to the lender and deposits to the sinking fund? (b) Compare to the amortization method of repaying the loan; with the same term and the same annual outlay, what interest rate r is implied?

Solution. (a) Reasoning as above, the annual outlay is:

$$\text{sinking fund payment} = Lr + \frac{L}{s_{\overline{n}|j}} = \$10,000(.08) + \frac{\$10,000}{s_{\overline{5}|.06}} = \$2573.96.$$

(b) If the $2573.96 were the annual payment direct to the lender under amortization, we can set up an equation of value for the implied amortization rate r by:

$$\$10,000 = \$2573.96 \cdot a_{\overline{5}|r} = \$2573.96 \cdot \frac{1 - (1+r)^{-5}}{r} \implies r \approx .09.$$

The solution of the equation needs numerical methods. The implication is that setting up the loan this way is advantageous to the lender; the sinking fund payments actually imply an effective per period interest rate of about 9% as opposed to the stated 8%. The main reason for this is that, since the borrower is getting a lower rate of $j = 6\%$ on his sinking fund savings, he must pay more in order to bring his future value up to $10,000. Ultimately, the lender receives all of this money. Had the $10,000 loan been amortized at 8%, the payments by the borrower would have been about $70 less:

$$P = \frac{\$10,000}{a_{\overline{5}|.08}} = \$2504.56.$$

(Exercise 23 asks under what conditions the two forms of loan repayment are equivalent.) ∎

Example 7. A *sinking fund schedule* is a table in which is displayed the per-period interest and sinking fund deposit, the interest and balance in the sinking fund, and the balance of the loan principle, that is, the initial loan amount minus the sinking fund balance. It is easy to implement such a schedule of sinking fund payments and balances on a spreadsheet. Consider a $5000 loan repayed by a sinking fund for 10 years at effective interest rate 10% per year, with a sinking fund interest rate of 7% per year. The sinking fund payments are:

$$Lr + \frac{L}{s_{\overline{n}|j}} = \$5000(.10) + \frac{\$5000}{s_{\overline{10}|.07}} = \$861.89.$$

The sinking fund schedule is a table that looks like the one in Figure 1.15. The first line is the initial situation at time 0. Payments of interest and payments by the borrower to the sinking fund occur at the end of each interval, so at time 1 there is interest on the loan to pay of $5000(.10) = \$500$, the first sinking fund deposit comes in but has earned no interest yet, the balance in the sinking fund is the $861.89 deposit minus the $500 that goes to the loan interest, and the net balance of the loan is the original loan balance of $5000 minus the current sinking fund balance of $361.89. In general, the loan interest and sinking fund deposit columns are constantly $500 and $861.89; the sinking fund interest is .07 times the previous sinking fund balance; the new sinking fund balance is the old sinking fund balance plus the sinking fund interest plus the difference $361.89 between the sinking fund deposit and the loan interest; and the net loan balance is $5000 minus the sinking fund balance.

Years	Loan Interest	Sink Fund Payment	Sink Fund Interest	Sink Fund Balance	Net Loan Balance
0	0	0	0	0	5000
1	500.	861.89	0.	361.89	4638.11
2	500.	861.89	25.33	749.11	4250.89
3	500.	861.89	52.44	1163.44	3836.56
4	500.	861.89	81.44	1606.77	3393.23
5	500.	861.89	112.47	2081.13	2918.87
6	500.	861.89	145.68	2588.7	2411.3
7	500.	861.89	181.21	3131.8	1868.2
8	500.	861.89	219.23	3712.92	1287.08
9	500.	861.89	259.9	4334.71	665.29
10	500.	861.89	303.43	5000.03	−0.03

Figure 1.15: Sinking fund schedule for a $5000 loan, 10% loan interest, 7% sinking fund interest.

As with amortization schedules, spreadsheets are well suited for producing sinking fund schedules. A template for a sinking fund schedule in a spreadsheet is shown in Figure 1.16. The problem parameters are input by the user into the upper left block of cells: the initial loan amount L, the per-period rate on the loan r (this time we assume that the user has already calculated this), the per-period sinking fund interest rate j, and the term of the loan n. The actuarial function $s_{n]j}$ is computed in cell C4 and used to find the total loan interest plus sinking fund deposit as in (1.56) in cell B5. After header and initialization information in rows 7 and 8, we have two constant columns for the loan interest payments $L \cdot r$ from cells B1 and B2 and the total sinking fund payment in cell B5. In cell D9 the sinking fund hasn't had a chance to earn interest yet, but in general the contents of a column D entry are the interest rate from cell B3 times the previous balance in the sinking fund in column E in the row above (see the formula in cell D10). The sinking fund balance in column E will always be the entering balance from the row above in column E, plus interest from column D, plus the difference between the payment in column C and the loan interest in column B for that row, which explains the formulas in cells E9 and E10. In column F, the current loan balance is the original balance from cell B1 minus the current sinking fund balance from column E. All of the formulas in row 10 can now be copied and pasted into subsequent rows to complete the table.

	A	B	C	D	E	F
1	init balance	L				
2	loan rate	r				
3	sink fund rate	j				
4	term	n	$\frac{(1+\$B\$3)^{\$B\$4}-1}{\$B\$3}$			
5	payment	$\$B\$1 * \$B\$2$ $+\$B\$1/\$C\$4$				
6						
7	period	loan interest	sink fund payment	sink fund interest	sink fund balance	loan balance
8	0	0	0	0	0	$= \$B\$1$
9	A8 + 1	$\$B\$1 * \$B\$2$	$\$B\$5$	0	E8 + D9 +C9 − B9	$\$B\$1 − E9$
10	A9 + 1	$\$B\$1 * \$B\$2$	$\$B\$5$	$\$B\$3 * E9$	E9 + D10 +C10 − B10	$\$B\$1 − E10$
11	⋮	⋮	⋮	⋮	⋮	⋮

Figure 1.16: Template sinking fund schedule spreadsheet.

Example 8. A sinking fund is set up to pay off a debt of $10,000 by paying interest only at a nominal yearly rate of 6% convertible monthly. The sinking fund can earn interest at 7% nominal yearly, also convertible monthly. It is desired to retire this debt in 5 years. At what point in time could half of the original debt be paid off?

Solution. Each month, interest of $10,000(.06/12) = \$50$ must be paid, and to pay off the principal in 5 years, or 60 months, a sinking fund deposit of

$$\frac{L}{s_{\overline{n}|j}} = \frac{\$10,000}{s_{\overline{60}|.07/12}} = \$139.68$$

must be made. This gives us an annuity future value problem, in which we would like to solve for the month m such that an annuity with a $139.68 monthly payment at the end of each month reaches at least a value of $5000, with monthly effective interest rate .07/12. The inequality is:

$$\$139.68 \cdot s_{\overline{m}|.07/12} \ \geq \ \$5000$$

$$\Longrightarrow s_{\overline{m}|.07/12} \ \geq \ \frac{\$5000}{\$139.68}$$

$$\Longrightarrow \frac{(1+.07/12)^m - 1}{.07/12} \ \geq \ \frac{\$5000}{\$139.68}$$

$$\Longrightarrow (1+.07/12)^m - 1 \ \geq \ \frac{(.07)5000}{12(139.68)}$$

$$\Longrightarrow m \ \geq \ \frac{\log\left(1 + \dfrac{(.07)5000}{12(139.68)}\right)}{\log(1+.07/12)} = 32.6039.$$

So it will need 33 months for the sinking fund to build up to at least half of the original debt. ■

Important Terms and Concepts

Amortization of a loan - A process of repayment of a loan in which in each period is a payment that offsets the interest for that period and reduces the principal by the amount of the payment that remains after paying interest.

Amortization table - A table that gives initial balances, interest, payments, and final balances for each period in an amortized loan.

Outstanding balance - At any point in time in an amortized loan, the amount remaining to be paid.

Retrospective form - An expression for the outstanding balance of a loan obtained by finding the future value of the principal of the loan at that time, minus the total future values of the loan payments to that time.

Prospective form - An expression for the outstanding balance of a loan obtained by finding the present value at that time of all future payments.

Sinking fund - A loan repayment strategy in which only interest is paid each period, but an additional amount is deposited into an annuity, not necessarily at the same rate of interest, set up to pay the principal of the loan in full at the end of the loan.

Sinking fund schedule - A table in which the loan interest, payment, sinking fund deposit, sinking fund balance, and loan amount minus sinking fund balance are displayed for each period.

Exercises 1.4

1. What nominal yearly interest rate is necessary to keep the monthly payments under $400 for a 5-year auto loan of $20,000 compounded monthly? (Assume end of the month payments. You will need to be able to solve an equation of value numerically.)

2. For the 5-year $20,000 auto loan in Exercise 1, suppose that the nominal yearly rate of interest is 6.99% compounded monthly, the car is purchased on December 15, and the dealer allows the purchaser to make beginning of the month payments with the first due on February 1. Find the monthly payments.

3. A family owes $200,000 on a home mortgage, which has 25 years remaining on the loan and a nominal interest rate of 6% per year. By how much is the monthly payment reduced if the interest rate declines to 5% nominal per year? Assume end-of-the-month payments. How much total interest is saved in the process?

4. Suppose that a credit card company has a policy that states that the minimum payment on a balance is the larger of $10 and a percentage q of the outstanding balance. David receives a bill with the following information and pays $100 on it. If he makes no purchases in the next month, what will next month's bill say?

Previous Balance:	$234.90
Payments:	$20
Other Credits:	$0.00
Purchases, Balance Transfers, & Other Charges:	$8.00
Interest Charged:	$2.72
New Balance:	$225.62
Minimum Payment:	$12

5. In the section we computed that an $80,000 30-year mortgage on a house at nominal yearly interest rate 7.5% would have monthly payments of $559.37. Write a function that gives the monthly payment in terms of the nominal yearly interest rate. What rate is necessary to bring the payments down to $530? (You will need to solve the appropriate equation numerically.)

6. How much interest is saved if a 30-year mortgage taken out for $100,000 at nominal rate 8% per year payable at the end of each month is renegotiated after the 6th year of the mortgage to a rate of 7% and a term of 20 years?

7. Use a spreadsheet to make an amortization table for a 5-year auto loan of $20,000 at 8% compounded monthly.

8. Compute without the use of a spreadsheet or computer algebra system the first three lines of an amortization table for the following loan: $10,000 for 36

months at nominal rate 5% per year convertible monthly, with payments scheduled at the end of the month.

9. Given the first two lines of the amortization table shown below, in how many
time periods is the loan paid off?

period	balance	interest	new balance	payment	final balance
1	$25,000	$125	$25,125	$587.13	$24,537.87
2	$24,537.87	$122.69	$24,660.56	$587.13	$24,073.43
$\vdots$	$\vdots$	$\vdots$	$\vdots$	$\vdots$	$\vdots$

10. Ward and June take out a 30-year mortgage for $100,000 payable monthly
at the end of the month with nominal yearly interest rate 6.25%. What are their
monthly payments? If they pay an additional $100 per month every month, how
soon can they pay off the mortgage?

11. How much interest is paid during the entire first year of the loan in Example
5?

12. A student loan originally for $45,000 is due to be paid off in equal end-of-
month installments over 10 years. What are those payments, assuming interest
at a nominal yearly rate of 4% compounded monthly? How much of the principal
is paid off in the first 5 years? After the 5th year through the 7th year?

13. Sid borrows $5000 from his local bank at a nominal rate of 5% annually,
with payments at the end of each month. The term of the loan is 5 years. Find
the balance of the loan at the end of the first year and the total interest paid
during the first year.

14. A car loan of $20,000 is taken out for 6 years at 3% nominal yearly interest
compounded monthly, with payments at the end of each month. What is the
balance of the loan at the 3-year point? How much interest and how much loan
principal is paid during the 4th year?

15. With beginning of the month payments, find both prospective and retrospective formulas for the outstanding balance at the end of each month (i.e., before
the next payment).

16. Show algebraically that the retrospective and prospective forms for the
outstanding balance of a loan are equivalent, i.e., show that:

$$L(1+r)^k - P \cdot s_{k\rceil r} = P \cdot a_{n-k\rceil r}.$$

17. MaryAnn can buy a new washer-dryer combo for $2500 at her local big box
home improvement store. She can use store credit to buy the machines for no

payments and no interest for 1 year, but she must pay the balance by the end of the year or else monthly interest at nominal yearly rate 22% will be applied retroactively. But she could also charge the machines on a bank credit card whose nominal yearly rate is 14.9%. What would her monthly payments be if she chooses the second alternative and does pay off the debt by the end of the year? Under the first alternative, what would she owe at the end of the year if she cannot pay the $2500? If there is an even chance that she will or will not be able to afford to pay the $2500 at the end of the year, which payment plan is better for her?

18. Dennis has a $15,000 bill for the replacement of his roof. The roofing company is letting him backload his repayments over a period of 3 years, paying each month $20 more than the month before. Assuming end-of-month payments and a nominal yearly interest rate of 10% compounded monthly, what are his first three payments?

19. Repeat the previous problem about Dennis and his $15,000 roof, except that each month he pays 4% more than the previous month.

20. Wilma agrees to lend Fred $150 for a fancy new bowling ball. Wilma will accept weekly payments of interest only (at the end of each week), at nominal yearly interest rate 6%, as long as the original balance of the loan will be paid off in a year. How much will Fred have to pay each week if he can invest money with nominal yearly rate 4% compounding weekly? Including interest, how much did his bowling ball really cost him?

21. A college borrows $2,000,000 for a capital project, with the full balance due at the end of 5 years, and interest payments at an effective yearly rate of 6% due at the end of each year. If the college will set aside funds in an interest bearing account at the end of each year with effective yearly rate 4% to provide for the payment of the $2,000,000 balance in 5 years, what is the total that it pays, including the interest payments, at the end of each year?

22. Use a spreadsheet to make a sinking fund schedule for the problem in Example 6.

23. Under what conditions is the payment that a borrower must make for a sinking fund loan the same as for a standard amortized loan?

24. In this section we have mainly considered the case where the interest rates and payment amounts are constant, but these assumptions can be relaxed. Suppose that an amount L is borrowed, the interest rate for period k is r_k, and the payment amount at the end of period k is P_k. Let B_k denote the balance of the loan at time k after the payment has been made.

(a) Construct and justify a recursive equation for B_k in terms of B_{k-1}.

(b) Suppose that a 1-year loan of $1000 is made with payments scheduled at the end of each quarter. If the nominal yearly rates are 10%, 8%, 6%, and 8% in the four quarters, and payments of $250, $300, and $275 are made in the first three quarters, what payment in the last quarter finishes paying off the loan?

1.5   Measuring Rate of Return

A financial transaction can involve a series of payments, which may be deposits or withdrawals, investments or consumptions. We will denote these payments by $P_0, P_1, P_2, \ldots, P_n$, made, respectively, at times $0, 1, 2, \ldots, n$. By convention, a deposit is a positive contribution $P_j > 0$ and a withdrawal is negative, and also if no payment is made at time j, then P_j is taken to be 0. Let B denote the final balance, if any, of the investment at time n.

To give a measure of the performance of a transaction, or to compare two transactions to each other, we need a notion of rate of return that is more general than we had earlier. We will illustrate three such measures in this section: the *internal rate of return*, the *approximate dollar-weighted rate of return*, and the *time-weighted rate of return*.

1.5.1   Internal Rate of Return on a Transaction

Earlier we looked at rates of return on transactions involving a single payment and two times, "now" and "later," and the rate of return r satisfied the equation:

$$r = \frac{\text{value later} - \text{value now}}{\text{value now}}$$

$$\implies \text{value later} = (1 + r) \cdot \text{value now}$$

$$\implies \text{value now} = (1 + r)^{-1} \cdot \text{value later}. \tag{1.57}$$

Hence the rate of return was the interest rate that made the present value of the final balance of our investment equal to the amount we invested at the start. One straightforward way of generalizing the idea to the case where there are multiple payments, some investments, and some consumptions is to say that the rate of return is the rate such that the present value of the final balance equals the net present value of all earlier payments. This observation gives rise to the following definition.

Definition 1. The *internal rate of return* on the transaction is the interest rate i using which the net present value of all amounts invested P_j equals the present value of the final balance B, that is:

$$P_0 + P_1(1 + i)^{-1} + \cdots + P_n(1 + i)^{-n} = B(1 + i)^{-n}. \tag{1.58}$$

To see that this new version of the rate of return idea actually specializes to what we know about in simpler cases, consider the next two examples.

Example 1. In a straightforward problem of compound interest, in which P_0 is invested at effective rate r per period and allowed to accrue interest for n periods, what is the internal rate of return?

Solution. The final balance is $B = (1+r)^n P_0$. In the notation of this section, there is only one positive payment P_0, so Equation (1.58) reduces to:

$$P_0 + P_1(1+i)^{-1} + \cdots + P_n(1+i)^{-n} = B(1+i)^{-n}$$
$$\Longrightarrow P_0 = B(1+i)^{-n} = (1+r)^n P_0 \cdot (1+i)^{-n}.$$

Thus,

$$1 = (1+r)^n(1+i)^{-n} \Longrightarrow (1+r)^n = (1+i)^n.$$

For non-negative values of x, the function x^n is 1-1, so under the restriction that $i \geq -1$, the last equation implies $1 + r = 1 + i$, which means that the internal rate of return i is just equal to the effective rate r. ∎

Example 2. Audrey pays \$100 per month into an ordinary annuity with effective monthly rate $r = .005$ for 25 years. Find Audrey's internal rate of return.

Solution. For this annuity the future value at time $25 \cdot 12 = 300$ months is

$$B = \$100\frac{(1.005)^{300} - 1}{.005} = \$69,299.40.$$

The internal rate of return i satisfies the equation:

$$\$100(1+i)^{-1} + \cdots + \$100(1+i)^{-300} = B(1+i)^{-300}$$
$$\$100(1+i)^{-1} \cdot \frac{\left((1+i)^{-1}\right)^{300} - 1}{(1+i)^{-1} - 1} = B(1+i)^{-300}$$
$$\$100(1+i)^{-1} \cdot \frac{1 - (1+i)^{300}}{(1+i)^{-1} - 1} = B$$
$$\$100 \cdot \frac{(1+i)^{300} - 1}{i} = \$100\frac{(1.005)^{300} - 1}{.005}$$
$$\frac{(1+i)^{300} - 1}{i} = \frac{(1.005)^{300} - 1}{.005}.$$

The value $i = .005$, which is the same as the monthly effective rate, is certainly a solution of the last, revised equation. It is not obvious that the function $f(x) = ((1+x)^n - 1)/x$ is 1-1 on a reasonable domain for x (such as the non-negative reals in $[0,1]$), which would be enough to show that $i = .005$ is the

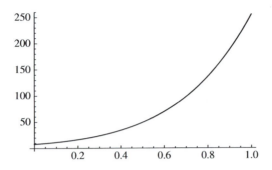

Figure 1.17: The function $f(x) = \big((1+x)^8 - 1\big)\big/x$.

unique internal rate of return. The graph in Figure 1.17 for the case $n = 8$ does seem to have the proper behavior.

We can deduce this fact analytically by recalling that the binomial theorem allows us to write:

$$(1+x)^n = 1 + \sum_{k=1}^{n} \binom{n}{k} x^k 1^{n-k} \implies f(x) = \frac{(1+x)^n - 1}{x} = \sum_{k=1}^{n} \binom{n}{k} x^{k-1}.$$

This is a polynomial with all positive coefficients, which clearly has a positive derivative on $(0, 1]$, so f is increasing and hence 1-1. Therefore, $i = r = .005$ is the IRR assuming that i is between 0 and 1. ∎

There was nothing special about the problem data in Example 2. Exercise 6 asks you to show in general that the internal rate of return on an annuity is the same as the effective rate per period.

Remark. In the definition of internal rate of return, we could have replaced "net present value" by "net future value." To see this, multiply both sides of Equation (1.58) by $(1 + i)^n$ to obtain:

$$(1+i)^n \left(P_0 + P_1(1+i)^{-1} + \cdots + P_n(1+i)^{-n} \right) = (1+i)^n \cdot B(1+i)^{-n}.$$

$$\implies P_0(1+i)^n + P_1(1+i)^{n-1} + \cdots + P_n = B \qquad (1.59)$$

The left side is the net future value of the payments at time n, and B is the future final balance. Equation (1.59) can be used as an alternate method for finding an internal rate of return.

The next example illustrates how complex it can be to measure the performance of investments when many payments are involved.

Example 3. Trader Bob begins with $10,000, with which he buys 500 shares of stock priced at $20 per share. One month later he notices that the price per share has gone down to $18, and he purchases another 100 shares, paying $1800. Three months after that, he sells half of his shares, now priced at $21 each. Finally, at the end of a year he sells off all existing shares at $20 each. Did Bob do better with all this manipulation than he could have done if he had invested his original $10,000 in an account bearing 5.5% nominal yearly interest compounded monthly?

Solution. To answer the question we will compare internal rates of return, but we have to adjust for the differences in period length. For the second, safe investment, we know that the IRR is $i_2 = r_2 = .055/12 = .00458$ expressed on a monthly basis. The effective rate of return for a time period of 12 months is:

$$\left(1 + \frac{.055}{12}\right)^{12} - 1 = .0564.$$

We will compare this to the complicated stock transaction by computing its IRR, first monthly, then converted to an effective yearly rate.

Clearly $P_0 = \$10,000$. At time 1 he invests $P_1 = \$1800$, obtaining 100 more shares for a total of 600 shares. At time 4, Bob withdraws money through the sale, in the amount $\$21 \cdot 300 = \$6300$, so we set $P_4 = -\$6300$. Finally, at time 12 he receives the proceeds of the sale of 300 shares, for a final balance $B = \$20 \cdot 300 = \$6000$. (Equivalently, you could say that $P_{12} = -\$6000$ and $B = 0$.) The monthly internal rate of return i solves the future value equation:

$$\$10,000(1 + i)^{12} + \$1800(1 + i)^{11} - \$6300(1 + i)^8 - \$6000 = 0.$$

Numerical equation-solving methods are necessary, which yield that the only positive real solution of this equation is $i = .00540$. In yearly terms, this corresponds to an effective rate of return of $(1 + .00540)^{12} - 1 = .0667$, so Bob did do better with his stock trading than he could have done by putting his money into a safe account. ∎

Although the construction of the internal rate of return makes intuitive sense, unfortunately there are examples in which the IRR cannot be solved for at all, or there are multiple possible values. Consider the next example, in which an entity pays at time 0 for the privilege of borrowing at time 1 and then settling the debt at time 2. The rate of return for the lender may not be well determined.

Example 4. Harry knows that he will need to have landscaping done in his yard in 1 year, and that it will cost $3000 to do it. Since he also knows that he won't have that money then, he contracts with a lawn care company to do a down payment of $1000 this year in order for them to do the work next year and give him another year's grace on the repayment of the balance. The company will ask him to repay $2200 at that time. Does the internal rate of return for the company exist, and is it unique? What if the repayment amount were $2300?

Can you say anything about the solution(s) as a function of the final repayment p at time 2?

Solution. From the company's point of view, it is investing $P_0 = -\$1000$ at time 0, $P_1 = \$3000$ at time 1, and $P_2 = -\$2200$ at time 2 with final balance 0, or alternatively, $P_2 = 0$ and the final balance is \$2200. Either way, we arrive at the future value equation for the internal rate of return i:

$$-1000(1+i)^2 + 3000(1+i) = 2200.$$

Dividing by 200 on both sides and letting $x = 1+i$, we have the equation:

$$-5x^2 + 15x = 11 \Longrightarrow 5x^2 - 15x + 11 = 0.$$

For this quadratic equation, the discriminant is $b^2 - 4ac = 225 - 4(5)11 = 5$. There are two solutions:

$$x = (1+i) = \frac{15 + \sqrt{5}}{10} = 1.72361,$$

$$x = (1+i) = \frac{15 - \sqrt{5}}{10} = 1.27639.$$

It is not clear whether we should report $i = 72\%$ or $i = 28\%$ as the IRR. So, the defining equation (1.58) or its future value equivalent (1.59) may not give a unique answer.

If the repayment amount is \$2300 instead, then, using the same approach, we are led to the equation:

$$-1000(1+i)^2 + 3000(1+i) = 2300$$

$$\Longrightarrow -5x^2 + 15x = 11.5$$

$$\Longrightarrow 5x^2 - 15x + 11.5 = 0.$$

But this means that the discriminant is $b^2 - 4ac = 225 - 20 \cdot 11.5 = -5$. So there would be no solution in this case.

The nature of the solution appears to be very sensitive to the final repayment amount. In general, if the final repayment is p, then the equation of value would look like:

$$-1000(1+i)^2 + 3000(1+i) = p$$

$$\Longrightarrow -5x^2 + 15x = \frac{p}{200}$$

$$\Longrightarrow 5x^2 - 15x + \frac{p}{200} = 0,$$

whose discriminant is $b^2 - 4ac = 225 - 20 \cdot \frac{p}{200} = 225 - \frac{p}{10}$. This is positive when $p < \$2250$, it equals 0 when $p = \$2250$, and it is negative when $p > \$2250$.

So we could have two solutions, we could have none, or in a very special case $p = \$2250$ we could have the unique solution:

$$x = 1 + i = \frac{15 + \sqrt{0}}{10} = 1.5,$$

in other words, when $p = \$2250$, the internal rate of return is 50%. ∎

In view of Example 4, the internal rate of return is somewhat flawed as a description of investment performance. But the next theorem redeems it in a more restrictive case in which there are only investments and no consumptions. A more general theorem is in Exercise 9.

Theorem 1. If all payments P_j are greater than or equal to zero, if $P_n \leq B$, and if at least one P_j, $j < n$ is strictly greater than zero, then there is a unique solution $i \geq -1$ to (1.58).

Proof. We choose to look at the equivalent future value form (1.59):

$$P_0(1 + i)^n + P_1(1 + i)^{n-1} + \cdots + P_n = B.$$

Set $x = 1 + i \geq 0$. The equation can then be written as:

$$P_0 x^n + P_1 x^{n-1} + \cdots + P_{n-1} x = B - P_n.$$

Since the coefficients of the polynomial on the left side are non-negative and at least one is positive, the polynomial tends to infinity as $x \longrightarrow \infty$. Also the value of the polynomial at $x = 0$ is 0, and its derivative is positive for $x > 0$, so the function is monotonically increasing and continuous. By the Intermediate Value Theorem, it must pass through the point $B - P_n \geq 0$ for some $x \geq 0$. It can only do so once, by monotonicity, so the solution is unique in the domain $x \geq 0$, or $i \geq -1$. ∎

1.5.2 Approximate Dollar-Weighted Rate of Return

There is another way of measuring investment performance that comes at a cost of an assumption that interest accrues as simple interest over a unit time interval, but has the advantage of allowing us to solve explicitly for a unique value of the rate of return (except in a rare case of division by 0). To set up the idea, let the notations P_k and B be as in the last subsection, where we think of P_0 as the initial balance of an investment and B as the final balance. The internal rate of return, which is also called the ***dollar-weighted rate of return***, satisfies the equation:

$$P_0(1 + i)^n + P_1(1 + i)^{n-1} + \cdots + P_n = B.$$

The terms $(1+i)^k$ in the expression are, of course, compound interest multipliers for rate i to convert the payments into their appropriate future values at time n. But by the binomial theorem,

$$(1+i)^k = 1 + k \cdot i + \binom{k}{2}i^2 + \binom{k}{3}i^3 + \cdots + \binom{k}{k}i^k,$$

so if k is not too large and i is near zero, all terms except $1 + k \cdot i$ on the right are small. (For instance, when $k = 4$, $i = .01$, $\binom{k}{2}i^2 = 6(.01)^2 = .0006$.) We approximate the dollar-weighted rate of return by approximating $(1+i)^k$ by $1 + k \cdot i$, which is the simple interest factor converting a payment into a value k time units into the future. A new equation results, and to keep this approximation distinct from the IRR, we will use a new notation j for the rate of return that satisfies the equation.

Definition 2. The ***approximate dollar-weighted rate of return*** is the simple interest rate j for which the total future value of all net deposits accumulating at simple interest rate j equals the final balance:

$$P_0(1 + n \cdot j) + P_1(1 + (n-1)j) + \cdots + P_{n-1}(1 + 1 \cdot j) + P_n = B. \qquad (1.60)$$

The approximate dollar-weighted rate of return can be written in closed form, which shows immediately that it exists, except in one coincidental case, and is unique.

Theorem 2. The approximate dollar-weighted rate of return is equal to the following, when the denominator does not equal zero:

$$j = \frac{B - \sum\limits_{k=0}^{n} P_k}{\sum\limits_{k=0}^{n} P_k(n-k)} = \frac{I}{\sum\limits_{k=0}^{n} P_k(n-k)}, \qquad (1.61)$$

where $I = B - \sum\limits_{k=0}^{n} P_k$ is the net interest earned during time interval $[0, n]$.

Proof. By Equation (1.60),

$$P_0(1 + n \cdot j) + P_1(1 + (n-1)j) + \cdots + P_{n-1}(1 + 1 \cdot j) + P_n = B$$

$$\implies \sum_{k=0}^{n} P_k + \sum_{k=0}^{n} j \cdot P_k(n-k) = B$$

$$\implies j \cdot \sum_{k=0}^{n} P_k(n-k) = B - \sum_{k=0}^{n} P_k.$$

Dividing both sides of the last equation by $\sum\limits_{k=0}^{n} P_k(n-k)$ yields the first equation in (1.61). ∎

Example 5. Susannah has an investment in a mutual fund that is liquid, so that she sometimes puts more money into the fund and sometimes withdraws money from the fund when she has an emergency. The fund begins at $100,000 on January 1, and the table below lists all of the end-of-month balances for the year. In addition, Susannah made a withdrawal of $3000 on May 1 for some medical expenses, and she made deposits of $2000 each on September 1 and December 1. Find the approximate dollar-weighted rate of return on her mutual fund.

Fund values	
Month	Amount
	$100,000
1/31	$100,520
2/28	$100,980
3/31	$101,110
4/30	$100,860
5/31	$98,440
6/30	$99,080
7/31	$100,350
8/31	$101,000
9/30	$104,870
10/31	$105,120
11/30	$105,330
12/31	$107,100

Solution. The time units are months; at time 0 there is an initial deposit of $P_0 = \$100,000$, and the final balance is $B = \$107,100$. There are only three non-zero payments: $P_4 = -\$3000$, $P_8 = \$2000$, and $P_{11} = \$2000$. Note that the time subscripts are referring to the ends of months, and we treat, say, the end of April and the beginning of May as the same. This labeling strategy allows us to subtract the subscript from 12 in order to find the numbers of months that the payment was active. By formula (1.61), the approximate dollar-weighted rate of return j is:

$$
\begin{aligned}
j &= \frac{B - \sum\limits_{k=0}^{n} P_k}{\sum\limits_{k=0}^{n} P_k(n-k)} \\
&= \frac{\$107,100 - (\$100,000 - \$3000 + \$2000 + \$2000)}{\$100,000 \cdot 12 - \$3000 \cdot 8 + \$2000 \cdot 4 + \$2000 \cdot 1} = .00514.
\end{aligned}
$$

So the return on a monthly basis was about .514%, and multiplying by 12 gives about 6.17% expressed as a yearly rate of return. It should be clear to you why the expression in the numerator above represents interest earned by the fund, correcting Susannah's account activity. It is the difference between the final and initial balance, less the net deposits into the account of $-\$3000 + \$2000 + \$2000 = \$1000$. This difference can only result from money that the fund is earning for her. ∎

Remark. We have chosen to introduce the approximate dollar-weighted rate of return in this way in order for the definition to be as consistent as possible with the definition of the internal rate of return. It is easy to see why the adjective "approximate" is applied, since we are approximating compound interest by simple interest. Other sources in financial mathematics restrict the problem to a real time interval $[0, 1]$, in which there are times of transaction $t_0 = 0 < t_1 < t_2 < \cdots < t_n \leq 1$, beginning with an initial balance of some value $P_0 = A$ and leading to a final balance of B. The main difference is that, in our world, time values are discrete, non-negative integers, and in the other construction time values could be real, and also unevenly spaced. The approximate dollar-weighted rate is defined verbally just as in Definition 2, but the computational formula (1.61) would change to:

$$j = \frac{B - \sum\limits_{k=0}^{n} P_k}{\sum\limits_{k=0}^{n} P_k \left(1 - t_k\right)} = \frac{I}{\sum\limits_{k=0}^{n} P_k \left(1 - t_k\right)}. \tag{1.62}$$

(You are asked to show this in Exercise 13.) This formula also makes clear an intuitive interpretation of j, namely, that it is the *ratio of the interest earned during the investment period and the average amount on deposit*, where, in the averaging process, the weight $(1 - t_k)$ is used to express the fact that the kth deposit P_k made at time t_k is only on hand for $1 - t_k$ time units.

Example 6. Redo Example 4, in which there were two IRRs and no IRR respectively in the two cases, using the approximate dollar-weighted rate of return.

Solution. Remember that the company is receiving a payment originally from Harry, making the balance of his account (i.e., what he owes) $P_0 = -\$1000$ at time 0. Then a payment occurs to Harry in the form of lawn care service of $P_1 = \$3000$ at time 1, and the final balance B on his account in case 1 is $\$2200$ at time 2. The approximate dollar-weighted rate is:

$$j = \frac{B - \sum\limits_{k=0}^{n} P_k}{\sum\limits_{k=0}^{n} P_k(n - k)} = \frac{\$2200 - (-\$1000 + \$3000)}{-\$1000 \cdot 2 + \$3000 \cdot 1} = \frac{\$200}{\$1000} = .20.$$

This rate of return is expressed on a per-year basis. Similarly, in case 2, where the final balance on Harry's account is $\$2300$, we have:

$$j = \frac{\$2300 - (-\$1000 + \$3000)}{-\$1000 \cdot 2 + \$3000 \cdot 1} = \frac{\$300}{\$1000} = .30.$$

It is satisfying to get unique solutions in both these cases, and it is also satisfying that the profit that the company made from the deal, $200 and $300, respectively, figures prominently into the calculation. The methodology gives us a denominator to which to compare the profit in order to compute a rate of return.

Before leaving this example, let us show how to rescale the times and work according to Equation (1.62) in the Remark above. Just consider the first case. Now the payment of $3000 is made at time $t_1 = .5$, and the final balance of $2200 is observed at time 1. We recompute:

$$j = \frac{B - \sum\limits_{k=0}^{n} P_k}{\sum\limits_{k=0}^{n} P_k (1 - t_k)} = \frac{\$2200 - (-\$1000 + \$3000)}{-\$1000 \cdot 1 + \$3000 \cdot (.5)} = \frac{\$200}{\$500} = .40.$$

This is actually consistent with the answer above, because it expresses a rate of return on the basis of the new time unit of 2 years; split it in half to produce a per-year rate and we get the same result as before. ■

1.5.3   Time-Weighted Rate of Return

A third method to compute a rate of return on a transaction with multiple payments plays on some earlier, fundamental ideas of the meaning of the term. To begin with an easy case, suppose that there are two time periods involved, and beginning with a balance of B_0 in an investment, the balance grows to B_1 at time n_1 and to B_2 at time $n_2 > n_1$, where the time intervals n_1 and $n_2 - n_1$ are not necessarily equal to 1, nor even equal to each other. Let r be the rate of return on the entire interval $[0, n_2]$, let r_1 be the rate of return on $[0, n_1]$, and let r_2 be the rate of return on $[n_1, n_2]$. Then r is:

$$
\begin{aligned}
r &= \frac{B_2 - B_0}{B_0} \\
&= \frac{B_2 - B_1 + B_1 - B_0}{B_0} \\
&= \frac{B_2 - B_1}{B_0} + \frac{B_1 - B_0}{B_0} \\
&= \frac{B_1}{B_0} \cdot \frac{B_2 - B_1}{B_1} + \frac{B_1 - B_0}{B_0} \\
&= \frac{B_1}{B_0} \cdot r_2 + r_1 \\
&= \frac{B_1 - B_0 + B_0}{B_0} \cdot r_2 + r_1 \\
&= (r_1 + 1) \cdot r_2 + r_1 \\
&= (1 + r_1)(1 + r_2) - 1.
\end{aligned} \tag{1.63}
$$

This can be generalized to more than two time intervals in a straightforward way. Now we have been considering cases in which the investment is not simply left to its own devices, but rather payments are made to it and withdrawals are taken from it. From our experience with the approximate dollar-weighted rate of return, it is logical in these problems to correct the balance difference $B_i - B_{i-1}$ in the formula for the rate of return on a time interval $[n_{i-1}, n_i]$ by subtracting away P_i, the net payment to the investment account during the time interval.

The **time-weighted rate of return** compounds the deposit-adjusted rates of return

$$
\frac{\text{closing balance} - \text{initial balance} - \text{net deposit}}{\text{initial balance}}
$$

between successive times to determine an overall rate of return for the full time interval. The length of the time intervals between the transactions doesn't matter. We summarize the important notation below, and then give the formal definition.

$$
0 = n_0 < n_1 < n_2 < \cdots < n_k \text{ the times of transaction}
$$

$$
B_0 = \text{initial investment balance}
$$

$$
B_i = \text{investment balance at time } n_i
$$

$$
P_i = \text{net deposit during time interval } (n_{i-1}, n_i]
$$

$$
\begin{aligned}
r_i &= \text{deposit-adjusted rate of return over interval } (n_{i-1}, n_i] \\
&= \frac{B_i - B_{i-1} - P_i}{B_{i-1}} \text{ for } 1 \leq i \leq k
\end{aligned}
$$

Definition 3. The **time-weighted rate of return** is:

$$r = (1 + r_1)(1 + r_2) \cdots (1 + r_k) - 1. \tag{1.64}$$

Example 7. Find the time-weighted rate of return for Susannah's mutual fund in Example 5.

Solution. The key time intervals during the year for which we have the information are January 1 through April 30 ($n_0 = 0, n_1 = 4$), May 1 through August 31 ($n_2 = 8$), September 1 through November 30 ($n_3 = 11$), and December 1 through December 31 ($n_4 = 12$). The opening balance is $B_0 = \$100,000$, and the closing balance at time n_1 is $B_1 = \$100,860$. There is no account activity in this interval, so $P_1 = 0$. Susannah's withdrawal of \$3000 (or deposit of $P_2 = -\$3000$) occurs on the second interval $(n_1, n_2]$, and the closing balance is $B_2 = \$101,000$. Reasoning similarly, we have $P_3 = \$2000$ for interval $(n_2, n_3]$, and $B_3 = \$105,330$. Also, $P_4 = \$2000$ occurs on the fourth interval, and the ending balance is $B_4 = \$107,100$. The deposit-adjusted rates of return are:

$$r_1 = \frac{B_1 - B_0 - 0}{B_0} = \frac{\$100,860 - \$100,000}{\$100,000} = .0086;$$

$$r_2 = \frac{B_2 - B_1 - P_2}{B_1} = \frac{\$101,000 - \$100,860 - (-\$3000)}{\$100,860} = .0311;$$

$$r_3 = \frac{B_3 - B_2 - P_3}{B_2} = \frac{\$105,330 - \$101,000 - (\$2000)}{\$101,000} = .0230;$$

$$r_4 = \frac{B_4 - B_3 - P_4}{B_3} = \frac{\$107,100 - \$105,330 - (\$2000)}{\$105,330} = -.0022.$$

So we compute the time-weighted rate as:

$$\begin{aligned} r &= (1 + r_1)(1 + r_2)(1 + r_3)(1 + r_4) - 1 \\ &= (1.0086)(1.0311)(1.0230)(.9978) - 1 = .0615. \end{aligned}$$

This is quite close to the answer we found earlier, 6.17%, for the approximate dollar-weighted rate of return. ■

Example 8. Find the time-weighted rate of return for a simple investment of P_0 at time 0 for a period of 1 year, at nominal rate y per year convertible quarterly. Use $n_0 = 0$, $n_1 = 1/4$, $n_2 = 1/2$, $n_3 = 3/4$, and $n_4 = 1$ as the time points (in units of years) in the computation. Is the result different if you use $n_0 = 0$, $n_1 = 1/2$, and $n_2 = 1$ for the time points?

Solution. For the first question we have an effective rate of $y/4$ per quarter, and there is just one non-zero payment P_0 at time 0. Therefore the balances at the time points are:

$$B_0 = P_0(\text{after deposit}),$$

$$B_1 = P_0\left(1 + \frac{y}{4}\right), B_2 = P_0\left(1 + \frac{y}{4}\right)^2,$$

$$B_3 = P_0\left(1 + \frac{y}{4}\right)^3, B_4 = P_0\left(1 + \frac{y}{4}\right)^4.$$

The rates of return in the four time intervals are:

$$r_1 = \frac{P_0\left(1 + \frac{y}{4}\right) - P_0}{P_0} = \left(1 + \frac{y}{4}\right) - 1 = \frac{y}{4},$$

$$r_2 = \frac{P_0\left(1 + \frac{y}{4}\right)^2 - P_0\left(1 + \frac{y}{4}\right)}{P_0\left(1 + \frac{y}{4}\right)} = \left(1 + \frac{y}{4}\right) - 1 = \frac{y}{4},$$

and similarly $r_3 = y/4$ and $r_4 = y/4$. Thus, the time-weighted rate of return is:

$$(1 + r_1)(1 + r_2)(1 + r_3)(1 + r_4) - 1 = \left(1 + \frac{y}{4}\right)^4 - 1,$$

which we recognize as being identical to the effective rate of return for compound interest in which compounding is quarterly at a nominal rate y.

If the times are 0, 1/2, and 1, then the rates of return in the two time intervals are:

$$r_1 = \frac{P_0\left(1 + \frac{y}{4}\right)^2 - P_0}{P_0} = \left(1 + \frac{y}{4}\right)^2 - 1,$$

$$r_2 = \frac{P_0\left(1 + \frac{y}{4}\right)^4 - P_0\left(1 + \frac{y}{4}\right)^2}{P_0\left(1 + \frac{y}{4}\right)^2} = \left(1 + \frac{y}{4}\right)^2 - 1.$$

The time-weighted rate of return becomes:

$$(1 + r_1)(1 + r_2) - 1 = \left(\left(1 + \frac{y}{4}\right)^2\right)^2 - 1 = \left(1 + \frac{y}{4}\right)^4 - 1,$$

which is the same result. ■

Example 8 illustrates that the time-weighted rate of return does generalize our simpler notion of rate of return for a single investment earning compound interest. See Exercise 20 for a similar problem involving annuities.

Important Terms and Concepts

Internal rate of return - The interest rate under which the present value of the final balance in a multi-payment transaction equals the net present value of all payments. The phrase "present value" may be replaced by "future value."

Dollar-weighted rate of return - Same as internal rate of return.

Approximate dollar-weighted rate of return - In the defining equation for internal rate of return, replace compound interest terms $(1 + i)^k$ by simple interest terms $(1 + k \cdot i)$. The approximate dollar-weighted rate of return is the solution j of the equation that sets the future value of all payments equal to the future value of the final balance.

Time-weighted rate of return - The quantity $r = (1 + r_1)(1 + r_2) \cdots (1 + r_k) - 1$, where the r_i are rates of return on subintervals of time, adjusted for net deposits that occurred during those intervals.

Exercises 1.5

1. Bruce invests an amount of $1000 at the end of each year to save for his daughter's college education. If the effective yearly rate r remains the same each year and we consider the end of time to be the fourth year, find the final value B. What is the internal rate of return?

2. Dave deposited $1000 into a savings account on Jan. 1. At that point the account earned 2% nominal yearly interest, compounded monthly. On April 1 the interest rate increased to 2.5%. Then on July 1 it increased to 3% and remained there for the rest of the year. Dave made no deposits or withdrawals. Find the internal rate of return, expressed as a yearly effective interest rate, on Dave's investment.

3. Suppose that Burt deposits $1000 in an interest-bearing account with nominal rate of return 5% convertible monthly, and at the end of 1 year the nominal rate changes to 5.5% and he deposits another $500. The rate stays at 5.5% for the next year, and Burt makes no further deposits or withdrawals. What is the internal rate of return for the first 2 years on this investment, expressed as an effective rate per year?

4. Julie starts up an annuity in which she pays $1000 at the start of the first year and increases that by 5% each year. Assume that the effective yearly rate of interest is 4%. What is the internal rate of return on this investment for the first 3 years?

5. At the beginning of a particular year, Tom deposits $1000 in a certificate of

deposit earning yearly interest at 3% compounded semiannually, and also deposits $50, $60, $70, and $80, respectively, at the beginning of each quarter in a simple savings account earning interest at yearly nominal rate 2%, compounding monthly. At the end of the tenth month, he withdraws $40 from the savings account. Find the amount of money the entire investment is worth at the end of the year, and use numerical techniques to find Tom's internal rate of return on the entire transaction, expressed on a monthly basis.

6. Show in general that the internal rate of return on an ordinary annuity with payment P and time horizon n is the same as the effective rate per period r.

7. David pays $1000 to a loan shark this month in order to gain credit for a $4000 bet on the Super Bowl next month. The deal is that he will pay back the $4000 in the following month. Also, he will pay another $100 two more months after he repays the $4000 to renew his relationship with the shark for the next year, at which point both sides call it even. Is the internal rate of return well determined on this transaction?

8. Steve makes a deal with his brother Bob, who will need money in the next 2 years for a club membership, to receive $1000 from Bob now, if he pays the membership fees of $1200 and $1210 in the next 2 years, after which Bob will pay Steve back $1452 the following year. Find, if possible, the internal rate of return to Steve.

9. Prove that the uniqueness part of Theorem 1 can be generalized to the case where the outstanding balance of the investment remains positive at all times, even if some of its terms may be negative.

10. Argue that the expression $j = \dfrac{I}{\sum\limits_{k=0}^{n} P_k(n-k)}$ for the approximate dollar-weighted rate of return in (1.61) represents the average interest per period divided by the average amount on deposit per period. (Hint: What do you get when you divide the numerator and denominator by n?)

11. Sue has a speculative investment in a startup company run by a college friend. Assume that its initial value is $1000 for a particular year, and the values at the end of months 2, 4, 6, 8, 10, and 12 are $1050, $1080, $1150, $1275, $1400, and $1450. Suppose that she withdrew $40 at the end of month 2, invested $55 at the end of month 6, and invested $75 at the end of month 8 and month 10. Find the approximate dollar-weighted rate of return on the investment.

12. Compute the approximate dollar-weighted rate of return using (a) Equation (1.61) and (b) Equation (1.62) for the following scenario, in which contributions to an interest-earning fund, withdrawals, the dates of transactions, and balances are listed at key times during a 1-year span. In both cases, express the rate of

return on a nominal yearly basis.

date	contribution	withdrawal	balance
1/1			$20,000
2/28	$1,000		$23,500
6/30		$1,500	$23,000
9/30		$2,000	$22,500
12/31			$23,000

13. Show that, if payments P_k are made at real-valued times $t_0 = 0 < t_1 < t_2 < \cdots < t_n \leq 1$, then the approximate dollar-weighted rate of return is:

$$j = \frac{B - \sum\limits_{k=0}^{n} P_k}{\sum\limits_{k=0}^{n} P_k (1 - t_k)} = \frac{I}{\sum\limits_{k=0}^{n} P_k (1 - t_k)}.$$

14. Find the approximate dollar-weighted rate of return for Steve in Exercise 8.

15. A pension fund begins the year with $2,000,000. At the end of the first quarter, the fund has balance $2,225,000. At the end of the second quarter, an addition to the fund of $50,000 is made, which brings the fund's balance to $2,400,000. At the end of the third quarter a withdrawal of $500,000 is made, and the fund balance becomes $2,200,000. Finally, at the end of the year the fund balance is $2,300,000. Find the approximate dollar-weighted rate of return earned by the pension fund using Equation (1.62).

16. Find the time-weighted rate of return on Tom's savings account (excluding the CD) whose activity is described in Exercise 5.

17. Find the time-weighted rate of return for the year of the investment portfolio described as follows. The portfolio balance initially is $50,000. The investor adds $4000 of value to it at the end of each of the first, second, third, and fourth quarters. During the second quarter, he consumes $5000 of value. The balance at the end of the first quarter is $55,000, at the end of the second is $53,000, at the end of the third is $58,000, and at the end of the year is $62,500.

18. Find the time-weighted rate of return for the lender in the transaction described as follows. Pedro has a $5000 balance on his credit card on January 1. The account has an annual nominal rate of 13.9% convertible monthly. To keep his business, the credit card company offers him the chance to skip payments during the first 6 months of the year and a reduced rate of 5.9% nominal on all new purchases made from the beginning of April through the end of June. Pedro takes advantage of this offer and makes a $100 purchase in April and a $200 purchase in May. Assume that the new purchases only are charged interest beginning the month after they occurred.

19. Using the data in Example 5,

(a) Redo the computation of the approximate dollar-weighted rate of return on a per-year basis using the time model expressed in (1.62);

(b) Recompute the time-weighted rate of return, assuming that the three transactions that Susannah made were on April 30, August 31, and November 30 after the final balances were reported as in the table, instead of May 1, September 1, and December 1, respectively.

20. Find the time-weighted rate of return for an ordinary annuity with initial balance $B_0 > 0$ and constant payment P occurring at the end of each month for a period of 1 year, at nominal rate y per year convertible monthly. Use the multiples of $1/12$ as the time points (in units of years) in the computation. (Hint: instead of using the closed formula for the future value, consider setting up a recurrence relation for the sequence of balances B_k.)

1.6 Continuous Time Interest Theory

Our study of interest is more or less complete for the case of discrete time periods. But occasionally a situation is able to be modeled with a continuous time framework, and then single-variable calculus becomes an important tool. The idea is that interest is compounded at a particular rate at every instant in real time. This is more of a theoretical, approximate construct than a real one, but it has the additional benefit of setting up later work on modeling risky asset price processes.

1.6.1 Continuous Compounding: Effective Rate and Present Value

Let us begin by generalizing the most basic ideas of compound interest and present value. Recall the formula for the future value of an investment of \$$K$ earning compound interest in discrete time, with per-period rate r.

$$\text{Future value of investment after } n \text{ compoundings} = K(1 + r)^n. \qquad (1.65)$$

If we compound m times per year, the nominal yearly rate is y, and we express the future value in t years, then we get:

$$\text{Future value of investment after } t \text{ years } = A(t) = K\left(1 + \frac{y}{m}\right)^{mt}. \qquad (1.66)$$

We would like to study what happens to this future value as the number of compoundings per year approaches infinity.

Example 1. (a) Compute the future value of an investment of $2000 earning 6% nominal yearly interest after 10 years if interest is compounded semi-annually, quarterly, monthly, weekly, daily. (b) Approximate the limit of the future value as the number of compoundings approaches infinity.

Solution. (a) By Equation (1.66), since the number of compoundings would be 20, 40, 120, 520, and 3650, respectively, in the five cases, we can compute the future values as follows:

$$\text{semi-annually: } \$2000 \left(1 + \frac{.06}{2}\right)^{20} = \$3612.22,$$

$$\text{quarterly: } \$2000 \left(1 + \frac{.06}{4}\right)^{40} = \$3628.04,$$

$$\text{monthly: } \$2000 \left(1 + \frac{.06}{12}\right)^{120} = \$3638.79,$$

$$\text{weekly: } \$2000 \left(1 + \frac{.06}{52}\right)^{520} = \$3642.98,$$

$$\text{daily: } \$2000 \left(1 + \frac{.06}{365}\right)^{3650} = \$3644.06.$$

(b) The numerical results in part (a) indicate that the future value doesn't grow indefinitely as we let the number of compoundings m become very large. With $m = 1000$ and $m = 2000$, we get, respectively:

$$\text{1000 compoundings: } \$2000 \left(1 + \frac{.06}{1000}\right)^{10000} = \$3644.17.$$

$$\text{2000 compoundings: } \$2000 \left(1 + \frac{.06}{2000}\right)^{20000} = \$3644.20.$$

It seems as if the limiting value is just over $3644.20. ∎

Example 1 suggested that the future value expression $K\left(1 + \frac{y}{m}\right)^{mt}$ does reach a limit as $m \longrightarrow \infty$. We will determine that limit below, but we can now state what we mean by **continuous compounding**.

Definition 1. The **future value of an investment** under continuous compounding is the limit as the number of compounding periods approaches infinity of the future value under discrete compounding.

To find this limit, look again at the future value expression. Its logarithm is:

$$\log(A(t)) = \log(K) + mt \log\left(1 + \frac{y}{m}\right).$$

The limit of this log as $m \to \infty$ is found by applying L'Hopital's rule.

$$
\begin{aligned}
\lim_{m \to \infty} \log(A(t)) &= \lim_{m \to \infty} \log(K) + mt \log\left(1 + \frac{y}{m}\right) \\
&= \log(K) + t \cdot \lim_{m \to \infty} \frac{\log\left(1 + \frac{y}{m}\right)}{m^{-1}} \\
&= \log(K) + t \cdot \lim_{m \to \infty} \frac{\left(1 + \frac{y}{m}\right)^{-1} \cdot \left(\frac{-y}{m^2}\right)}{-m^{-2}} \\
&= \log(K) + t \cdot \lim_{m \to \infty} y \cdot \left(1 + \frac{y}{m}\right)^{-1} \\
&= \log(K) + t \cdot y.
\end{aligned}
$$

Raising e to both sides gives the following result.

Theorem 1. Let $A(t)$ be the future value at time t of an initial investment of $A(0) = K$ under continuous compounding at nominal rate y. Then,

$$A(t) = Ke^{y \cdot t}. \tag{1.67}$$

For instance, using the data in Example 1, the future value of the investment at time 10 years is:

$$\$2000 \cdot e^{.06 \cdot 10} = \$2000 \cdot e^{.6} = \$3644.24.$$

Example 2. How long does it take an amount of money earning interest continuously at rate 5% per year to double? What interest rate is necessary for the doubling time to be exactly 10 years? Do either of these results depend on the initial investment?

Solution. Denote t_2 as the time required to double the initial value. We demand that:

$$A(t_2) = 2K = Ke^{.05t_2}.$$

Solving for t, we get:

$$2 = e^{.05t_2} \implies \log(2) = .05t_2 \implies t_2 = \frac{\log(2)}{.05} = 13.86.$$

So it takes nearly 14 years, and this value clearly does not depend on K, since K divided away in the equation of value.

For the second question, we can set up an equation of value in which the interest rate is unknown and the doubling time is 10 years:

$$A(10) = 2K = Ke^{y \cdot 10}.$$

Solving for y this time, we obtain:

$$2 = e^{y \cdot 10} \implies \log(2) = 10y \implies y = \frac{\log(2)}{10} = .0693.$$

So we need an interest rate of nearly 7% to double in 10 years, and again this doesn't depend on K. ∎

Generalizing Example 2, Exercise 2 asks you to show that, if t_n is the time required for an investment earning interest at rate y to multiply by a factor of n, then

$$t_n = \frac{\log(n)}{y}. \tag{1.68}$$

It might have surprised you to see that it is the nominal rate of interest that has the primary role in the formula for the future value of an investment under continuous compounding. We can nevertheless define a notion of effective rate of interest.

Definition 2. The **effective annual rate of interest** r under continuous compounding is the rate of return of an investment over a year using nominal rate y, that is:

$$r = e^y - 1. \tag{1.69}$$

To verify that Equation (1.69) makes sense as a definition of the effective yearly rate, the rate of return on a $\$K$ investment for a 1-year time horizon is:

$$r = \frac{A(1) - A(0)}{A(0)} = \frac{Ke^{y \cdot 1} - Ke^{y \cdot 0}}{Ke^{y \cdot 0}} = \frac{e^{y \cdot 1} - e^{y \cdot 0}}{e^{y \cdot 0}} = \frac{e^y - 1}{1} = e^y - 1.$$

Example 3. Is there a 1-1 correspondence between effective and nominal rates? Is the effective rate always greater than the nominal rate?

Solution. The relationship $r = e^y - 1$ between the effective yearly rate and the nominal yearly rate is clearly 1-1, and the inverse relationship is $y = \log(1 + r)$. The graph in Figure 1.18 gives evidence that, even for negative values of the nominal rate, the effective rate is higher. You are asked for an analytical proof in Exercise 7. ∎

Example 4. (a) What is the effective rate if the nominal rate is 8%? (b) What nominal rate corresponds to an effective rate of 10%?

Solution. (a) The corresponding effective rate is $r = e^{.08} - 1 = .0833$.

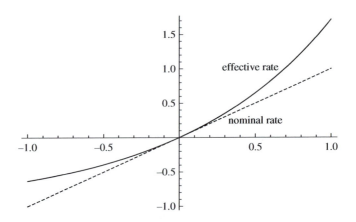

Figure 1.18: Effective yearly rate is at least as large as nominal yearly rate.

(b) The corresponding nominal rate is $y = \log(1 + .1) = .0953$. ■

As in the discrete case, there is an obvious notion of the **present value** of a future amount, obtained by inverting the future value expression $A(t) = Ke^{yt}$.

Definition 3. The **present value** of an amount A at time t under continuous compounding at nominal rate y is:

$$PV = Ae^{-yt}. \tag{1.70}$$

Example 5. If an investor has a savings goal to reach \$25,000 in 20 years using continuous compounding at nominal annual rate 7%, how much should be deposited?

Solution. This is the present value of \$25,000, namely:

$$\$25,000e^{-.07 \cdot 20} = \$6164.92. \quad ■$$

As before, we can also compute net present values of streams of earnings. If payments $P_1, P_2, ..., P_n$ occur at real times $t_1, t_2, ..., t_n$, then the net present value is:

$$\text{NPV} = \sum_{i=1}^{n} P_i \cdot e^{-yt_i}. \tag{1.71}$$

A similar formula holds if the sequence of payments is infinite. But a new possibility arises, as we will explore further when we study continuous annuities: payments may occur continuously in some interval $[a, b]$ of real-valued times at a rate $P(t)$. The amount of the payment in a short time interval $[t_i, t_i + \Delta t]$ is approximately $P(t_i)\Delta t$, and the net present value of the payments is approximately:

$$\text{NPV} = \sum_{i=1}^{n} P\left(t_i\right) \Delta t \cdot e^{-yt_i} \longrightarrow \int_{a}^{b} e^{-yt} \cdot P(t)dt \text{ as } n \to \infty. \qquad (1.72)$$

The integral in (1.72) is the **continuous net present value** of payments made at rate $P(t)$ during time interval $[a, b]$ under nominal rate y.

Example 6. What was the nominal rate of interest if the net present value of payments of \$1000 at time 1.5 and \$2000 at time 3.0 is \$1800?

Solution. The nominal rate y satisfies the equation of value:

$$\$1800 = \$1000e^{-1.5y} + \$2000e^{-3.0y} \implies 9 = 5e^{-1.5y} + 10e^{-3.0y}.$$

Let $x = e^{-1.5y}$, so that $y = \log(x)/(-1.5)$. Then x satisfies the quadratic equation:

$$9 = 5x + 10x^2 \implies 10x^2 + 5x - 9 = 0 \implies x = \frac{-5 \pm \sqrt{25 + 360}}{20}.$$

Only the positive solution makes economic sense, and this comes out to about $x = .731$. The nominal rate is therefore $y = \log(.731)/(-1.5) = .209$. ∎

Example 7. Suppose that payments are made continuously at a constant rate of \$200 per year to a savings account with nominal yearly continuous interest rate 4% during a particular year, then a year is skipped, then more payments are made continuously at a rate of \$100 per year for 1 more year. What is the net present value of these payments?

Solution. This would be the total of the present values of the two continuous streams:

$$\int_0^1 e^{-.04t} \cdot 200dt + \int_2^3 e^{-.04t} \cdot 100dt = \left(200 \cdot \frac{e^{-.04t}}{-.04}\right)\Big|_0^1 + \left(100 \cdot \frac{e^{-.04t}}{-.04}\right)\Big|_2^3$$

$$= 286.54. \blacksquare$$

1.6.2 Force of Interest

There is another way to look at how continuous compounding can be defined, which gives rise to a new concept.

Denote by $A(t)$ the total accumulated value of an investment earning interest continuously at nominal yearly rate y at time t, and for the moment suppose that y is constant. We could choose to interpret the concept of continuous compounding as meaning that the rate of growth of the accumulated value is proportional to the current accumulated value. Notice that, in the sense of average rate of change, we already have this in discrete compounding, because, if r is the per-period interest rate, we have:

$$A(n+1) = (1+r)A(n) \implies \frac{A(n+1) - A(n)}{1} = r \cdot A(n). \qquad (1.73)$$

Working by analogy to (1.73) in the continuous time world, we could say that the investment earns interest continuously at rate y if the accumulated value function satisfies the differential equation:

$$\frac{dA}{dt} = y \cdot A(t). \qquad (1.74)$$

It is well known that the solution of this equation is:

$$A(t) = A(0)e^{y \cdot t}, \qquad (1.75)$$

so that we end up in the same place even though we started with a different definition of continuous interest.

What is gained by thinking this way? Just this: there is no reason why the growth rate y in Equation (1.74) must be constant. The rate of change of A at time t might be a function $y = \delta(t)$ times the current value $A(t)$, which allows us to model more general situations. So we are led to the following definition.

Definition 4. If an investment has accumulated value $A(t)$ at time $t \geq 0$ satisfying:

$$\delta(t) = \frac{A'(t)}{A(t)}, \qquad (1.76)$$

then we call $\delta(t)$ the **force of interest** for the investment.

The defining differential equation (1.76) allows us to interpret the force of interest as the rate at which an investment is growing, per dollar invested.

The accumulated value can be solved for in terms of the force of interest by integrating both sides of the equation as follows.

$$\delta(t) = \frac{A'(t)}{A(t)} \implies \int_0^t \delta(s)ds = \int_0^t \frac{A'(s)}{A(s)} ds$$

$$\implies \int_0^t \delta(s)ds = \int_0^t \frac{d\log(A(s))}{ds} ds$$

$$\implies \int_0^t \delta(s)ds = \log(A(s))\Big]_0^t = \log(A(t)) - \log(A(0)).$$

Raising e to both sides of the equation and solving for $A(t)$ gives:

$$A(t) = A(0)e^{\int_0^t \delta(s)ds}. \qquad (1.77)$$

Example 8. Suppose that the continuous force of interest over a time period is composed of a linearly increasing trend, with a periodic component superimposed, specifically $\delta(t) = .002t + .05\cos(6t)$. What are the rates of change of accumulated value per dollar when $t = 0, \pi/12$? Can the force of interest ever be negative, and what does that imply? Find an expression for the accumulated value for an initial investment of $100. Does this value reach an absolute minimum, and, if so, when?

Solution. The rate of change of the accumulated value $A(t)$ per dollar is just $\delta(t)$. Then we have:

$$\delta(0) = 0 + .05 \cdot 1 = .05,$$

$$\delta(\pi/12) = .002(\pi/12) + .05 \cdot \cos(\pi/2) = .002(\pi/12) + 0 = .00524.$$

The table of values below shows us that this force of interest can be negative, which says intuitively that the rate of change of value per dollar invested is negative, that is, the value of the investment is declining.

time	force of interest
0	0.05
$\pi/36$	0.0434758
$\pi/18$	0.0253491
$\pi/12$	0.000523599
$\pi/9$	-0.0243019
$5\pi/36$	-0.0424286
$\pi/6$	-0.0489528
$7\pi/36$	-0.0420795
$2\pi/9$	-0.0236037
$\pi/4$	0.0015708

The graph in Figure 1.19(a) shows that the force of interest function has a slight upward trend that is dominated by a periodic wave of relatively high amplitude. This force of interest function is simple enough that we can find the accumulated value function in closed form. For an initial investment of $100, it is:

$$A(t) = \$100e^{\int_0^t .002s+.05\cos(6s)ds} = \$100e^{.001t^2 + \frac{.05}{6}\sin(6t)}.$$

The function $A(t)$ is graphed in part (b) of Figure 1.19, and we see as well in this function that there is an upward trend starting from the initial value of $100 subject to a periodic wave. The accumulated value function reaches its lowest value when:

$$A'(t) = 0 \implies 100e^{.001t^2 + \frac{.05}{6}\sin(6t)} \cdot (.002t + .05\cos(6t)) = 0.$$

force of interest

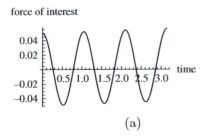

accumulated value

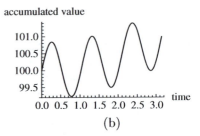

(a)   (b)

Figure 1.19: (a) Force of interest $\delta(t) = .002t + .05\cos(6t)$. (b) Accumulated value for $A(0) = \$100$.

Notice that this happens when $\delta(t) = 0$, and from the graph, we are looking for the second non-negative time at which this happens, somewhere between .5 and 1, at which $\delta(t) = .002t + .05\cos(6t) = 0$. The equation can be solved numerically to find $t \approx .78$. At this value of time, the accumulated value is $A(.78) = \$99.23$. ∎

1.6.3   Continuous Annuities

In a simple **continuous annuity**, payments are made at a constant rate p per unit time at every time. Let T be a time at which we want to compute the future value of the annuity.

In a short time interval $[t, t + dt]$, a payment of approximately $p \cdot dt$ is made, which accumulates interest for approximately $T - t$ time units, hence its value at time T is:

$$p \cdot dt \cdot e^{y(T-t)}.$$

The **future value of the continuous annuity** is the sum of all these separate contributions, in the limit as $dt \to 0$, which gives rise to the following.

$$\text{Future value of continuous annuity at } T = \int_0^T pe^{y(T-t)}dt = p \cdot \frac{e^{yT} - 1}{y}. \quad (1.78)$$

Similarly, the present value of the payment of $p \cdot dt$ during $[t, t + dt]$ is:

$$p \cdot dt \cdot e^{-yt}.$$

Hence the **present value of a continuous annuity** with payments from time 0 to time T at rate p is:

$$\text{Present value of continuous annuity} = \int_0^T pe^{-yt}dt = p \cdot \frac{1 - e^{-yT}}{y}. \quad (1.79)$$

You can also derive this formula for the future value by considering the limit of the future value of a discrete annuity with more and more periods. Recall that the latter is:

$$\text{FV} = P\frac{(1 + y/m)^{m \cdot T} - 1}{y/m}, \qquad (1.80)$$

where P is the payment in each period, and y is the nominal yearly rate. To translate, since there are $\$P$ payments in each of m periods in a year, payments are made at a rate of $p = Pm$ per year, hence $P = p/m$. Then we must find:

$$\lim_{m \to \infty} \frac{p}{m}\frac{(1 + y/m)^{m \cdot T} - 1}{y/m} = \lim_{m \to \infty} p \cdot \frac{(1 + y/m)^{m \cdot T} - 1}{y}.$$

The exponential term in the numerator is well known to converge to $e^{y \cdot T}$, which agrees with (1.78).

Example 9. In 2012, Lyle deposits $5 per day into an account compounding daily at 4% nominal annual interest. In the year 2013 he increases the deposit to $8 per day, but the interest rate falls to 3% for that year. Using (a) the discrete annuity formulation with daily compounding, and (b) a continuous approximation, find the value of this annuity at the end of 2013.

Solution. (a) In the discrete case, the accumulated value through the end of 2012 is:

$$\text{FV} = P\frac{(1 + r)^n - 1}{r} = \$5\frac{(1 + .04/365)^{365} - 1}{.04/365} = \$1861.89.$$

That amount earns compound interest for 1 more year at nominal rate 3% per year. For the deposits beginning in 2013, the accumulated value is:

$$\$8\frac{(1 + .03/365)^{365} - 1}{.03/365} = \$2964.12,$$

for a total of:

$$\$1861.89\left(1 + \frac{.03}{365}\right)^{365} + \$2964.12 = \$4882.71.$$

The value of the action of ignoring interest considerations and putting $5 per day in a shoebox for a year and $8 for the next year is:

$$\$5 \cdot 365 + \$8 \cdot 365 = \$4745,$$

so about $138 of interest was earned.

(b) In a continuous approximation, he is putting money at a rate $\$5(365) = \$1825$ per year into the account for the first year. Using the continuous annuity formula $p \cdot \frac{e^{yT} - 1}{y}$, we have a value after the first year of:

$$\$1825 \cdot \frac{e^{.04} - 1}{.04} = \$1861.99.$$

That amount earns continuous interest at nominal rate .03 for a year, in addition to which is the value of the third year of payments, at a rate of $\$8(365) = \$2920$ per year. The latter is:

$$\$2920 \cdot \frac{e^{.03 \cdot 1} - 1}{.03} = \$2964.24.$$

Thus, the total future value under continuous compounding is:

$$\$1861.99 \cdot e^{.03 \cdot 1} + \$2964.24 = \$4882.94.$$

The result is around \$.23 greater than for daily compounding. ■

Annuities need not always be paid at a constant rate. The exercise set for this section examines some variations. If the payment rate p is actually a function of time t, then in the integrals in formulas (1.78) and (1.79) we simply replace the constant p by $p(t)$. The previous closed forms no longer apply; the availability of a closed form is dependent on our ability to calculate the integral. Our last example shows one case in which a calculation is possible.

Example 10. A company has studied the pattern of its earnings in time and found that they begin at a rate of \$500,000 per year, rise roughly linearly to \$600,000 in the middle of the year, then fall back down roughly linearly to \$500,000 at the end of the year. Using a nominal yearly interest rate of $y = .04$, compute the present value of the earnings for the year.

Solution. The earnings form a continuous annuity with a non-constant payment function. We need to find this earnings rate function first, and we will use monetary units of \$100 thousand in order to simplify. A line that goes through the (time, earnings) pairs $(0, 5)$ and $(.5, 6)$ has slope $1/.5 = 2$ and intercept 5, hence the equation of the earnings rate function $p(t)$ is $p(t) = 2t + 5$ on the time interval $[0, .5]$, From time .5 to time 1 year, we have a line through $(.5, 6)$ and $(1, 5)$, whose slope is $1/(-.5) = -2$. The equation turns out to be $p(t) = -2t + 7$ on this interval. In summary, the earnings rate function is:

$$p(t) = \begin{cases} 2t + 5 & \text{if } 0 \le t < .5 \\ -2t + 7 & \text{if } .5 \le t \le 1. \end{cases}$$

The present value of the earnings would be the integral:

$$\int_0^T p(t)e^{-yt}dt = \int_0^{.5} (2t + 5)e^{-.04t}dt + \int_{.5}^1 (-2t + 7)e^{-.04t}dt.$$

Integration by parts on these integrals (with $u = p(t)$ and $dv = e^{-.04t}$) can be carried out to give a present value of \$539,144. ■

1.6.4   Continuous Loans

All of the ideas about loan amortization that we saw in Section 1.4 carry over to the continuous time world, and our experience with continous annuities points out the direction for us. Though it gives us more good practice in financial thinking, this area of investigation may be a little less applicable than others in the sense that most real loans are made under discrete conditions, so let us just look briefly at how this goes.

Example 11. This example reprises Example 1 of Section 1.4. A college borrows \$2 million on a bond issue, which is due to be paid back with interest in 20 years. At what continuous rate p per year should they pay, if they can get a 7% effective annual interest rate? How much total interest is paid? If they are able to repay \$.5 million each year, in how many years can the bond be retired?

Solution. Since 7% is the effective annual rate, before we do anything we should convert that to the nominal annual rate. The formula $y = \log(1 + r)$, derived earlier, gives us a nominal rate of $y = \log(1.07) \approx .0677$.

Now to answer the first question, let us find an expression for the payment rate p in general for a loan of $\$L$ at nominal rate y per year and time horizon T years. The ideas we have used before still help us. The lender should be indifferent between keeping the amount $\$L$ now and receiving the continuous annuity of loan payments at rate p. This produces the present value equation:

$$L = p \cdot \frac{1 - e^{-yT}}{y} \implies p = L \cdot \frac{y}{1 - e^{-yT}}. \tag{1.81}$$

(Compare to the formula $P = L \cdot \frac{y/m}{1-(1+y/m)^{-mT}}$ in the discrete case; note $p = mP$.) For our data, the repayment rate is:

$$p = \$2 \text{ million} \cdot \frac{.0677}{1 - e^{-.0677 \cdot 20}} = \$182,530/\text{year}.$$

The interest paid is $20(\$182,530) - \$2,000,000 = \$1,650,600$.

If the college makes payments at the rate $p = \$.5$ million per year, then the present value equation in (1.81) can be employed to solve for the term T of the loan:

$$L = p \cdot \frac{1 - e^{-yT}}{y} \implies \$2 \text{ million} = \$.5 \text{ million} \cdot \frac{1 - e^{-.0677T}}{.0677}$$
$$\implies 4(.0677) = 1 - e^{-.0677T}$$
$$\implies e^{-.0677T} = 1 - 4(.0677)$$
$$\implies T = -\log(1 - 4(.0677))/.0677 \approx 4.66 \text{ years}. ■$$

Example 12. Since time is continuous rather than discrete, and payments are made continuously, the device of the amortization table is not quite as appropriate for continuous time loans, but it is rather easy to find an outstanding balance function that could be computed and tabulated at various discrete instants to produce the same information as an amortization table. Let us find a formula for the outstanding balance at a given time t of a 30-year loan of $80,000 at nominal rate 7.5% per year, as in Example 4 of Section 1.4.

Solution. We need to find the payment rate p that will pay off the loan in 30 years. By Equation (1.81),

$$p = L \cdot \frac{y}{1 - e^{-yT}} = \$80,000 \cdot \frac{.075}{1 - e^{-.075 \cdot 30}} = \$6706.90.$$

This corresponds to a rate of payment per month of about $558.91, similar to the result of the earlier example $559.37.

Thinking prospectively, the outstanding balance at time t should be the total time t present value of all payments yet to be made from time t through time 30. Taking time t as the time origin, there are $T - t$ time units left during which payment takes place. By Equation (1.79), the outstanding balance at time t would be:

$$\text{Outstanding balance (prospective form)} \ = \ \int_0^{T-t} pe^{-ys}ds = p \cdot \frac{1 - e^{-y(T-t)}}{y}.$$

$$(1.82)$$

For our problem data, the outstanding balance would be:

$$\text{OB}(t) = \$6706.90 \cdot \frac{1 - e^{-.075 \cdot (30-t)}}{.075}.$$

In Section 1.4 we produced an amortization table for the first 20 months. The kth month would correspond to time $k/12$ in years. Sampling out the first 20 outstanding balances gives a table like the following:

month	outstanding balance
0	80000.
1	79940.9
2	79881.4
3	79821.6
4	79761.4
5	79700.8
6	79639.8
7	79578.5
8	79516.7
9	79454.6
10	79392.1
11	79329.2
12	79265.9
13	79202.2
14	79138.1
15	79073.6
16	79008.7
17	78943.4
18	78877.7
19	78811.5
20	78745.

A graph of the outstanding balance function appears in Figure 1.20, and its shape is nearly identical to the point graph in Figure 1.14 for the discrete case, which we have superimposed on the continuous graph. ■

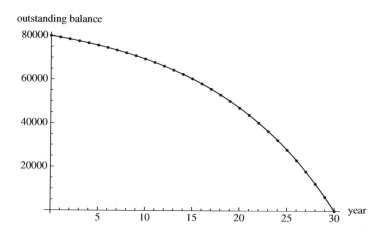

Figure 1.20: End-of-year outstanding balance on $80,000 loan.

The highlight of the last example is the argument that led to the prospective form of the outstanding balance in (1.82). In Exercise 24 you will hone your financial reasoning skills by deriving a retrospective form for the outstanding balance.

Important Terms and Concepts

Continuous compounding - Interest is compounded at every real instant of time with rate of change equal to the nominal yearly rate. The future value $A(t)$ of the investment satisfies the differential equation $dA/dt = y \cdot A$ with solution $A(0)e^{yt}$.

Effective annual rate of interest - The effective rate of return earned by an investment under continuous compounding for a 1-year period $r = e^y - 1$.

Continuous present value - The amount Ae^{-yt} that must be deposited at continuous interest with nominal rate y in order to reach a value of A at time t.

Continuous net present value - The total $\sum_{i=1}^{n} P_i \cdot e^{-yt_i}$ of continuous present values with nominal rate y of discrete payments; in the case of continuous payments at rate $P(t)$ on time interval $[a, b]$, this is $\int_a^b e^{-yt} \cdot P(t)\, dt$.

Continuous effective rate of interest - The quantity $r = e^y - 1$, which is the 1-year rate of return on an investment earning continuous interest at nominal rate y.

Force of interest - The rate $\delta(t) = \frac{A'(t)}{A(t)}$ at which an investment is changing value, per dollar invested.

Continuous annuity - An investment situation in which payments are made at a constant rate p per unit time at every time in a continuous interval. The future and present values are integrals $\int_0^T pe^{y(T-t)}dt$ and $\int_0^T pe^{-yt}dt$, respectively. These generalize to non-constant payment rate functions $p(t)$.

Continuous loan - A continuous annuity of payments made by the borrower to the lender at a given rate p per unit time.

Exercises 1.6

1. How long does it take an investment of $1000 earning interest continuously at yearly nominal rate 5% to reach a value of $1500?

2. Prove that the time t_n required for the initial value of an investment earning interest at rate y to be multiplied by n is:

$$t_n = \frac{\log(n)}{y}.$$

3. (a) If an investment earns interest continuously at nominal rate 5%, what is the effective rate of interest? (b) An investment earning continuous interest at effective rate 8% corresponds to what nominal rate?

4. How often must an investment of $1000 be compounded at nominal yearly rate 4% in order to be worth within $1 of an investment compounding continuously at the same nominal rate?

5. What amount of money, deposited in a continuously compounding account at nominal yearly rate 3.5%, will accumulate to a value of $4000 in 2 years?

6. At least how large must the effective yearly rate of interest be on a continuously compounding investment to make $3000 grow to $5000 in 10 years?

7. Prove that, for continuous compounding, the effective yearly rate is at least as large as the nominal yearly rate.

8. Suppose that an investment of $500 earns interest continuously at nominal rate 3% for 6 months, at which point the rate changes to 3.5% for the next year, and then decreases to 3.2% for the next 6 months after that. What is the value of the investment at the end of 2 years?

9. A bond is scheduled to make payments of $100 at the end of each of the next 5 years to its holder, and at the end of the full 5 years can be cashed in for $2000. Using a continuous nominal yearly rate of 3%, what is the total present value of the bond's payments?

10. Suppose as in Example 7 that payments are made continuously at a constant rate $200 per year to a savings account with nominal yearly continuous interest rate 4% during the first year, then more payments are made continuously at a rate $100 per year during the third year, but during this year the nominal rate changes to 3%. Find the net present value of the payments.

11. A record label collects royalties for an artist on a particular album. The artist earns 10% of the total sales, and proceeds from sales of the album begin at a rate of $350,000 per year on the release of the album, and decay exponentially with time according to the equation $dP/dt = -.05P$. Find the present value of the artist's share of the first 2 years of sales, discounting under the assumption of an annual nominal interest rate of 3%.

12. Find the accumulated value function for an investment of initial value $100 such that the force of interest is $\delta(s) = .01 + .02s$. At what time does the value exceed $120?

13. An investment in continuous time has a constant force of interest $\delta(t) = i$ for all $t \geq 0$. What value of i allows a $2000 investment to grow to $2500 in 3

years?

14. An investment has a linear force of interest $\delta(t) = kt$. At least how large must k be in order that the time that it takes to double in value is no more than 8 years?

15. Explain why the accumulated value function of an investment with positive initial value and continuous force of interest function $\delta(t)$ can have local maxima and minima only for values of t such that $\delta(t) = 0$.

16. A town wants to end its commitment to funding the retirement of its unionized city employees by paying off the union, which will then divide the funds fairly among its members. Given the current salary structure of its workforce, the town expects to have to pay out continuously at rate $5,000,000 per year for 20 years. Using a discount factor of 4%, what lump sum should the town pay the union?

17. A particular continuous annuity pays at a rate of $2000 per year with a nominal yearly rate of 5% for a period of 20 years.
(a) Find the future and present value of this annuity;
(b) A *continuous perpetuity* pays forever. Using the same payment and interest rate as in (a), find the present value of the perpetuity.

18. If a continuous annuity has nominal yearly rate y and payment rate p, show that its future value at time T can also be written:

$$p \cdot \frac{(1+r)^T - 1}{\log(1+r)},$$

where r is the effective yearly rate of interest.

19. Find the present value of the continuous annuity in Example 9(b).

20. As illustrated in Example 10, in a non-constant continuous annuity, the payment rate may change with time. For such an annuity of duration 10 years with payment rate equal to $1500e^{.05t}$ and yearly nominal interest rate .04, find the present and future value.

21. For discrete annuities in which payments were in arithmetic progression $P_k = P + k \cdot \Delta P$, it was rather difficult to find a formula for the future value of the annuity. In the continuous time world, what is the analog of payments in arithmetic progression? Derive a general formula for the future value at time T for such an annuity.

22. A 5-year auto loan for $25,000 is taken out with interest compounded continuously at nominal yearly rate y. At most how high can y be to keep the loan

payments under $450 per month? (You will need numerical techniques.)

23. John and Mary have a 30-year mortgage whose outstanding balance is $41,000 and they are 18 years into the loan. The terms of the mortgage calculate payments via continuous compound interest with nominal rate 5.9%. They have the opportunity to renegotiate the loan at a lower rate of 4.7%, but will incur closing costs of $1500 to do it. By how much will their monthly payments be reduced if they take the new deal, financing the closing costs?

24. Derive a retrospective form for the outstanding balance of a continuous loan, and show that it gives the same result as the prospective form in (1.82).

Chapter 2

Bonds

One of the primary goals of financial mathematics is to give fair values to financial securities that trade in markets, such as futures, options, and, in this chapter, bonds.

A **bond** is a contract between its issuer and its holder that specifies a series of payments called **coupons** (usually fixed and regular) that will be paid by the issuer to the holder, plus a lump payment of principal, its **face value**, at the end of the term of the bond. In essence, the issuer of the bond is borrowing money from the holder, in the form of the initial price of the bond paid by the bond holder. The coupon payments are determined as a fixed perentage of the face value, and two of the key quantities to relate are the price of the bond and its rate of return or **yield rate**, which is the internal rate of return on the whole transaction.

Bonds are issued by local municipalities as well as the federal government in order to raise money. But they are not limited to the public financial sphere; they are also issued by private firms for the same reason (as an alternative to selling off ownership of the company by issuing stock). Individuals, banks, insurance companies, and pension plans own bonds in their asset portfolios, and bonds can change hands between owners. So it is of some importance to be able to give value to bonds depending on their specific attributes and on the current time.

Other issues related to bonds that we will introduce in this chapter include the **amortization** of bonds (thinking of them as loans made to the issuer) and revisions to the basic valuation formulae in the case that the bond is **callable**, that is, redeemable at the discretion of the issuer prior to its maturity. We will also consider in depth the market relationships between bond yield rates and maturities, a problem referred to in the literature as the **term structure of interest rates**.

2.1   Bond Valuation

There is a wealth of notation and terminology that is necessary to understand bonds. Below is a glossary in which we elaborate on the attributes described above.

coupons - regular payments to bond holder prior to expiration of the bond;

face value - stated amount of loan upon which the coupons are calculated; notation: F;

coupon rate (or *interest rate*) - rate of interest per coupon period used to calculate coupon payments; notation: r;

bond period - time between coupon payments (often semi-annual);

maturity date (or *term to maturity*) - when the bond is scheduled to be fully paid; notation $n =$ number of periods to maturity;

redemption value (or *claim value*) - what the holder receives at the end, notation C, usually the same as face amount;

initial value of the bond - the value when the bond is issued of all payments; notation P;

yield rate - rate of return j that makes the initial value of the bond equal to the present value of the stream of payments;

bond bought at a premium - bond price P is greater than face value F;

bond bought at par - bond price P equals face value F;

bond bought at a discount - bond price P is less than face value F.

Among the most common examples of bonds are U.S. **Treasury bills**, or "T-bills," which are typically issued with short maturities such as 1 month (4 weeks), 3 months (13 weeks), or 6 months (26 weeks). These are debt obligations on the part of the U.S. government that are bought at a discount and redeemable at maturity for their face value, and they are examples of zero-coupon bonds in the sense that there are no coupon payments between purchase and maturity ($r = 0$ in our notation). **Treasury notes**, or "T-notes" have longer maturities, such as 2, 3, 5, 7, or 10 years and pay coupons every semi-annum. **Treasury bonds** are similar to T-notes, but have even longer maturities, typically 20 or 30 years.

Information on U.S. Treasury securities can be found at the U.S. Treasury website *http://www.treasurydirect.gov*, among other places. The table in Figure

type	issue date	maturity date	interest rate(%)	yield rate(%)	price per $100
13 WK. BILL	05/02/13	08/01/13	0	0.051	99.9874
26 WK. BILL	05/02/13	10/31/13	0	0.081	99.9596
52 WK. BILL	05/02/13	05/01/14	0	0.107	99.8938
2 YR. NOTE	04/30/13	04/30/15	0.125	0.233	99.7846
5 YR. NOTE	04/30/13	04/30/18	0.625	0.710	99.5832
9 YR.,10 MO. NOTE	04/15/13	02/15/23	2.000	1.795	101.8392

Figure 2.1: Some U.S. bond quotations.

2.1 gives some data from bills and notes issued in Spring 2013. These securities are sold in $100 increments, and their prices are quoted in dollars per $100 in face value in the last column. The "interest rate" column refers to the yearly coupon rate used to calculate semi-annual coupon payments to the bond holder, and the "yield rate" column gives the associated nominal yearly value of j, compounded semi-annually.

Our first task will be to find the initial bond price in terms of the yield rate.

2.1.1 Bond Value at Issue Date

Consider first the case of a zero-coupon bond such as a T-bill. The holder buys the bond for P and receives the face value of F at the end of the term of the bond. No other payments are made to the bond holder. So, the price should be the present value, using the yield rate j, of the face amount. In other words,

$$\text{price of zero coupon bond} = P = F(1 + j)^{-n}. \tag{2.1}$$

For example, look at the 52 week T-bill in Figure 2.1. Since the nominal yearly yield rate is $y = .00107$, the semi-annual effective yield rate is $j = .00107/2$, hence a single $100 T-bill should cost:

$$P = \$100 \left(1 + \frac{.00107}{2}\right)^{-2} = \$99.8931$$

on its issue date, which is in line with the figure reported in the table. (Rounding errors in the yield rate calculation can result in minor departures from the listed price per $100.)

Now let us think about how to value a coupon paying bond. If the face value is F and the coupon rate is r, then the bond holder receives payments of Fr at each of times $1, 2, ..., n$, and also at time n receives the redemption value C. The price of the bond should be the present value of this stream of payments, using j to discount. Figure 2.2 illustrates the situation and easily leads to the next theorem.

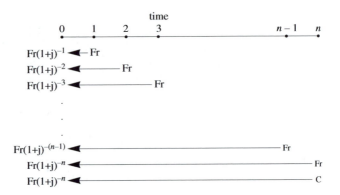

Figure 2.2: Present value stream for $\$Fr$ coupon payments with n periods, rate r per period.

Theorem 1. The price P at the issue date of a bond with face value F, redemption value C, coupon rate r, yield rate j, and maturity n is:

$$P = \frac{C}{(1+j)^n} + Fra_{n\rceil j}. \tag{2.2}$$

When the redemption value equals the face value, the initial price becomes:

$$P = \frac{F}{(1+j)^n} + Fra_{n\rceil j}. \tag{2.3}$$

Proof. As Figure 2.2 indicates, the present value of the stream of payments is the sum of the present value of the redemption amount C earned at time n, plus the total value of an ordinary annuity with payments of $\$Fr$ at the end of each period.

$$
\begin{aligned}
P &= \frac{C}{(1+j)^n} + \left(\frac{Fr}{(1+j)^1} + \frac{Fr}{(1+j)^2} + \cdots + \frac{Fr}{(1+j)^n} \right) \\
&= \frac{C}{(1+j)^n} + Fra_{n\rceil j}. \quad \blacksquare
\end{aligned}
$$

Unless otherwise noted, we will assume that $C = F$ and so formula (2.3) is the main tool for valuing bonds at the time of issue.

Notice that (2.3) also nicely covers the case of a zero-coupon bond, since $r = 0$ would imply that $P = F/(1+j)^n$, which agrees with (2.1).

Two alternative forms of the bond price P can be found as follows. First, recall that $a_{n\rceil j} = \frac{1-(1+j)^{-n}}{j}$. Thus,

$$(1+j)^{-n} = 1 - ja_{n\rceil j}.$$

Using this, we can rewrite formula (2.3) as:

$$P = F\left(1 - ja_{n\rceil j}\right) + Fra_{n\rceil j}$$
$$= F + F(r - j)a_{n\rceil j}. \tag{2.4}$$

This form is sometimes called the **premium discount form** of the initial price. From this form, since $a_{n\rceil j} > 0$, we can see how the relationship between r and j determines whether the bond price is at a premium, at par, or at a discount. If $r > j$, then $P > F$ and the bond is bought at a premium. The economic intuition is that, if r is large, then the coupons provide most of the value to the bondholder in this investment and the bondholder is inclined to pay a high price for the bond. In the case that $r = j$, the second term in (2.4) disappears, resulting in $P = F$, so the bond is bought at par. If $r < j$, then $P < F$ and the bond is bought at a discount. Economically, the coupons are not valuable enough for the holder to pay more than the face value of the bond in order to own it. This is the case for the 2-year and 5-year Treasury notes in Figure 2.1, whereas the long 9-year, 10-month note has $r > j$ and is bought at a premium.

A third form can be obtained by rewriting (2.3) by substituting the explicit expression for $a_{n\rceil j}$ and denoting the present value $\frac{F}{(1+j)^n} = F(1 + j)^{-n}$ of the face amount by K:

$$P = K + Fr\frac{1 - (1 + j)^{-n}}{j}$$
$$= K + \frac{r}{j} \cdot F\left(1 - (1 + j)^{-n}\right) = K + \frac{r}{j}(F - K). \tag{2.5}$$

Equation (2.5) is called **Makeham's formula**, which we will see is useful when we consider portfolios of bonds.

Example 1. Let us finish confirming the prices of the T-bills in Figure 2.1. For the 13-week bill with nominal yearly yield rate $j = .051$, we must discount for $1/4$ year or $1/2$ of a semi-annum, hence the present value of the bond is:

$$P = \$100\left(1 + \frac{.00051}{2}\right)^{-1/2} = \$99.9873.$$

Also, for the 26-week bill whose yield rate is .081, we discount for one semi-annual period to find the present value as

$$P = \$100\left(1 + \frac{.00081}{2}\right)^{-1} = \$99.9595.$$

These results are in accordance with the U.S. Treasury report. ∎

Example 2. For the T-notes in Figure 2.1, we can use any of forms (2.3), (2.4), or (2.5) to find the price. First, for the 2-year note with coupon rate .00125 per year and $y = .00233$ nominal yield per year, we split both of these rates in half for semi-annual periods, and so formula (2.3) implies that:

$$
\begin{aligned}
P &= \frac{F}{(1+j)^n} + Fra_{n\rceil j} \\
&= \frac{\$100}{\left(1 + \dfrac{.00233}{2}\right)^4} + \$100\left(\frac{.00125}{2}\right) a_{4\rceil .00233/2} \\
&= \$99.7846.
\end{aligned}
$$

The 5-year note has yearly coupon rate .00625 and yearly nominal yield rate $y = .00710$. Formula (2.4) yields:

$$
\begin{aligned}
P &= F + F(r - j)a_{n\rceil j} \\
&= \$100 + \$100\left(\frac{.00625}{2} - \frac{.00710}{2}\right) a_{10\rceil .00710/2} \\
&= \$99.5832.
\end{aligned}
$$

Let us try Makeham's formula (2.5) on the 9-year, 10-month note $\left(n = 2\left(9 + \frac{10}{12}\right)\right)$. Here, $r = .02/2$ and $j = .01795/2$, and we get:

$$
K = \$100\left(1 + \frac{.01795}{2}\right)^{-2(9+10/12)} = \$83.8852;
$$

$$
P = K + \frac{r}{j}(F - K) = \$83.8852 + \frac{.02/2}{.01795/2}(\$100 - \$83.8852) = \$101.84.
$$

All of these results are quite close to the table values. ∎

Example 3. Suppose that a 3-year Treasury note with face value $1000 and semi-annual coupons at yearly rate 2.4% is purchased for $1020. What is the nominal annual yield rate received by the purchaser? (Numerical techniques will be needed.)

Solution. We can use Equation (2.4) to set up an equation for j as an effective semi-annual rate of return, and then convert j to a nominal yearly rate. We are given that the bond extends for six semi-annual periods, $F = \$1000$, $P = \$1020$, and $r = .024/2 = .012$. Then, by the premium-discount form (2.4),

$$
P = F + F(r - j)a_{n\rceil j}
$$

$$
\Longrightarrow \$1020 = \$1000 + \$1000(.012 - j) \cdot \frac{1 - (1+j)^{-6}}{j}.
$$

This non-linear equation can be solved to find $j = .00857$. Note that this bond was sold at a premium, and correspondingly the coupon rate $r = .012$ is more than the semi-annual yield rate. The nominal yearly rate is $y = 2j = .01714$. ∎

Remark. Clearly the yield rate and the purchase price are strongly dependent on one another. You can derive some intuition by looking at a graph of the price P as a function of j. Using the data in Example 3, $F = \$1000$, $r = .012$ and $n = 6$, the specific function is $P = \$1000 + \$1000(.012 - j)\frac{1-(1+j)^{-6}}{j}$, whose graph is in Figure 2.3. We see that, for low yields in comparison to the semi-annual coupon rate of 1.2%, the bond is bought at a price higher than the face value of \$1000, the price is exactly \$1000 when the yield rate is the same as the coupon rate, and the price continues to fall as the yield increases beyond 1.2%. If the face value and coupon rate are fixed, lower price is associated with higher yield, and higher price with lower yield.

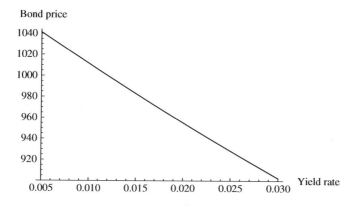

Figure 2.3: Bond price as a function of yield rate.

By returning to first principles, we can price bonds in which the coupon rates are not constant, as in the following example.

Example 4. Suppose that a 5-year \$200 municipal bond with monthly coupons has nominal yearly coupon rates in arithmetical progression: 1% for the first year, 2% for the second, etc. If the nominal annual yield rate is $y = 2\%$, what is the initial price of the bond?

Solution. The first 12 monthly coupons will be for $\$200 (.01/12) = \$ .167$, the next 12 for $\$200 (.02/12) = \$ .333$, and similarly the coupons in the third, fourth, and fifth years will be \$ .50, \$ .667, and \$ .833, respectively. Taking the time period as months, the yield rate j per month is $2\%/12 = .167\%$. We find the value of this bond as the present value of the coupons plus the face amount of \$200 collected at the end of 5 years:

$$\begin{aligned}
\text{initial price} \quad = \quad & \$.167\left((1.00167)^{-1} + \cdots + (1.00167)^{-12}\right) \\
& +\$.333\left((1.00167)^{-13} + \cdots + (1.00167)^{-24}\right) \\
& +\$.50\left((1.00167)^{-25} + \cdots + (1.00167)^{-36}\right) \\
& +\$.667\left((1.00167)^{-37} + \cdots + (1.00167)^{-48}\right) \\
& +\$.833\left((1.00167)^{-49} + \cdots + (1.00167)^{-60}\right) \\
& +\$200(1.00167)^{-60}
\end{aligned}$$

We can recognize each of the five terms prior to the face value term as the present value of an annuity, with the last four being deferred; note that, to convert to present value at time 0, we must factor $(1.00167)^{-12}$ out of the second term, $(1.00167)^{-24}$ out of the third term, etc. Therefore, the sum can be computed as:

$$\begin{aligned}
\$.167 \cdot a_{\overline{12}|.00167} \quad & + \quad \$.333(1.00167)^{-12} \cdot a_{\overline{12}|.00167} \\
& + \quad \$.50(1.00167)^{-24} \cdot a_{\overline{12}|.00167} \\
& + \quad \$.667(1.00167)^{-36} \cdot a_{\overline{12}|.00167} \\
& + \quad \$.833(1.00167)^{-48} \cdot a_{\overline{12}|.00167} \\
& + \quad \$200(1.00167)^{-60} \\
& = \quad \$209.129. \quad \blacksquare
\end{aligned}$$

Example 5. A **serial bond** is used by some bond issuers to spread out their obligations. Specifically, a serial bond with m components has face amounts $F_1, F_2, ..., F_m$, maturing at times $n_1, n_2, ...n_m$, respectively, with coupon rates $r_1, r_2, ..., r_m$ and yield rates $j_1, j_2, ..., j_m$. Suppose that a particular serial bond has five components with face amounts of \$1000, \$2000, \$3000, \$2000, and \$1000, maturity dates 1, 2, 3, 4, and 5 years, with semi-annual coupons at semi-annual rate .03 each, and effective semi-annual yield rates all equal to .025. Use Makeham's formula to derive the total present value of this serial bond issue.

Solution. It makes sense to value the whole serial bond as the sum of the present values of its components. For the ith component of the serial bond, Makeham's formula gives a value:

$$P_i = K_i + \frac{r_i}{j_i}\left(F_i - K_i\right),$$

where $K_i = (1 + j_i)^{-n_i} F_i$. In terms of semi-annual periods, the maturities are $n_1 = 2$, $n_2 = 4$, $n_3 = 6$, $n_4 = 8$, and $n_5 = 10$. Also, the coupon rate $r_i = .03$ for each component, and each yield rate $j_i = .025$. Then, by the given data, the total value of the serial bond is:

$$\begin{aligned}
P = \sum_{i=1}^{5} P_i \quad & = \quad \sum_{i=1}^{5}\left(K_i + \frac{r_i}{j_i}\left(F_i - K_i\right)\right) \\
& = \quad \sum_{i=1}^{5} K_i + \frac{.03}{.025}\sum_{i=1}^{5}\left(F_i - K_i\right) \\
& = \quad \sum_{i=1}^{5} K_i + \frac{.03}{.025}\left(\sum_{i=1}^{5} F_i - \sum_{i=1}^{5} K_i\right).
\end{aligned}$$

The sum of the face values in the middle term is: $1000+$2000+$3000+$2000+$1000 = $9000. The sum of present values is:

$$
\begin{aligned}
\sum_{i=1}^{5} K_i &= \$1000(1.025)^{-2} + \$2000(1.025)^{-4} + \$3000(1.025)^{-6} \\
&\quad + \$2000(1.025)^{-8} + \$1000(1.025)^{-10} \\
&= \$7773.30.
\end{aligned}
$$

Therefore the value of the serial bond is:

$$
\$7773.30 + \frac{.03}{.025}(\$9000 - \$7773.30) = \$9245.34. \quad \blacksquare
$$

2.1.2   Bond Value at Coupon Date

So far we have only considered the value of a bond at its time of issue. But bonds can change hands from an original holder to a new one, and so we should look at the value of a bond in the midst of its lifetime. The most basic assumption would be that the yield rate of the bond is the same for the buyer as for the seller. However, it might be that the market yield rates of similar bond instruments have changed since the bond was originally issued, so that the new buyer might not expect the same yield as the issue price of the bond implies. For example, if the bond had a yield of 2% nominal yearly when it was issued, but since then market yield rates have declined to 1%, a buyer might be very happy to purchase the bond from its original holder at a price that implies a 1.5% yield. But then what is the yield for the seller, and would the seller be happy with that?

Assume that we have a bond of maturity n, coupon rate r, and face value F. Any of formulas (2.3), (2.4), or (2.5) gives the relationship between the price at issue P and the yield rate j. Suppose this bond is sold at time k just after the coupon is collected. We ask what the **book value** B_k is that makes the yield rate equal to j for both the seller of the bond and the new buyer. Taking the buyer's point of view, this value should be the present value at time k of the future payments, of which there are $n - k$ coupons, each in the amount Fr, plus the redemption value F at time n. This reasoning shows that:

$$
B_k = \frac{F}{(1+j)^{n-k}} + Fr a_{n-k\rceil j}.
$$

The analysis begs the question: is this the same as the price the seller would demand at time k to sell the bond for the original rate of return j? Essentially, the seller of the bond wants to set up a new bond whose initial price was P, whose maturity is k, whose redemption value is some unknown price C, for which the associated yield rate is j. The first k coupons of Fr each have been received. The seller's desired price C would satisfy the present value equation:

$$
P = \frac{C}{(1+j)^k} + Fr a_{k\rceil j},
$$

which is just an adaptation of Equation (2.3) for a bond with revised conditions. Then, using Equation (2.3) for the original issue price P of the bond, we can derive as follows:

$$
\begin{aligned}
C &= P(1+j)^k - Fr(1+j)^k a_{k\rceil j} \\
&= \left(\frac{F}{(1+j)^n} + Fra_{n\rceil j} \right) (1+j)^k - Fr(1+j)^k a_{k\rceil j} \\
&= \frac{F}{(1+j)^{n-k}} + Fr \left((1+j)^k a_{n\rceil j} - (1+j)^k a_{k\rceil j} \right) \\
&= \frac{F}{(1+j)^{n-k}} + Fr \left((1+j)^k \frac{1-(1+j)^{-n}}{j} - (1+j)^k \frac{1-(1+j)^{-k}}{j} \right) \\
&= \frac{F}{(1+j)^{n-k}} + Fr \left(\frac{(1+j)^k - (1+j)^{-(n-k)}}{j} - \frac{(1+j)^k - 1}{j} \right) \\
&= \frac{F}{(1+j)^{n-k}} + Fr \cdot \frac{1-(1+j)^{-(n-k)}}{j} \\
&= \frac{F}{(1+j)^{n-k}} + Fra_{n-k\rceil j} = B_k.
\end{aligned}
$$

Thus, if the seller sells the bond at the book value B_k at the time k at which the kth coupon is issued, then both the seller and the buyer realize the same rate of return j. The definition below summarizes what we have done.

Definition 1. The **book value** of a bond at coupon time k is the time k present value of the remaining payments just after the kth coupon has been received. If the bond parameters are F, r, j, and n, this is:

$$
\text{book value of bond at time } k \equiv B_k = \frac{F}{(1+j)^{n-k}} + Fra_{n-k\rceil j}. \tag{2.6}
$$

The book value is also the price at which the bond can be sold at time k in order that both the buyer and seller receive yield rate j.

The derivation above verifies in general our intended meaning for the term book value, but it is perhaps useful to see how it plays out in a particular situation.

Example 6. An investor purchases a $2000, 20-year, 2% per period coupon rate bond with semi-annual coupons. The purchase price will give a nominal annual yield of 5%, compounded semi-annually. After the 12th coupon, he sells the bond. At what price did he sell it if his actual nominal annual yield is 5%? Compute the semi-annual yield rate for the buyer of the bond.

Solution. Since the investor is attempting to match the original semi-annual yield rate of $5\%/2 = 2.5\%$, we would anticipate that the selling price is the book value at time $k = 12$ semi-annual periods. We would also anticipate that the buyer's yield is 2.5% semi-annually. But we should verify these guesses.

We first compute the original purchase price of the bond. The problem states that $F = \$2000, r = .02, n = 40, j = .05/2 = .025$. Using formula (2.3),

$$P = \frac{F}{(1+j)^n} + Fra_{\overline{n}\rceil j} = \frac{\$2000}{(1.025)^{40}} + \$2000 \cdot (.02) \cdot a_{\overline{40}\rceil.025} = \$1748.97.$$

To answer the question, we observe that the sale of the bond at unknown price C sets up a new redemption value for the bond of C, as well as a new maturity time of 12. Using this, we can equate the initial price to the present value of coupon payments, each of which is $Fr = \$2000 \cdot .02 = \$40$, plus the present value of redemption at time 12 half-years.

$$\$1748.97 = C(1.025)^{-12} + \$40 \cdot a_{\overline{12}\rceil.025} \implies C = \$1800.35.$$

Comparing this to formula (2.6) for the book value B_{12} at time 12,

$$B_{12} = \frac{F}{(1+j)^{40-12}} + Fra_{\overline{40-12}\rceil j} = \frac{\$2000}{(1.025)^{28}} + \$40a_{\overline{28}\rceil.025} = \$1800.35,$$

which conforms to what we expected.

As for the buyer, since there will be $40 - 12 = 28$ coupon payments of $\$40$ each and a redemption value of $\$2000$ in another $40 - 12$ time periods, the yield rate j_0 satisfies the time $k = 12$ present value equation:

$$\$1800.35 = \frac{\$2000}{(1+j_0)^{28}} + \$40a_{\overline{28}\rceil j_0}.$$

The solution can be found numerically to be $j_0 = .025$, which is the same as the seller's semi-annual yield. ■

As we pointed out above, there may be cases in which a bondholder is inclined to sell a bond prior to its maturity, but, due to changes in market yield rates, the buyer will not have any motivation to buy the bond according to its original bond provisions. We illustrate this situation in the next example.

Example 7. Grumpy owns a $\$10,000$, 10-year mining company bond with quarterly coupons at 1% per coupon period, which he originally bought at a price of $\$9420$. Owing to his need to pay off a gambling debt after holding this bond for 5 years, he decides to sell the bond to Sleepy. But yield rates have risen to 1.5% in the meantime for bonds with 5-year maturities, so Sleepy offers a price consistent with this per quarter yield. What is that price, and by how much does it differ from the book value? If Grumpy accepts Sleepy's offer, what is Grumpy's yield rate? (Numerical equation solving will be necessary.)

Solution. We first compute the original yield rate of the bond. The problem states that $F = \$10,000, r = .01, n = 40, P = \$9560$. We must numerically solve for j in the equation:

$$P = \frac{F}{(1+j)^n} + Fra_{\overline{n}|j} \implies \$9420 = \frac{\$10,000}{(1+j)^{40}} + \$100 \cdot \frac{1-(1+j)^{-40}}{j}$$

$$\implies j = .0118.$$

Using this yield rate, the book value of Grumpy's bond at time $k = 20$ (5 years) is:

$$
\begin{aligned}
B_k &= \frac{F}{(1+j)^{n-k}} + Fra_{\overline{n-k}|j} \\
&= \frac{\$10,000}{(1.0118)^{20}} + \$100 \cdot \frac{1-(1.0118)^{-20}}{.0118} \\
&= \$9680.99.
\end{aligned}
$$

Next, we must compute Sleepy's offering price, if the bond gives him 20 more periods of coupon payments and is valued according to yield rate 1.5%. This price turns out to be:

$$P = \frac{\$10,000}{(1.015)^{20}} + \$100a_{\overline{20}|.015} = \$9141.57.$$

This is lower than the book value of $\$9680.99$ by $\$539.42$. If Grumpy sells the bond at Sleepy's price, then his yield rate solves the equation:

$$\$9420 = \frac{\$9141.57}{(1+j)^{20}} + \$100a_{\overline{20}|j} \implies j = .00926. \blacksquare$$

2.1.3 Recursive Approach: Bond Amortization Table

As we know, bonds may be considered as loans made to the issuer by the holder of the bond. In each period the issuer is charged a new amount of interest at rate j, where j is the effective yield rate per coupon period. Coupon payments are then the analogs of loan installment payments, but in this case there is also a lump sum payment to the holder at the end to redeem the bond at its redemption value. The coupons serve to transition the outstanding balance of the loan from the original value of P, the initial bond price, to the final value of C, the redemption value. We may look at the outstanding balance, that is, the book value of the bond at each coupon time, and make note of interesting behaviors depending on whether the bond was bought at a discount, par, or at a premium.

The book value B_0 of a bond at the beginning is the initial price P of the bond, and the final book value B_n after the last coupon is the redemption amount (again for simplicity we assume that the redemption amount is the face value F).

Formula (2.6) gives an explicit expression for the book values. In between times 0 and n, the book values $B_1, B_2, ...$ either increase to F if the bond was bought at a discount (called **writing up** the bond) or decrease to F if it was bought at a premium (called **writing down** the bond). If the bond was bought at a discount, the coupons were too small to cover the interest on the loan for the period, which accounts for the increase in outstanding balance for the period. If the bond was bought at a premium, the coupons more than cover the per-period interest; thus the outstanding balance is reduced on receipt of a coupon.

A **bond amortization table** is like a loan amortization table in that it shows for each period the outstanding balance (book value) for the period, the payment (coupon), and the interest paid. Optionally, the table can show the amount of principal paid during the period, which is negative in the case of a bond bought at a discount. It is relatively easy to form recursive relationships that generate an amortization table. In addition to the usual notation, denote:

$$I_k = \text{interest paid in period } k;$$

$$P_k = \text{principal adjustment amount after coupon } k \text{ paid.}$$

Since the coupons have value Fr and they are divided between interest jB_{k-1} for the period and adjustment to principal, we have the obvious recursive relationships:

$$I_k = jB_{k-1}, \ \ P_k = Fr - I_k, \ \ B_k = B_{k-1} - P_k, k = 1, 2, ..., n. \qquad (2.7)$$

(Make sure you understand why these equations hold.)

The recursive approach to computing the outstanding balances is well suited to spreadsheet implementation. In Figure 2.4 we illustrate a possible layout for a bond amortization table including the appropriate formulae. The face value, yield rate per period, coupon rate per period, and maturity time of the bond are input parameters in the upper left of the spreadsheet. The coupon value Fr is computed in cell B4, and formula (2.3) for the initial bond price is implemented in cell B6. After a header in row 8, we construct a column to show the period, starting at 0. In line 0 we just initialize the book value $B_0 = P$ that is computed in cell B6, and leave the rest of the entries blank. In column B from lines 10 onward the coupon value from cell B4, like a constant loan repayment amount, is displayed. The interest paid in each row is the book value in column E of the previous row times the yield rate j in cell B2. The amount of principle paid is displayed in column D, and it is the coupon payment in column B minus the interest in column C for each row. Then the book value in column E is computed as the previous book value minus the principal paid in column D.

The result of doing these computations looks something like the table in Figure 2.5. This bond has face value \$1000, yield rate $j = .01$, coupon rate $r = .005$, and maturity 8 periods. It was bought at a discount for \$961.74, which you can verify with any of formulas (2.3)–(2.5). The coupon payments are only $Fr = \$5$ per period, and the interest $B_{k-1}j$ always exceeds this, so additional

	A	B	C	D	E
1	face value	F			
2	yield rate	j			
3	coupon rate	r			
4	coupon	$B\$1 * \$B\$3$			
5	maturity	n			
6	initial price	$\frac{\$B\$1}{(1+\$B\$2)^{(\$B\$5)}} + \$B\$4$ $* \frac{1-(1+\$B\$2)^{(-\$B\$5)}}{\$B\$2}$			
7				principal	book
8	period	coupon	interest	paid	value
9	0	0	0	0	$\$B\$6$
10	1	$B\$4$	E9 * $B\$2$	= B10 − C10	E9 − D10
11	A10 + 1	$B\$4$	E10 * $B\$2$	B11 − C11	E10 − D11
12	⋮	⋮	⋮	⋮	⋮

Figure 2.4: Template bond amortization table spreadsheet.

principal is added as this bond is written up. At maturity, the outstanding balance is the face value of $1000, which is paid off in a lump sum by the issuer of the bond.

Figure 2.6 shows the case where we interchange the roles of j and r, that is, we set $j = .005$ and $r = .01$, hence the bond is bought at a premium. The other bond parameters are the same. The initial price of the bond turns out to be $P = B_0 = \$1039.11$, and the coupon payments of $\$1000 \cdot .01 = \$10$ are more than enough to account for the interest for each period, hence some principal is paid and the book value decreases to its ultimate value of $F = \$1000$ at time 8. This illustrates the process of "writing down" a bond.

Example 8. Consider a $10,000 10-year bond with semi-annual coupons at rate 2.5% per semi-annum and yield rate 2% per semi-annum. How much interest is paid from the second through the third year of the bond's existence, and what is the reduction in principal in that time?

Solution. We could do a complete amortization table (see Exercise 16), but if we just compute the first six values of B_k and use the relationships $I_k = jB_{k-1}$, $P_k = Fr - I_k$, $B_k = B_{k-1} - P_k$, it will be enough. We will add I_3 through I_6 and P_3 through P_6.

First, the intial price of the bond is:

period	coupon	interest	paid	book
0	0.	0.	0.	961.74
1	5.	9.62	−4.62	966.36
2	5.	9.66	−4.66	971.02
3	5.	9.71	−4.71	975.73
4	5.	9.76	−4.76	980.49
5	5.	9.8	−4.8	985.3
6	5.	9.85	−4.85	990.15
7	5.	9.9	−4.9	995.05
8	5.	9.95	−4.95	1000.

Figure 2.5: Amortization table for bond bought at a discount.

period	coupon	interest	paid	book
0	0.	0.	0.	1039.11
1	10.	5.2	4.8	1034.31
2	10.	5.17	4.83	1029.48
3	10.	5.15	4.85	1024.63
4	10.	5.12	4.88	1019.75
5	10.	5.1	4.9	1014.85
6	10.	5.07	4.93	1009.93
7	10.	5.05	4.95	1004.98
8	10.	5.02	4.98	1000.

Figure 2.6: Amortization table for bond bought at a premium.

$$
\begin{aligned}
P = B_0 &= \frac{F}{(1+j)^n} + Fr a_{n]j} \\
&= \frac{\$10,000}{(1.02)^{20}} + \$10,000 \cdot (.025) \cdot a_{20].02} \\
&= \$10,817.60.
\end{aligned}
$$

We need the balance for the first two semi-annual periods in order to find the values for years 2 and 3:

$$
I_1 = .02(\$10,817.60) = \$216.35; \quad P_1 = \$10,000(.025) - I_1 = \$33.65;
$$

$$
B_1 = \$10,817.60 - \$33.65 = \$10,783.95.
$$

$$
I_2 = .02(\$10,783.95) = \$215.68; \quad P_2 = \$10,000(.025) - I_2 = \$34.32;
$$

$$
B_2 = \$10,783.95 - \$34.32 = \$10,749.63.
$$

Now we can compute the ensuing four interests, principals paid, and book values as follows:

$$I_3 = .02(\$10,749.63) = \$214.99; \quad P_3 = \$10,000(.025) - I_3 = \$35.01;$$

$$B_3 = \$10,749.63 - \$35.01 = \$10,714.62.$$

$$I_4 = .02(\$10,714.62) = \$214.29; \quad P_4 = \$10,000(.025) - I_4 = \$35.71;$$

$$B_4 = \$10,714.62 - \$35.71 = \$10,678.91.$$

$$I_5 = .02(\$10,678.91) = \$213.58; \quad P_5 = \$10,000(.025) - I_5 = \$36.42;$$

$$B_5 = \$10,678.91 - \$36.42 = \$10,642.49.$$

$$I_6 = .02(\$10,642.49) = \$212.85; \quad P_6 = \$10,000(.025) - I_6 = \$37.15;$$

$$B_6 = \$10,642.49 - \$37.15 = \$10,605.34.$$

The interest paid from the second through third year is:

$$I_3 + I_4 + I_5 + I_6 = \$214.99 + \$214.29 + \$213.58 + \$212.85 = \$855.71,$$

and the principal paid is:

$$P_3 + P_4 + P_5 + P_6 = \$35.01 + \$35.71 + \$36.42 + \$37.15 = \$144.29. \blacksquare$$

The arithmetic involved in generating interest and principal values recursively can be somewhat tedious, as shown by the previous example. So the result of the next example is useful.

Example 9. Find closed formulas for the interest and principal paid in each period.

Solution. We have already derived the formula for the book value of the bond at time k:

$$B_k = \frac{F}{(1+j)^{n-k}} + Fra_{\overline{n-k}|j}.$$

Therefore,

$$I_k = jB_{k-1} = j \cdot \left(\frac{F}{(1+j)^{n-(k-1)}} + Fra_{n-(k-1)\rceil j} \right), \qquad (2.8)$$

and

$$P_k = Fr - I_k = Fr - j \cdot \left(\frac{F}{(1+j)^{n-(k-1)}} + Fra_{n-(k-1)\rceil j} \right). \qquad (2.9)$$

(See Exercise 20 for an alternative form of the principal paid.) In Example 8, for instance,

$$I_3 = .02 \cdot \left(\frac{\$10,000}{(1.02)^{20-(3-1)}} + \$10,000 \left(.025a_{20-(3-1)\rceil.02} \right) \right) = \$214.99,$$

and

$$\begin{aligned} P_3 &= \$10,000(.025) - .02 \cdot \left(\frac{\$10,000}{(1.02)^{20-(3-1)}} + \$10,000 \left(.025a_{20-(3-1)\rceil.02} \right) \right) \\ &= \$35.01. \blacksquare \end{aligned}$$

Example 10. For a \$2000 5-year bond with semi-annual coupons at 2% nominal yearly rate and yield rate 3% nominal yearly, by how much is the balance increased in the first 2 years, and what is the book value at that time? Sketch a graph of the book values for the bond's duration.

Solution. We have $n = 10$, $j = .015$, and $r = .01$. Each coupon is in the amount $\$2000(.01) = \$20$. We can either compute the total of the amounts $P_1 + P_2 + P_3 + P_4$, or we can just compute $B_4 - B_0$ in order to answer the first question. The initial price of the bond is

$$B_0 = \frac{F}{(1+j)^n} + Fra_{n\rceil j} = \frac{\$2000}{(1.015)^{10}} + \$20a_{10\rceil.015} = \$1907.78.$$

In general, after the kth coupon is received the book value of this bond is:

$$B_k = \frac{F}{(1+j)^{n-k}} + Fra_{n-k\rceil j},$$

hence:

$$B_4 = \frac{\$2000}{(1.015)^6} + \$20a_{6\rceil.015} = \$1943.03.$$

The increase in balance is $\$1943.03 - \$1907.78 = \$35.25$. The table below shows the book values, and the graph is in Figure 2.7.

period	0	1	2	3	4	5
book value	$1907.78	$1916.39	$1925.14	$1934.02	$1943.03	$1952.17

period	6	7	8	9	10
book value	$1961.46	$1970.88	$1980.44	$1990.15	$2000

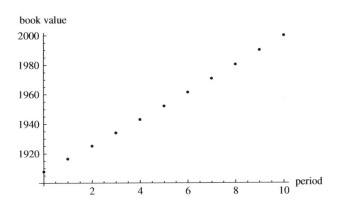

Figure 2.7: Book values of a $2000 5-year semi-annual coupon bond with $r = .01$, $j = .015$.

Important Terms and Concepts

Bond coupons - Regular payments to bondholder prior to expiration of bond.

Face amount - Stated amount of loan upon which coupons are calculated.

Coupon rate or interest rate - Rate of interest used to calculate coupon payments.

Bond period - Time between coupon payments.

Maturity date (*term to maturity*) - When the bond is scheduled to be redeemed.

Redemption amount - What the holder receives when the bond is redeemed, usually the same as face amount.

Current bond price - What the bond is selling for on the market.

Yield rate - Rate of return j that makes the current price equal to the present value of the stream of payments.

Bond bought at a premium - Bond price exceeds face value.

Bond bought at par - Bond price equals face value.

Bond bought at a discount - Bond price is less than face value.

Treasury bills - Short-term, zero-coupon U.S. debt obligations.

Treasury notes - U.S. coupon bonds with maturities ranging from 2 to 10 years.

Treasury bonds - Long-term coupon bonds.

Premium discount form - The formula $F + F(r - j)a_{n\rceil j}$ for the intial bond price that shows how the relation between the coupon rate and the yield rate determines whether the bond is sold at a discount or not.

Makeham's formula - The issue price of the bond in the form $K + \frac{r}{j}(F - K)$, where K is the present value of the redemption amount.

Book value of a bond (at coupon time) - The price at which the bond would sell just after the coupon is collected, so that the seller and the buyer earn the same yield rate.

Bond amortization table - A table listing coupons, interest paid, principal paid, and bond value at each coupon period.

Writing up (down) a bond -The process of increasing (resp. decreasing) the book value of a bond from its original price to its face value through coupon payments.

Exercises 2.1

1. What is the initial value of a 6-year semi-annual bond with coupon rate 4% nominal yearly, yield rate 5% nominal yearly, and face value $1000?

2. Suppose that a $1000 bond with quarterly coupons, maturity 2 years, and yield rate 5% nominal yearly is purchased for $950. What must be the bond's nominal yearly coupon rate?

3. Use each of the three bond price formulas to compute the initial price of a bond of face value $5000, with maturity 10 years, quarterly compounding and coupon issue, coupon rate 2% nominal yearly, and yield rate 4% nominal yearly.

4. A startup Internet company issues an 8-year bond of face value $1,000,000 and semi-annual coupons with coupon rate 5%. (a) If an investor wishes a 4%

nominal yearly rate of return on the bond, what price should the investor pay?
(b) If the price of the bond is $950,000, what is the nominal yearly yield rate?
(Numerical solution techniques will be necessary.)

5. In Example 4, arrive at the same answer by a strategy in which you value
the bond as if coupons were constantly 5%, but subtract the value of coupons
at rates lower than 5%.

6. If, in the serial bond of Example 5, the bond again has five components with
face amounts of $1000, $2000, $3000, $2000, and $1000, maturity dates 1, 2, 3,
4, and 5 years, and effective semi-annual yield rates all equal to .025, but we
assume that the total initial value of the bond issue is $9516.00, then what must
have been the semi-annual coupon rate?

7. A serial bond has four components with maturities 1, 2, 3, and 4 periods.
The yield rates of the bond are all equal to 3% per period, and the coupon rates
are all 2% per period. If the first three face values are $1000 and the initial price
of the bond is $4868.53, what is the face value of the fourth bond component?

8. Derive a formula for the present value of a bond with a non-level coupon
rate that begins at rate r and such that the coupons increase by a fixed multiple
$(1 + i)$ at each period.

9. Consider a five-period bond in which the yield rate for the first period is 1%
and the yields are not constant, but instead grow by an increment of .1% each
period. If the face value of the bond is $1000 and the per period coupon rate
remains constant at 2%, find the initial price of the bond.

10. Rederive Makeham's formula (2.5) if the redemption value C is not necessarily the same as the face value F.

11. A particular bond portfolio consists of a zero-coupon bond with face value
$2000, effective yield rate 2% per semi-annum, and maturity 6 years, and a
coupon bond of face value $1000, nominal yearly coupon rate 1%, semi-annual
coupons, nominal yearly yield rate 4%, and maturity 8 years. Assume that the
zero-coupon bond can be held indefinitely, earning compound interest at its effective yield rate. Set up and numerically solve an equation for the internal rate
of return on this bond portfolio.

12. Gladys purchases a $100 bond with maturity 8 years and semi-annual
coupons at nominal yearly rate 3%. The yield rate on the bond is listed at
6% nominal yearly. After 4 years she then sells the bond to Hilda, who holds it
until maturity, and earns an effective yield of 2.5% per half-year. For what price
did Gladys sell the bond to Hilda? What was Gladys' yield rate per semi-annual
period? (You will need numerical techniques.)

13. Sylvia buys a 4-year bond of face value $1000, coupon rate 2% for each semi-annual period, and nominal yield rate 5% annual, compounded semi-annually. One year later she decides to sell the bond to Blaine. At what price should it be sold, if Sylvia wants her original yield rate of 5%?

14. Theresa buys a $1000 bond with annual coupon rate 4% paid semi-annually and term 5 years at a discounted value of $975. What is the yield rate of the bond? If she decides to sell the bond to her sister Kate after 2 years and wants a yield rate that is .1% higher per period than her original yield, what must she charge Kate? In that case, what will Kate's yield rate be? (You will need numerical techniques for the yield rate calculations.)

15. If a bond is bought at par, what would the amortization table look like?

16. Produce a bond amortization table for Example 8.

17. A $10,000 10-year bond has yield rate 6% nominal per year, with semi-annual compounding. The nominal yearly coupon rate is 4%, with semi-annual coupons. Use a bond amortization table to find the time it takes for the bond to be paid up to at least $9,880 in value.

18. Consider a $20,000 5-year bond with semi-annual coupons at 3% nominal yearly rate and an initial price of $19,000. How much interest is paid during the first two years? (Use the recursive approach.) Use the closed form for the interest in period k to sketch a graph of the interest paid in each period for the duration of the bond. (Numerical techniques will be needed.)

19. For a $50,000 10-year bond with semi-annual coupons at 6% nominal yearly rate and an initial price of $48,000, how much interest is paid in the fourth year, and what is the book value of the bond at the end of that year? Sketch a graph of the book values for the duration of the bond. (You will need numerical techniques.)

20. An alternative, simpler form of the principal paid in period k is:

$$P_k = B_{k-1} - B_k = F(r - j) \left(\frac{1}{1+j} \right)^{n-k+1}.$$

Show this using a premium discount form of the book value. Use the defining equation for $a_{n|i}$ to show that our form is equivalent to this.

21. With F and r as usual, and C being the redemption value (not necessarily equal to F), define the **modified coupon rate g** by:

$$g = \frac{F \cdot r}{C}.$$

(a) Rederive the three expressions (2.3)-(2.5) for the initial value of a bond with redemption value C using g.

(b) Derive an expression using the modified coupon rate for the initial value of a serial bond with m components of face amounts $F_1, F_2, ..., F_m$, redemption values $C_1, C_2, ..., C_m$ maturing at times $n_1, n_2, ...n_m$, respectively, with common coupon rates r and yield rates j for each component.

2.2   More on Bonds

There are still a few important issues and ideas related to bonds that we have not touched on. It is the purpose of this section to expose you to some of these, which include the problem of valuing bonds between coupon dates; a bond modification where either the holder or the issuer puts a condition on the contract that allows for early redemption; and an interesting numerical measure that turns out to be simultaneously a measure of the average time at which bond payments are made and also the sensitivity of the price of the bond to changes in the yield rate. We begin with the valuation of a bond between the times at which coupons are paid.

2.2.1   Value of a Bond between Coupons

In the last section we put some effort into finding the value of a bond at the coupon dates. Obviously, bondholders may choose to make transactions at dates other than coupon dates, which leads to the question of how to value bonds at arbitrary times prior to maturity.

For simplicity, we will consider only the case where the redemption value C equals the face value F, and we will choose the coupon period to be the time unit, so coupons will be paid at times $k = 1, 2, ..., n$. The bond's book values at successive coupon times k and $k + 1$ are:

$$B_k = \frac{F}{(1+j)^{n-k}} + Fra_{\overline{n-k}|j}, \text{ and} \tag{2.10}$$

$$B_{k+1} = \frac{F}{(1+j)^{n-(k+1)}} + Fra_{\overline{n-(k+1)}|j}. \tag{2.11}$$

There are several approaches that we can take to value the bond at a time $t \in (k, k+1)$. Linear interpolation would be an easy approach, so that we would take the value at t to be:

$$B_t^i = ((k+1) - t)B_k + (t - k)B_{k+1}. \tag{2.12}$$

but this method lacks much economic justification. Besides simplicity though, the benefit that linear interpolation carries is that the value of the bond is a continuous function of time; note that, as $t \searrow k$, $B_t^i \longrightarrow B_k$, and as $t \nearrow (k+1)$,

$B_t^i \longrightarrow B_{k+1}$. You can try out this approach in Exercises 1 and 2, but we would like to be more sophisticated in our discussion and so we leave it here.

Because the yield rate of the bond is j, another approach is to let the bond gain value for $t - k$ time units at rate j, which results in the so-called **price plus accrued value** formula:

$$\text{price plus accrued value at } t \in (k, k+1) = B_k(1+j)^{t-k}. \qquad (2.13)$$

But there is a problem with this arising from the fact that coupons are paid discretely at integer times. Let us compare the limit of the price plus accrued value as time $t \nearrow (k+1)$ to the formula for B_{k+1}.

$\lim_{t \to k+1} B_k(1+j)^{t-k} - B_{k+1}$

$$= \lim_{t \to k+1}(1+j)^{t-k}\left(\frac{F}{(1+j)^{n-k}} + Fra_{n-k\rceil j}\right)$$
$$- \left(\frac{F}{(1+j)^{n-(k+1)}} + Fra_{n-(k+1)\rceil j}\right)$$
$$= (1+j)^1\left(\frac{F}{(1+j)^{n-k}} + Fra_{n-k\rceil j}\right) - \left(\frac{F}{(1+j)^{n-(k+1)}} + Fra_{n-(k+1)\rceil j}\right)$$
$$= (1+j)Fra_{n-k\rceil j} - Fra_{n-(k+1)\rceil j}$$
$$= Fr\left((1+j)a_{n-k\rceil j} - a_{n-(k+1)\rceil j}\right)$$
$$= Fr\left((1+j)\frac{1-(1+j)^{-(n-k)}}{j} - \frac{1-(1+j)^{-(n-k-1)}}{j}\right)$$
$$= Fr\left(\frac{(1+j)-(1+j)^{-(n-k-1)}}{j} - \frac{1-(1+j)^{-(n-k-1)}}{j}\right)$$
$$= Fr\frac{(1+j)-1}{j} = Fr.$$

So there is a downward jump in the price plus accrued value of exactly the coupon value Fr at the time of the next coupon. For example, Figure 2.8 is the graph of the price plus accrued value function for a \$2000 bond of eight periods with coupons at rate 2% per period and yield rate 3% per period. Its value at the kth coupon time is B_k for all k, and in between coupon times the value is governed by Equation (2.13). The dots are the book values at the coupon times, and the vertical drops represent the discontinuities in the function that occur when a new coupon is received.

We look instead for a way to give value to the bond between coupon times that results in a continuous bond value function, and does so with reasonable economic intuition. What we have done so far does point the way to a correction: we ought to subtract from the price plus accrued value the value at time t of the portion of the next coupon that has been earned. Once time $k + 1$ is reached, the full value of the coupon will have been subtracted, thus taking up the gap between the limit of the price plus accrued value and the book value at time $k + 1$.

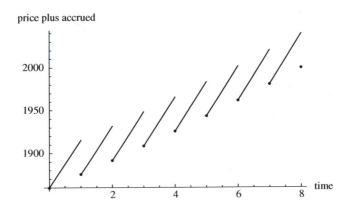

Figure 2.8: Price plus accrued values of a $2000 eight-period coupon bond with $r = .02$, $j = .03$.

But still there is not a unique way to do this. In the exercises we expose you to the method that some call the **practical bond value**, in which the portion of the coupon earned is approximated linearly (as if simple interest held):

$$\text{practical value at } t \in (k, k+1) = B_t^p = B_k(1+j)^{t-k} - (t-k)Fr. \qquad (2.14)$$

Like linear interpolation, this strategy creates a continuous function for the value of the bond.

What we will do instead is similar. From the price plus accrued value we will subtract a term that supposes partial compound interest applies to the next coupon value Fr. We must do this carefully in order to preserve continuity of the bond value function. The idea is to approximate the actual bond by one in which coupons are being paid out as a continuous annuity during each period at a rate of $p = Fr$ per period. Recall the formula for a continuous annuity that we derived in Chapter 1:

$$\text{Future value of continuous annuity at } T = \int_0^T pe^{y(T-t)}dt = p \cdot \frac{e^{yT} - 1}{y}, \qquad (2.15)$$

where y is the nominal rate of interest, which relates to the effective rate j by the equation $j = e^y - 1$. Then,

$$e^y = 1 + j, \ y = \log(1 + j),$$

and, setting the time origin as k and setting $T = t-k$, the value of the continuous coupon payment at time $t \in (k, k+1)$ becomes:

$$Fr \cdot \frac{(1+j)^{t-k} - 1}{\log(1 + j)}.$$

Unfortunately, it turns out that we do not quite get the desired continuity with this formulation, but we can get it if we replace the $\log(1+j)$ term in the denominator by its linear approximation j. (This is equivalent to making an artificial assumption that coupons are paid at a continuous rate of $Fr \log(1+j)/j$.) We are led to the following as the definition of the book value of a bond between coupon times.

Definition 1. Let a bond have face value F, coupon rate r, and yield rate j, and let B_k be the book value given by (2.10) at coupon time k. Then we define:

$$B(t) = \text{book value at } t \in [k, k+1) = B_k(1+j)^{t-k} - Fr \cdot \frac{(1+j)^{t-k} - 1}{j}. \quad (2.16)$$

As in the practical value formula, when $t = k$, the book value is:

$$B_k(1+j)^0 - Fr \cdot \frac{1-1}{j} = B_k,$$

and, as $t \longrightarrow k+1$, the book value approaches $B_k(1+j) - Fr$, which equals B_{k+1}. This shows that $B(t)$ is a continuous function of t for $t \in [0, n]$.

It is amusing that the human preference for continuity shows itself in the common languages that are in use for the discontinuous price plus accrued value function and the continuous book value function, which are the "dirty" and "clean" values, respectively.

Exercise 9 asks you to use the greatest integer function $\text{Floor}[x] = \lfloor x \rfloor$ to express the piecewise formula (2.16) as a single formula:

$$B(t) = B(\text{Floor}[t])(1+j)^{t-\text{Floor}(t)} - Fr \cdot \frac{(1+j)^{t-\text{Floor}(t)} - 1}{j}, \quad t \in [0, n], \quad (2.17)$$

where, on the right side, the function $B(k)$ is the book value function at integer coupon times k, i.e., (2.10).

Example 1. A \$100 5-year bond has semi-annual coupons issued at the beginning of February and of August. The coupon rate is 5% per annum and the nominal yield rate is 3% per annum. The bond was purchased in February, 2013 (no coupon was issued then). Find the book value of the bond on March 1, 2015 and May 1, 2015.

Solution. The coupon rate is $r = 5\%/2 = .025$ per semi-annual period, and the yield rate is $j = 3\%/2 = .015$. Coupons have value \$100(.025) = \$2.50. There are coupons on August 1, 2013 (time 1), February 1, 2014, (time 2), August 1, 2014 (time 3), February 1, 2015 (time 4), August 1, 2015 (time 5), etc. up to February 1, 2018 (time 10). Ignoring the slight differences in the number of days per month, March 1, 2015, which is one month after the coupon at time 4, is time $t = 4 + \frac{1}{6}$ in units of semi-annual periods. Similarly, May 1, 2015 is time $t = 4 + \frac{3}{6} = 4 + \frac{1}{2}$. The book value of this bond at time 4 is:

$$B_4 = \frac{F}{(1+j)^{10-4}} + Fra_{\overline{10-4}|j} = \frac{\$100}{(1.015)^6} + \$2.50 \cdot a_{\overline{6}|.015} = \$105.70.$$

By formula (2.16),

$$B\left(4 + \frac{1}{6}\right) = B_4(1.015)^{1/6} - \$2.50 \cdot \frac{(1.015)^{1/6} - 1}{.015} = \$105.55,$$

and

$$B\left(4 + \frac{1}{2}\right) = B_4(1.015)^{1/2} - \$2.50 \cdot \frac{(1.015)^{1/2} - 1}{.015} = \$105.24.$$

The graphs of the price plus accrued and book value functions are in Figure 2.9. ∎

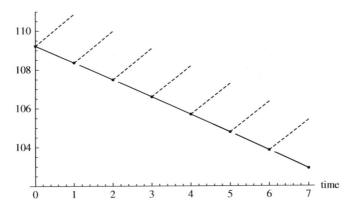

Figure 2.9: Price plus accrued (dashed) and book (solid) values of a $100 bond bought at a premium.

Example 2. Sarah buys a $1000 bond with maturity six periods, coupon rate $r = .04$ per period, and yield rate $j = .05$ per period. At time 2.25 periods, she sells the bond for its book value as specified by formula (2.16) to Ben, who holds it to maturity. Find the yield rates for both Ben and Sarah. Also, find their yield rates when the transaction price is determined by linear interpolation as in (2.12), and when the price is determined by the price plus accrued formula (2.13).

Solution. We need to compute the initial price of the bond, and the price at which it is sold, using formulas (2.10) and (2.16):

$$B_0 = \frac{F}{(1+j)^n} + Fra_{\overline{n}|j} = \frac{\$1000}{(1.05)^6} + \$40a_{\overline{6}|.05} = \$949.24;$$

$$B_2 = \frac{F}{(1+j)^{n-2}} + Fra_{n-2\rceil j} = \frac{\$1000}{(1.05)^4} + \$40a_{4\rceil.05} = \$964.54;$$

$$B(2.25) = B_2(1.05)^{.25} - \$40 \cdot \frac{(1.05)^{.25} - 1}{.05} = \$966.56.$$

From Sarah's point of view, she has purchased a bond for \$949.24, earned coupons of \$40 each at times 1 and 2, and redeemed the bond for \$966.56 at time 2.25. An equation of present values for her yield j_s is:

$$\$949.24 = \$40 (1 + j_s)^{-1} + \$40 (1 + j_s)^{-2} + \$966.56 (1 + j_s)^{-2.25}.$$

Solving numerically, we get $j_s \doteq .046$.

From Ben's perspective, he has bought a bond for \$966.56 at time 2.25, earned coupons of \$40 each at times 3, 4, 5, and 6, and redeemed the bond for \$1000 at time 6. Taking the time origin as 2.25, his yield rate j_b satisfies:

$$966.56 = 40 (1 + j_b)^{-(3-2.25)} + 40 (1 + j_b)^{-(4-2.25)} + 40 (1 + j_b)^{-(5-2.25)}$$
$$+ 40 (1 + j_b)^{-(6-2.25)} + 1000 (1 + j_b)^{-(6-2.25)}.$$

Numerical solution techniques give $j_b = .053$. So pricing the bond at time $t = 2.25$ according to formula (2.16) gives the advantage to the buyer.

The price at which the bond would be sold using linear interpolation is:

$$B_t^i = ((k + 1) - t)B_k + (t - k)B_{k+1} = (3 - 2.25)B_2 + (2.25 - 2)B_3,$$

where

$$B_3 = \frac{F}{(1+j)^{n-3}} + Fra_{n-3\rceil j} = \frac{\$1000}{(1.05)^3} + \$40a_{3\rceil.05} = \$972.77.$$

The interpolated value comes out to be $B_t^i = \$966.60$. This is only barely different from the book value, and resolving the two equations gives again $j_s = .046$ and $j_b = .053$ to the third decimal place.

If the transaction price is determined by the price plus accrued formula, then it becomes:

$$B_2(1.05)^{.25} = \$964.54(1.05)^{.25} = \$976.38.$$

This higher value could substantially decrease Ben's yield and increase Sarah's. The new equations to solve are:

$$949.24 = 40 (1 + j_s)^{-1} + 40 (1 + j_s)^{-2} + 976.38 (1 + j_s)^{-2.25},$$
$$976.38 = 40 (1 + j_b)^{-(3-2.25)} + 40 (1 + j_b)^{-(4-2.25)} + 40 (1 + j_b)^{-(5-2.25)}$$
$$+ 40 (1 + j_b)^{-(6-2.25)} + 1000 (1 + j_b)^{-(6-2.25)},$$

and we do get to a high degree of accuracy $j_b = j_s = .05$ for both the buyer and the seller. So we have found the interesting result that, between coupon times, the book value is not the fair transaction price at which the bond can be sold to give the original yield rate to both the seller and the buyer; instead it is the price plus accrued value that would achieve this goal. (See Exercise 11 for the general result.)   ∎

2.2.2   Callable and Putable Bonds

Some bonds may be redeemed at agreed upon coupon dates before the maturity date. In a **callable bond**, the bond issuer makes the choice to redeem the bond; in a **putable bond**, the bondholder makes the choice. The early redemption value, called in this context the **call price** (or **put price**) need not be the same as the face amount of the bond. The ability of one or the other party to decide whether to redeem the bond before maturity allows them to influence the yield rate. This means that the party who is not the decision maker may demand some compensation in the form of a price discount or premium, or a reduction or increase in the yield as compared with prevailing market conditions. There is much that can be said about these bonds with early redemption conditions, but we will just illustrate the main ideas with a few examples and state two theorems that establish rules of thumb about yields according to whether the bond was bought originally at a discount or at a premium.

The first example looks at the effect of call price on yield for a callable bond.

Example 3. A \$100,000 5-year callable bond of coupon rate 5% convertible semi-annually is issued by a town for improvements to its infrastructure. The bond can be called by the town at the end of the third year. The nominal annual yield rate for the bond at issue is 6%. What is the yield if the bond is called and the call price is \$100,000? What must the call price be so that the bond still yields 6% if called? What is the yield if the call price is \$3000 less than the face value?

Solution. There are to be $n = 10$ coupons, at rate $r = .05/2 = .025$ per semi-annum, hence the coupons are in the amount of \$2500. The yield rate per semi-annum is $j = .06/2 = .03$. We can compute the initial price of the bond using the premium discount version as:

$$F + F(r - j)a_{n\rceil j} = \$100,000 + \$100,000(.025 - .03)a_{10\rceil .03} = \$95,734.90.$$

Note that this bond is bought at a discount. If the call is made at the sixth period and the call price is the face value of \$100,000, then the yield per semi-annum j satisfies the equation:

$$\$95,734.90 = \frac{\$100,000}{(1 + j)^6} + \$2500 \cdot a_{6\rceil j} \implies j \approx .03295,$$

by solving numerically. So the nominal annual yield to the bond holder is about 6.59% as opposed to the 6% under which the bond was originally issued. It is good for the investor, and bad for the issuer, if the call occurs (at face value). Since coupons are relatively less valuable to the holder than the appreciation from purchase price to redemption value, the earlier the face value is paid, the better for the holder. The issuer who is in charge of making the choice would not be inclined to redeem early, unless external considerations made it desirable to clear the debt off the books.

To answer the next question, assume that j remains at 3% per semi-annum if the call is made. Then the call price C satisfies the equation of value:

$$\$95,734.90 = \frac{C}{(1.03)^6} + \$2500 \cdot a_{\overline{6}|.03}$$

$$\implies C = 1.03^6 \left(\$95,734.90 - \$2500 \cdot a_{\overline{6}|.03}\right) = \$98,141.50.$$

So the call price must be $98,141.50 to keep the nominal annual yield rate at 6%. If, on the other hand, the call price is $3000 less than the face value, namely, $97,000, then, if the call is made, the yield rate satisfies:

$$\$95,734.90 = \frac{\$97,000}{(1+j)^6} + \$2500 \cdot a_{\overline{6}|j}.$$

Numerical solution of the equation gives $j \approx .0282$, so that the nominal annual yield in this case drops to about 5.64%. As the call price falls, of course the benefit goes to the issuer and not the holder. ∎

The next example considers a callable bond bought at a premium and analyzes the dependence of yield on the time when the call is made.

Example 4. Suppose that a $10,000 10-year callable bond of coupon rate 5% paid annually is priced at $10,500 and can be called for the face value of $10,000 at each of the fourth, fifth, and sixth years. Find the effective annual yield rates in these three cases, and in the case that the bond is never called.

Solution. The coupons are in the amount of $500 annually. The bond is bought at a premium (so the investor is counting on the coupons). If the bond is never called, then the yearly effective yield rate j satisfies:

$$\$10,500 = \frac{\$10,000}{(1+j)^{10}} + \$500 \cdot a_{\overline{10}|j},$$

whose solution is about $j = 4.37\%$ if the bond is not called. If it is called at the end of years 4, 5, or 6, respectively, then the yields can be computed as follows:

$$\text{call at 4: } \$10,500 = \frac{\$10,000}{(1+j)^4} + \$500 \cdot a_{\overline{4}|j} \implies j = .0363;$$

$$\text{call at 5: } \$10,500 = \frac{\$10,000}{(1+j)^5} + \$500 \cdot a_{\overline{5}|j} \implies j = .0388;$$

$$\text{call at 6: } \$10,500 = \frac{\$10,000}{(1+j)^6} + \$500 \cdot a_{\overline{6}|j} \implies j = .0404.$$

So the yearly yield rates are increasing as the call time increases. The longer the investor is able to hold the bond, the better the overall yield rate, as a result of the comparatively high coupon value. The bond issuer, whose decision it is to make, is induced to call the bond as early as possible. So the holder, to be conservative, should consider this investment as yielding the lowest rate possible among the issuer's choices, namely, 3.63%. ∎

In the next example we look at a putable bond. If the bond holder can make the choice to redeem the bond early, when should the holder do it? How does the decision depend on whether the bond was bought at a discount or at a premium?

Example 5. A \$5000 4-year bond pays coupons semi-annually at rate 2% per period. At the discretion of the bond holder, the face value can be redeemed at any time after the midpoint of its lifetime. Find the possible yields of the bond if: (a) it was purchased at \$4800; (b) it was purchased at \$5200.

Solution. (a) There are 8 coupons of value $\$5,000 \cdot .02 = \$100$. The equations below, which require numerical solution, characterize the yields per semi-annum if the bond is redeemed by the holder at times 5, 6, 7, and 8:

$$\text{put at 5: } \$4800 = \frac{\$5000}{(1+j)^5} + \$100 \cdot a_{\overline{5}|j} \implies j = .0287;$$

$$\text{put at 6: } \$4800 = \frac{\$5000}{(1+j)^6} + \$100 \cdot a_{\overline{6}|j} \implies j = .0273;$$

$$\text{put at 7: } \$4800 = \frac{\$5000}{(1+j)^7} + \$100 \cdot a_{\overline{7}|j} \implies j = .0263;$$

$$\text{put at 8: } \$4800 = \frac{\$5000}{(1+j)^8} + \$100 \cdot a_{\overline{8}|j} \implies j = .0256.$$

The highest rate of return for the holder occurs when the bond is put back to the issuer at the earliest possible time, namely, after the fifth coupon is paid, at which the yield rate is about 2.87%. As in Example 3, it is better for the holder to choose to redeem the bond early when the lost coupons are relatively less important than claiming the face value.

(b) The solution is similar to part (a), but this time the bond is bought at a premium. The equations for the yields per semi-annum if the bond is put back to the issuer by the holder at times 5, 6, 7, and 8 are now:

$$\text{put at 5: } \$5200 = \frac{\$5000}{(1+j)^5} + \$100 \cdot a_{\overline{5}|j} \implies j = .0117;$$

$$\text{put at 6: } \$5200 = \frac{\$5000}{(1+j)^6} + \$100 \cdot a_{\overline{6}|j} \implies j = .0130;$$

$$\text{put at 7: } \$5200 = \frac{\$5000}{(1+j)^7} + \$100 \cdot a_{\overline{7}|j} \implies j = .0140;$$

$$\text{put at 8: } \$5200 = \frac{\$5000}{(1+j)^8} + \$100 \cdot a_{\overline{8}|j} \implies j = .0147.$$

So the holder is better advised to hold on to the bond as long as possible, that is, to its maturity at time 8, at which the yield rate is about 1.47%. The relative desirability of the coupons as compared to claiming the face value of the bond explains this result. ∎

The preceding examples seem to point at general principles about early redemption of bonds, when the bond contract permits it. When a bond is bought at a discount, yield rates decrease as time of redemption approaches the maturity of the bond, and when a bond is bought at a premium, yield rates increase with time. Thus, ignoring other side factors, for a callable bond, the issuer's preference would be to not call a bond bought at a discount until maturity, and to call a bond bought at a premium as soon as possible. For a putable bond, the holder's preference is to put the bond back to the issuer as soon as possible in the discount case, and to hold onto the bond until maturity in the case that the bond was bought at a premium. To end this subsection, let us state and prove the results on monotonicity of yield rates that lead to these conclusions. (We leave half of the proof to the reader as Exercise 18.)

Theorem 1. (a) Suppose that a bond is redeemable at its face value F, and the bond is bought at a premium so that the initial price $P > F$ and the coupon rate r exceeds the bond's initial yield rate j. Then, if coupon times $k < l$ are possible redemption times of the bond, the yield rates on redemption satisfy $j_k < j_l$.

(b) If the conditions of (a) hold, except that the bond is bought at a discount $P < F$, then $k < l \implies j_k > j_l$.

Proof. Note first that, by the premium discount form for the value of a bond, since redemption is at the face value even if it occurs early at time k, we have:

$$P = F + F(r - j_k) a_{\overline{k}|j_k},$$

hence the assumption that $P > F$ implies that $r > j_k$ for any of the possible pre-maturity redemption yield rates j_k.

Now, since the redemption value is F and the purchase price is P at both times k and l, we have both:

$$P = Fra_{k\rceil j_k} + F(1+j_k)^{-k}$$

and

$$P = Fra_{l\rceil j_l} + F(1+j_l)^{-l}.$$

The definition of $a_{l\rceil j_l}$ allows us to rewrite it in the following way:

$$
\begin{aligned}
a_{l\rceil j_l} &= \left(\frac{1}{(1+j_l)} + \cdots + \frac{1}{(1+j_l)^k} + \frac{1}{(1+j_l)^{k+1}} + \cdots + \frac{1}{(1+j_l)^l} \right) \\
&= \left(\frac{1}{(1+j_l)} + \cdots + \frac{1}{(1+j_l)^k} + \frac{1}{(1+j_l)^k} \left(\frac{1}{(1+j_l)} + \cdots + \frac{1}{(1+j_l)^{l-k}} \right) \right) \\
&= a_{k\rceil j_l} + \frac{1}{(1+j_l)^k} \left(\frac{1}{(1+j_l)} + \cdots + \frac{1}{(1+j_l)^{l-k}} \right) \\
&= a_{k\rceil j_l} + \frac{1}{(1+j_l)^k} \cdot \frac{1-(1+j_l)^{-(l-k)}}{j_l}.
\end{aligned}
$$

In the equation defining j_l, expand $(1+j_l)^{-l}$ as $(1+j_l)^{-(l-k)}(1+j_l)^{-k}$. Then, since $r > j_l$, we can derive:

$$Fra_{k\rceil j_k} + F(1+j_k)^{-k}$$

$$
\begin{aligned}
&= P \\
&= Fra_{l\rceil j_l} + F(1+j_l)^{-l} \\
&= Fra_{l\rceil j_l} + F(1+j_l)^{-k}(1+j_l)^{-(l-k)} \\
&= Fr\left(a_{k\rceil j_l} + \frac{1}{(1+j_l)^k} \cdot \frac{1-(1+j_l)^{-(l-k)}}{j_l} \right) + F(1+j_l)^{-k}(1+j_l)^{-(l-k)} \\
&= Fra_{k\rceil j_l} + F(1+j_l)^{-k} \left(r \cdot \frac{1-(1+j_l)^{-(l-k)}}{j_l} + (1+j_l)^{-(l-k)} \right) \\
&> Fra_{k\rceil j_l} + F(1+j_l)^{-k} \left(j_l \cdot \frac{1-(1+j_l)^{-(l-k)}}{j_l} + (1+j_l)^{-(l-k)} \right) \\
&= Fra_{k\rceil j_l} + F(1+j_l)^{-k}.
\end{aligned}
$$

Looking at the two sides of this inequality, we see that they are the present values of two bonds with the same coupons, redemption values, and maturity k, but different yield rates j_k and j_l. The value for the bond with yield j_k is larger, which can only be the case if $j_k < j_l$.

(b) See Exercise 18.

2.2.3   Bond Duration

We close this section by briefly considering the notion of the **Macaulay duration** of a bond, which is a measure of the weighted average of times at which payments are received, with the weight for a payment being determined as the portion of the overall present value of all payments made up by the present value of that payment. But the true importance of the notion emerges in an unexpected connection between this duration and a measure of the sensitivity of the price of a bond to changes in the yield rate. We will only be treating the subject of duration in the context of bonds, but it is actually a general concept that can be applied to an investment with any sequence of payments.

We know that the initial price of a bond is related to the yield rate j by:

$$
\begin{aligned}
P(j) &= Fr a_{n\rceil j} + F(1+j)^{-n} \\
&= \sum_{l=1}^{n} Fr \cdot \frac{1}{(1+j)^l} + F \cdot \frac{1}{(1+j)^n} \\
&\equiv \sum_{l=1}^{n} P_l \cdot \frac{1}{(1+j)^l},
\end{aligned}
\tag{2.18}
$$

where, for convenience, we denote:

$$
P_l = \begin{cases} Fr & \text{if } l = 1, 2, ..., n-1; \\ Fr + F & \text{if } i = n. \end{cases}
\tag{2.19}
$$

The portion of the overall present value $P(j)$ attributable to the ith payment is then:

$$
\frac{P_i \cdot \dfrac{1}{(1+j)^i}}{P(j)} = \frac{P_i \cdot \dfrac{1}{(1+j)^i}}{\sum_{l=1}^{n} P_l \cdot \dfrac{1}{(1+j)^l}}.
$$

which leads us to the following definition.

Definition 2. Let a bond have face value F, coupon rate r, and yield rate j. The *Macaulay duration* of the bond is defined by

$$
D = \sum_{i=1}^{n} i \cdot \left(\frac{P_i \cdot \dfrac{1}{(1+j)^i}}{\sum_{l=1}^{n} P_l \cdot \dfrac{1}{(1+j)^l}} \right),
\tag{2.20}
$$

where the bond payments P_i are as in (2.19). Thus, D is the weighed average of payment times, weighted by the relative sizes of the present values of the payments.

Example 6. Compute the Macaulay duration of a $1000 4-year bond with semi-annual coupons at rate 4% per year and yield rate 5% nominal yearly.

Solution. Since $n = 8$ semi-annual periods, $r = .02$, and $j = .025$, we have the initial price as:

$$P = Fra_{n\rceil j} + F(1 + j)^{-n} = \$1000(.02)a_{8\rceil.025} + \$1000(1.025)^{-8} = \$964.15.$$

Note that the coupon payments are $Fr = \$20$, and the final payment includes the coupon as well as the redemption value \$1000. Then the Macaulay duration is:

$$D = \sum_{i=1}^{7} i \cdot \frac{\$20(1.025)^{-i}}{\$964.15} + 8 \cdot \frac{\$1020(1.025)^{-8}}{\$964.15} = 7.46.$$

Despite the discounting, the payments are very much backloaded because of the redemption of \$1000 at time 8, pulling the weighted average duration of the bond up to nearly 7.5 periods. ■

The idea of weighted average time of payment is a fairly interesting quantity, but the importance of the Macaulay duration would not be nearly as great without the connection to a different question that we will look at now. You can view the initial value of the bond $P(j)$ as a function of the yield rate j, holding other parameters fixed. A first-order approximation to the change in P over a small interval of yields $[j, j + \Delta j]$ is:

$$\Delta P \approx \frac{dP}{dj} \cdot \Delta j,$$

which implies that the relative change in P is approximated by:

$$\frac{\Delta P}{P(j)} \approx \frac{\frac{dP}{dj} \cdot \Delta j}{P(j)} = \frac{dP/dj}{P(j)} \Delta j. \tag{2.21}$$

But by formula (2.18),

$$P(j) = \sum_{i=1}^{n} P_i \cdot \frac{1}{(1 + j)^i} \implies \frac{dP}{dj} = \sum_{i=1}^{n} P_i \cdot (-i)(1 + j)^{-i-1}.$$

Substituting into (2.21) yields that:

$$\frac{\Delta P}{P(j)} \approx \frac{\sum_{i=1}^{n} P_i \cdot (-i)(1 + j)^{-i-1}}{P(j)} \Delta j = \left(\frac{-1}{1 + j} \cdot \sum_{i=1}^{n} \frac{P_i(1 + j)^{-i}}{P(j)} \cdot i \right) \Delta j. \tag{2.22}$$

We recognize the sum on the right as the Macaulay duration D, so we have that:

$$\frac{\Delta P}{P(j)} \approx \left(\frac{-1}{(1 + j)} \cdot D \right) \Delta j. \tag{2.23}$$

This expression shows that the price decreases as the yield increases, and the greater is D, the greater is the relative change in the price of the bond.

Important Terms and Concepts

Price plus accrued value - The value at time t after coupon k and before the next coupon computed as the book value of the bond at time k with compound interest at rate j added on: $B_k(1+j)^{t-k}$.

Practical bond value - A linear approximation to the bond value at time t after coupon k and before the next coupon, which is the price plus accrued value, less the earned value of the next coupon: $B_k(1+j)^{t-k} - (t-k)Fr$.

Value of a bond between coupons - The bond value at time t computed as the price plus accrued value, less the earned value of the next coupon assuming continuous compound interest: $B_k(1+j)^{t-k} - Fr \cdot \frac{(1+j)^{t-k}-1}{j}$.

Callable bond - A bond that may be redeemed early at the discretion of the issuer.

Putable bond - A bond that may be redeemed early at the discretion of the holder.

Macaulay duration of a bond - An average time of payment for a bond, computed by weighting the times of payments by the proportion that the payment makes up of the present value of all payments; also a measure of the sensitivity of the relative change in price of a bond to changes in the yield rate.

Exercises 2.2

1. For a bond of face value $1000, coupon rate .03 per period, yield rate .04 per period, and maturity 8 periods, use the linear interpolation method to value the bond at times: (a) $t = 1.5$; (b) $t = 3.25$; (c) $t = 6.4$.

2. Suppose that the bond in Exercise 1 is sold at time $t = 2.8$ at the price determined by the linear interpolation method. Compute the rates of return for both the original bondholder and the new buyer.

3. Produce a graph of the price plus accrued function for a bond with face value $100, coupon rate .02 per period, yield rate .01 per period, and maturity 6 periods. Also, produce a graph of the practical bond value for this bond.

4. Show that the practical value in (2.14) produces a continuous function of t to approximate the bond value.

5. Consider a $500 bond with coupon rate .01 per period, yield rate .015 per period, and maturity 8 periods. At what time, using the practical value approximation, does the bond reach a value of $490? Answer the same question using formula (2.16) for the book value.

6. Derive the inequality below for the absolute difference between the interpolated bond value and the practical bond value.

$$\left|B_t^p - B_t^i\right| \le B_k \cdot \max_{s\in[0,1)} \left|(1+j)^s - (1+j\cdot s)\right|, t \in [k, k+1).$$

Note that it does not depend directly on the coupon rate or face value. (Hint: in the expression for B_t^i, use the recursive form for B_{k+1}.) When $j = .02$, find this bound.

7. Given a 10-year, $5000 bond with coupon rate 3% per year convertible semi-annually and nominal annual yield rate 4%, use the book value formula (2.16) to find the value of the bond: (a) at time $3\frac{1}{4}$ years; (b) at time $6\frac{3}{4}$ years.

8. The bond in Exercise 7 is to be sold at time $2\frac{1}{12}$ years at a price such that the seller and buyer both have nominal annual yield of 4%. What is that price?

9. Show that the book value can be written as:

$$B(t) = B(\text{Floor}(t))(1+j)^{t-\text{Floor}(t)} - Fr \cdot \frac{(1+j)^{t-\text{Floor}(t)} - 1}{j}, \ t \in [0, n],$$

where $B(k)$ is the book value function at integer coupon times k, and Floor of t is the greatest integer less than or equal to t.

10. A $1000 bond with coupons at the rate of 4% per period, yield rate 3% per period, and maturity 6 is to be sold by Ann to Tom at time 2.3. Find the yield rate of each of Ann and Tom if the transaction price is determined by (a) the book value; (b) the price plus accrued formula.

11. Show that, in general, if a bond with face value F, coupon rate r, yield rate j, and maturity n is sold between coupons at a time $t \in (k, k+1)$, then the price plus accrued value $B_k(1+j)^{t-k}$ is the price for which both the seller and buyer receive a yield rate of j. (You may assume that the yield rates for buyer and seller are unique.)

12. A $4000 bond with maturity four periods can be called in by its issuer after two periods. If the bond has stated coupon rate of 4.5% per period and yield rate 4% per period, what will be the yield rate if the issuer does make the call? What rate should the bondholder attribute to the investment, thinking conservatively?

13. A corporate bond of term 10 years has value \$10,000 and gives coupons to its holder at the rate of 3% nominal yearly, payable at the end of each quarter. The bond was purchased for \$9500, and the corporation has the option of calling it at face value at the end of the 7th, 8th, or 9th year. What are the possible yields for the bondholder, expressed as nominal yearly rates? If the chance that the bond will be called after year 7 is 1/2, after year 8 is 1/4, and after year 9 is 1/8, formulate and compute a notion of the average yield per quarter (note that the bond may not be called).

14. A \$5000 5-period bond has coupon rate 2% per period and is bought to give a yield rate of 4% per period. But the bond is redeemable at the discretion of the bondholder just after the third period and at no other time. Is it in the bondholder's interest to exercise this option? What is the actual rate of return on investment if the bond is redeemed?

15. Gus buys an 8 year, \$2000 bond with yearly coupons at rate 1.4% and effective yield rate .8% annually. He has the right to redeem the bond at \$2080 in 4 years, or at \$2060 in 6 years. By numerically computing yield rates, decide whether it is in his best interest to redeem the bond early.

16. A U.S. bond of face value \$100 and maturity 20 years is able to be redeemed in 10 years if the bond holder wishes. The coupon rate is 2.5% per year, and coupons are paid semi-annually. The original nominal yield rate of the bond was 3% annually. What redemption value at the 10-year mark produces a nominal annual yield of 3% if the bond is redeemed?

17. Oscar borrows \$1500 from his bookie to bet on the Super Bowl. The bookie requires 1% payments of interest every week, and wants the principal of the loan paid back in 6 weeks after the game. The bookie is looking for a 2% yield on his investment, but Oscar demands the right to pay off the principal of the loan early if he can, at the end of the fourth or fifth week. (a) Find the amount that Oscar must repay in 6 weeks if he does not prepay; (b) find the amount that Oscar repays in each of the cases where he pays in the fourth and fifth week, assuming that the bookie will still require a 2% yield.

18. Prove part (b) of Theorem 1.

19. What is the Macaulay duration of a zero-coupon bond?

20. Compute the Macaulay duration of the bond in Exercise 1.

21. Compute the Macaulay duration of the bond in Exercise 10.

22. For the bond in Exercise 7, approximate the change in its initial price if yield changes from 4% to 4.1% (nominal annual). Compute the exact value of this change and find how far off the approximation is from the exact value.

23. Derive the following closed form for the Macauley duration of a bond of face value F, coupon rate r, yield rate j, and maturity n:

$$D = \frac{Fr}{P}(1+j)^{-1} \cdot \frac{n(1+j)^{-(n+1)} - (n+1)(1+j)^{-n} + 1}{\left(1 - (1+j)^{-1}\right)^2} + \left(\frac{n \cdot F}{P(1+j)^n}\right),$$

where P is the initial price of the bond. Hint: Consider the series of coupon payments separately from the final payment of F, and sum the series by differentiating a geometric series.

2.3   Term Structure of Interest Rates

Our work so far has focused on relating the value of a bond to its yield rate. We have not addressed the question of where yield rates or bond prices come from. Market factors of several kinds are considerations. Essentially, the buyer of a bond is loaning money to the issuer. The buyer should expect a return on investment that is competitive with other returns on other investments with a similar level of risk. For bonds, the only risk is in the default of the issuer, that is, the issuer's inability to make one or more payments. So the assessment of default risk will contribute to the price that a buyer will be willing to pay for a bond, determining accordingly the yield rate. Bonds with higher default risk will require higher yields and lower prices for similar face values. Also, in hard economic times in which risky investments become less attractive, the market for less risky assets becomes better, and more investors will be willing to compete with one another to buy bonds, driving the price up and the yield down. For government bonds for which the risk of default is very small, a major concern of the investor is liquidity, that is, the ease of availability of money, and investing in a bond ties up money during the term of the bond, so the investor would demand compensation in terms of yield, with longer terms typically being associated with higher yields demanded by the investor.

We will not attempt to address here in detail the prevailing theories about the determinants of yield rates, but the last observation above points the way to what we want to do in this section, which will be more descriptive than predictive. We would like to study bond yield rates as a function of their terms to maturity, and what this relationship implies about the current view of the market about risk-free rates in the future. This maturity-yield relationship is called the **term structure of interest rates**. Since bond yield rates change daily as market conditions change, empirical data sets are just snapshots at a particular time of the maturity-yield relationship, which is subject to change.

To see this point, consider the data below, reported by the U.S. Treasury website *www.treasury.gov*, which are nominal annual yield rates (in %) for par value treasury bills, notes, and bonds of several maturities for two dates in May

2013:

date	3 mo.	6 mo.	1 yr	2 yrs	3 yrs	5 yrs	7 yrs	10 yrs	20 yrs
				maturity					
5/1/13	.06	.08	.11	.20	.30	.65	1.07	1.66	2.44
5/8/13	.04	.08	.11	.22	.35	.75	1.20	1.81	2.61

A graph of the yield as a function of the maturity, interpolating linearly between points, is called a **yield curve**. (Some more sophisticated yield curves use curve-fitting techniques like cubic splines to connect data points.) Yield curves for these 2 days are shown together in Figure 2.10. Notice that both graphs are increasing, reflecting the investor desire to be compensated more when money is tied up in the bond for a longer period of time. The U.S. government has a very low risk of defaulting on its debt obligations, so that default risk should not have much impact here on the relation between term and yield. But if you think of the corporate bond market, you would also expect long terms to be associated with higher yields due to the increased chance of default during the lifetime of the corporate bond, for which the investor would expect compensation. But the yield curves at different times, though similar, are not identical. Market attitudes toward the risk-free rate are variable, so we would not expect the curves to be rigid as time progresses.

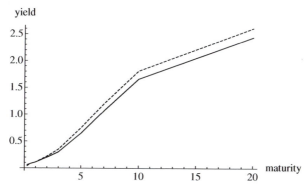

Figure 2.10: U.S. Treasury yield curves for 5/1/13 (solid) and 5/8/13 (dashed).

In fact, the differences between such yield curves can be clues to changes in investors' beliefs about the future performance of the economy, as we now discuss. Yield rates are one important measure of the demand for capital for investment or commercial purposes. As business booms, the demand for capital increases, driving up interest rates on loans. Hence high yield rates are associated with an expectation on the part of the market of a growth period for the economy. When the yield curve is very steep, it may be reflecting a belief that an economic recovery is beginning, because investors are bidding lower prices for relatively low-risk bonds with longer maturities, preferring instead to move to what they believe to be a healthy risky securities market. When might a yield curve be **inverted**, that is, have a range of maturities over which it decreases? Suppose, for instance, a decline in yields is observed between terms of 3 months and 1 year. Noting the inverse relation between yield rate and price of the security, investors are bidding less for 3-month bonds than for 1-year bonds. This indicates that they wish to move their money into the bond market and out of the risky securities market for the next year, thinking that during that time risky investments are going to decline in value. For these reasons, economists view inversions to the yield curve as signals of a downturn in the risky securities market, which may be an indicator of an upcoming recession.

2.3.1 Spot Rates, STRIPS, and Yield to Maturity

The discussion above was just a brief introduction to what we mean by the term structure of interest rates, and the reasons for the shape of a term structure graph and its change with time. Much more could be said, but for the rest of this section we choose to look at two alternative characterizations of the term structure, and their relationships with the yield to maturity. In particular, we will define and study **spot rates of interest** and **forward rates**.

It may surprise you to learn that financial institutions (not the U.S. Treasury itself) can actually split bond coupon payments into individual pieces called STRIPS (a descriptive acronym for Separate Trading of Registered Interest and Principal of Securities). This allows us to think of a single bond as a collection of zero-coupon bonds, with one bond at each coupon payment time of redemption value equal to the coupon payment. We could either lump the final coupon payment with the face value of the bond or split it off, but the former is usually done. Each of these STRIPS has a yield rate called the **spot rate**, and, in total, the spot rates must be consistent with the interest rate and price specifications of the original bond. We are led to the following definition.

Definition 1. The **spot rate of interest** is the yield rate of a zero-coupon bond of the given term, initial price, and face value.

We will use the symbol j as we have earlier in this chapter to indicate the per-period yield to maturity of a bond, and to emphasize its dependence on the maturity time n we will subscript it: j_n. For the spot rate, we will denote:

$$s_n = \text{effective yield rate per period for zero-coupon bond with maturity } n.$$
(2.24)

Thinking of a zero-coupon bond as a loan with a single repayment at the maturity time, essentially the spot rate is a measure of the time value of money, with the discounting factor $1/(1 + s_n)^n$ indicating the present value of \$1 paid at time n.

Using a little financial reasoning that we already know, let us see how to go back and forth between yields to maturity of bonds and spot rates.

Example 1. Suppose that a one-period zero-coupon \$100 bond can be bought at a yield rate of 1.5% per period, a two-period \$100 bond with coupons at .5%

has yield 2%, a three-period $100 bond with 1% coupons has yield 2.5%, and a four-period $100 bond with 1% coupons has yield 2.8%. Find each of the spot rates s_1, s_2, s_3, and s_4, and draw a graph of the spot rate as a function of maturity for maturities of 1 through 4.

Solution. Consider first the single-period bond. By Definition 1, the spot rate s_1 is the same as the yield rate $j_1 = .015$, since the one-period bond has no coupons. We can build up the other spot rates iteratively using the following idea. The two-period bond has coupons at times 1 and 2 of value $100(.005) = \$.50$, and pays $100 at time 2. The initial $.50 payment should be discounted back to a present value using rate s_1, but the second payment of $100 + \$.50$ should be discounted back using the unknown spot rate s_2 for an amount of money to be received two periods from now. Equate this total present value of the two STRIPS to the current price of the full bond, which is:

$$P = \frac{F}{(1+j_2)^n} + Fra_{\overline{n}|j_2} = \frac{\$100}{(1.02)^2} + \$.50a_{\overline{2}|.02} = \$97.09.$$

Thus, we have the equation of value:

$$\begin{aligned}
\$97.09 &= \frac{\$.50}{(1+s_1)} + \frac{\$100 + \$.50}{(1+s_2)^2} \\
&= \frac{\$.50}{(1.015)} + \frac{\$100.50}{(1+s_2)^2} \\
\implies \frac{\$100.50}{(1+s_2)^2} &= \$97.09 - \frac{\$.50}{(1.015)} = \$96.60 \\
\implies (1+s_2)^2 &= \frac{\$100.50}{\$96.60} \\
\implies s_2 &= \sqrt{\frac{\$100.50}{\$96.60}} - 1 \approx .02.
\end{aligned}$$

So, up to rounding errors, the spot rate for a two-period investment is the same as the yield to maturity. Next we set up a similar equation of value using the three-period bond, whose yield is $j_3 = .025$ and which pays three coupons of value $100(.01) = \$1$ at times 1, 2, and 3. First, its initial price is:

$$\begin{aligned}
P &= \frac{F}{(1+j_3)^n} + Fra_{\overline{n}|j_3} \\
&= \frac{\$100}{(1.025)^3} + \$1 \cdot a_{\overline{3}|.025} \\
&= \$95.72.
\end{aligned}$$

Then the spot rate s_3 satisfies:

$$\begin{aligned}
\$95.72 &= \frac{\$1}{(1+s_1)} + \frac{\$1}{(1+s_2)^2} + \frac{\$100 + \$1}{(1+s_3)^3} \\
&= \frac{\$1}{(1.015)} + \frac{\$1}{(1.02)^2} + \frac{\$101}{(1+s_3)^3}
\end{aligned}$$

$$\implies \quad \frac{\$101}{(1+s_3)^3} = \$95.72 - \frac{\$1}{(1.015)} - \frac{\$1}{(1.02)^2} = \$93.77$$

$$\implies \quad (1+s_3)^3 = \frac{\$101}{\$93.77}$$

$$\implies \quad s_3 = \sqrt[3]{\frac{\$101}{\$93.77}} - 1 \approx .0251.$$

Finally, the four-period bond also pays \$1 coupons, but has yield rate $j_4 = .028$. We compute its price and the spot rate s_4 in a similar way:

$$P = \frac{F}{(1+j_4)^n} + F r a_{\overline{n}|j_4} = \frac{\$100}{(1.028)^4} + \$1 \cdot a_{\overline{4}|.028} = \$93.28.$$

$$\begin{aligned}
\$93.28 &= \frac{\$1}{(1+s_1)} + \frac{\$1}{(1+s_2)^2} + \frac{\$1}{(1+s_3)^3} + \frac{\$100+\$1}{(1+s_4)^4} \\
&= \frac{\$1}{(1.015)} + \frac{\$1}{(1.02)^2} + \frac{\$1}{(1.0251)^3} + \frac{\$101}{(1+s_4)^4}
\end{aligned}$$

$$\implies \quad \frac{\$101}{(1+s_4)^4} = \$93.28 - \frac{\$1}{(1.015)} - \frac{\$1}{(1.02)^2} - \frac{\$1}{(1.0251)^3} = \$90.41$$

$$\implies \quad (1+s_4)^4 = \frac{\$101}{\$90.41}$$

$$\implies \quad s_4 = \sqrt[4]{\frac{\$101}{\$90.41}} - 1 \approx .0281.$$

In this case we have found very little difference between the spot rates and the yields to maturity. The graph of the spot rates is in Figure 2.11. ∎

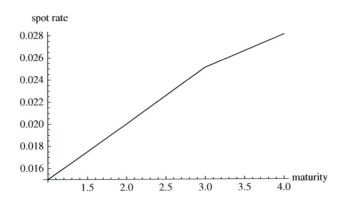

Figure 2.11: Spot rates as a function of maturity.

Given the spot rates of interest, we can also find the yields to maturity, as illustrated in the next example.

Example 2. Suppose that the spot rates for bonds of 1 and 2 periods are .01 and .015, respectively. Find the yield rates and prices of (a) a 2% coupon, $1000 bond with maturity 2 periods, and (b) a 1% coupon, $100 bond with maturity 2 periods.

Solution. (a) By definition of the spot rate, the yield for a single-period zero-coupon bond is .01, and for a two-period zero-coupon bond is .015. Split the 2% coupon bond into two strips: a single-period bond with payoff $1000(.02) = $20 at time 1, discounted by a factor of $1/1.01$, and a two-period bond with payoff $20 + $1000 with discount factor $1/1.015$. The present value of the total of the two strips is:

$$\$20 \cdot \frac{1}{1.01} + \$1020 \cdot \frac{1}{1.015} = \$1024.73.$$

We can determine the yield rate for the full bond by solving the equation:

$$P = \$1024.73 = \frac{F}{(1+j_2)^2} + Fra_{2|j_2} = \frac{\$1000}{(1+j_2)^2} + \$20 \cdot a_{2|j_2}.$$

Solving numerically, the solution is about $j_2 = .75\%$.

(b) Analyzing similarly to part (a), there is a coupon of $100(.01) = $1 at time 1 and a payment of $101 consisting of the face value plus coupon at time 2. The present value of these two payments is:

$$\$1 \cdot \frac{1}{1.01} + \$101 \cdot \frac{1}{1.015} = \$100.50.$$

The yield rate for the full bond satisfies the equation:

$$P = \$100.50 = \frac{F}{(1+j_2)^n} + Fra_{n|j_2} = \frac{\$100}{(1+j_2)^2} + \$1 \cdot a_{2|j_2}.$$

The solution is again about $j_2 = .75\%$. ∎

2.3.2 Forward Rates and Spot Rates

The spot rates of interest in the market carry information about what investors expect interest rates to be in the future. Specifically, if spot rates are known for two times, a specific interest rate is implied for a zero-coupon bond which will be bought at the earlier of these times and redeemed at the later time. Such an interest rate is called a *forward rate*.

Definition 2. The *forward rate of interest* $f_{m,n}$ corresponding to times $m < n$ is the yield rate for a zero-coupon bond that is purchased at the end of time period m and redeemed at the end of time period n.

To illustrate the idea, consider a single-term forward rate $f_{1,2}$ for a bond bought at time 1 and redeemed at time 2. Suppose the current spot rate s_1 for a one-period \$100 bond is 3%, and the spot rate s_2 for a two-period \$100 bond is 3.5%. This means that the one-period bond can be bought for a price of:

$$\$100 \cdot \frac{1}{1.03} = \$97.09,$$

and the two-period bond can be bought for a price of:

$$\$100 \cdot \frac{1}{1.035^2} = \$93.35.$$

Suppose for the sake of illustration that we can buy \$100 bonds in arbitrary quantities, not just positive integers. (See Exercise 11 for an argument that this assumption is not crucial.) Let us artificially construct an investment in which we borrow a one-period \$100 bond from someone now and sell it for \$97.09, using the proceeds to buy \$97.09/\$93.35 = 1.0401 units of two-period bonds. We must pay back the full accrued value of \$100 to the bondholder from whom we borrowed at time 1, which constitutes an investment of \$100 for us at that point in the future. But at time 2, we cash in the value of the two-period bonds, which is:

$$\frac{\$97.09}{\$93.35} \cdot \$100 = 1.0401 \cdot \$100 = \$104.01.$$

Then our \$100 investment at time 1 has grown to \$104.01 at time 2, which corresponds to a forward rate of:

$$f_{1,2} = \frac{\$104.01 - \$100}{\$100} = .0401.$$

The fact that the two spot rates of 3% for a 1-year investment and 3.5% for a 2-year investment are currently in place says that investors feel that the interest rate will be 4.01% for a 1-year investment beginning a year from now. Since the spot rates change often, corresponding changes will follow in the implied forward rate.

Example 3. If the yield rate on a 2-year zero-coupon bond is s_2 and the yield rate on a 5-year zero-coupon bond is s_5, find the yield rate that is implied on a 3-year zero-coupon bond purchased in 2 years.

Solution. The problem asks us to compute $f_{2,5}$. Let us assume a nominal face value of F for each bond, which will turn out not to be influential in the solution. Using the reasoning of the discussion above, the price of the 2-year bond is the present value:

$$P_2 = \frac{F}{(1 + s_2)^2},$$

and the price of the 5-year bond is:

$$P_5 = \frac{F}{(1+s_5)^5}.$$

Borrow a 2-year bond and cash it for its current value of P_2, purchasing P_2/P_5 units of the 5-year bond at its price of P_5. Note that the number of units of the 5-year bond is:

$$\frac{P_2}{P_5} = \frac{F \big/ (1+s_2)^2}{F \big/ (1+s_5)^5} = \frac{(1+s_5)^5}{(1+s_2)^2},$$

and if the term structure of spot rates is increasing, this will exceed 1. At time 2, the face value of F must be returned to the holder of the 2-year bond, which we view as an investment of F made at that time. At time 5 we receive the face value F of the 5-year bond, times the number of units of the 5-year bond that we bought. So the forward rate $f_{2,5}$, expressed as a per period effective rate, satisfies the equation of present values at time 2:

$$F = (1+f_{2,5})^{-3} \cdot \frac{(1+s_5)^5}{(1+s_2)^2} \cdot F$$

$$\implies \quad (1+f_{2,5})^3 \;=\; \frac{(1+s_5)^5}{(1+s_2)^2}$$

$$\implies \quad f_{2,5} \;=\; \sqrt[3]{\frac{(1+s_5)^5}{(1+s_2)^2}} - 1. \;\blacksquare$$

So far we have been computing forward rates given spot rates. The next example shows how to go backward in a specific numerical case; we will do a general derivation after the example.

Example 4. Suppose that we have two forward rates $f_{1,5} = 2\%$ and $f_{5,10} = 3\%$, and the current 1-year spot rate is $s_1 = 1\%$. Find s_5 and s_{10}.

Solution. All bonds in this discussion will have the same nominal face value F. First, to find s_5 let us synthesize a 5-year zero-coupon bond by buying a 1-year zero-coupon bond and redeeming it at its maturity for the expected price of a 4-year zero-coupon bond as determined by $f_{1,5}$. The time 0 prices of our synthetic bond and the real 5-year bond should agree.

Specifically, the time 1 price of a 4-year bond with forward rate $f_{1,5}$ is:

$$P_4 = \frac{F}{(1+f_{1,5})^4} = \frac{F}{(1.02)^4}.$$

Since the 1-year spot rate is $s_1 = .01$, the price of a 1-year bond is:

$$P_1 = \frac{F}{1+s_1} = \frac{F}{1.01},$$

and so, in order to have the amount of P_4 available at time 1 to purchase the 4-year bond, we must buy P_4/F units of the 1-year bond initially, which grow to $(P_4/F) \cdot F = P_4$ in value at time 1. The number of units of the 1-year bond to buy at time 0 is then:

$$\frac{P_4}{F} = \frac{1}{(1.02)^4},$$

and the total price of these 1-year bonds is:

$$P_1 \cdot \frac{P_4}{F} = \frac{P_1}{(1.02)^4} = \frac{F}{(1.01)(1.02)^4}.$$

This initial investment, with the transaction to the 4-year bond at time 1, yields a final value of F at time 5, just as the purchase of a 5-year bond initially does. So this initial value must equal the initial value of a 5-year bond. Thus,

$$\frac{F}{(1.01)(1.02)^4} = P_5 = \frac{F}{(1+s_5)^5}$$

$$\implies \quad (1.01)(1.02)^4 = (1+s_5)^5$$

$$\implies \quad s_5 = \sqrt[5]{(1.01)(1.02)^4} - 1 \approx .018.$$

We have attacked this problem with the synthesis method, but the details of the last computation suggest a valuable way of thinking about the relationship between spot and forward rates. Consider an initial investment of \$1. The per period spot rate s_5 for an investment of five periods multiplies the value of that dollar by the factor $(1+s_5)^5$. But we can also invest the dollar for a single period at rate $s_1 = .01$, and then four more periods at rate $f_{1,5} = .02$, multiplying the value of the dollar by the factor $(1.01)(1.02)^4$. Since these investments should give equivalent value at time 5, we find the equation $(1.01)(1.02)^4 = (1+s_5)^5$, from which s_5 can be determined. Adapting this approach to the determination of s_{10}, since we now have s_5, and we are also given $f_{5,10}$, the value of a dollar at time 10 that was invested at time 0 can be written in two ways:

$$(1+s_5)^5 (1+f_{5,10})^5 = (1+s_{10})^{10}$$

$$\implies \quad (1.018)^5(1.03)^5 = (1+s_{10})^{10}$$

$$\implies \quad s_{10} = \sqrt[10]{(1.018)^5(1.03)^5} - 1 \approx .0240. \quad \blacksquare$$

To compute spot rates in general given forward rates and an initial spot rate, consider a sequence of maturity times $n_1, n_2, ..., n_k$. A zero-coupon bond of face value F, maturity n_k, and spot rate s_{n_k} can be synthesized by buying an appropriate number of units N_0 of a zero-coupon bond with maturity n_{k-1} at spot rate $s_{n_{k-1}}$, and then rolling that over at maturity n_{k-1} to one zero-coupon bond of face value F, maturity n_k, and forward rate f_{n_{k-1},n_k}. The value P_{k-1} of the bond bought at time n_{k-1} in order to give a value of F at time n_k would be:

$$P_{k-1} = \frac{F}{\left(1 + f_{n_{k-1},n_k}\right)^{n_k - n_{k-1}}}.$$

The quantity N_0 of bonds of face value F that we would need to buy initially in order to achieve the value P_{k-1} at time n_{k-1} satisfies:

$$N_0 \cdot F = P_{k-1},$$

and under the spot rate $s_{n_{k-1}}$ the price per unit of these bonds is:

$$P_0 = \frac{F}{\left(1 + s_{n_{k-1}}\right)^{n_{k-1}}}.$$

So an initial investment of

$$
\begin{aligned}
N_0 P_0 &= \frac{P_{k-1}}{\left(1 + s_{n_{k-1}}\right)^{n_{k-1}}} \\
&= \frac{F}{\left(1 + f_{n_{k-1},n_k}\right)^{n_k - n_{k-1}} \left(1 + s_{n_{k-1}}\right)^{n_{k-1}}}
\end{aligned}
$$

produces a final value of F, just as an initial investment of one bond of face value F, maturity n_k, and spot rate s_{n_k} does. Therefore, the initial value of the latter bond must equal the initial value of our synthesized investment, which implies that:

$$\left(1 + f_{n_{k-1},n_k}\right)^{n_k - n_{k-1}} \left(1 + s_{n_{k-1}}\right)^{n_{k-1}} = \left(1 + s_{n_k}\right)^{n_k}. \tag{2.25}$$

Formula (2.25) is a general recursive formula using which we can build up spot rates from one given spot rate for the earliest time horizon available, if we know the appropriate forward rates. But the same formula rewritten gives us a way of computing forward rates directly from a sequence of spot rates:

$$\left(1 + f_{n_{k-1},n_k}\right)^{n_k - n_{k-1}} = \frac{\left(1 + s_{n_k}\right)^{n_k}}{\left(1 + s_{n_{k-1}}\right)^{n_{k-1}}}. \tag{2.26}$$

Notice that (2.25) represents the intuitive idea that \$1 invested for n_{k-1} time units at rate $s_{n_{k-1}}$ and then rolled over to another $n_k - n_{k-1}$ periods at rate f_{n_{k-1},n_k} yields the same as \$1 invested for a full n_k periods at rate s_{n_k}. Our next example illustrates the use of these two formulas.

Example 5. (a) If spot rates for 2-, 5-, and 10-year zero-coupon bonds are, respectively, 2.3%, 3.1%, and 3.5%, find forward interest rates for the time intervals $[2, 5]$ and $[5, 10]$.

(b) If the single-period spot rate for a zero coupon bond is .9%, and we know forward rates $f_{1,4} = 1.1\%$, $f_{4,8} = 1.6\%$, and $f_{8,10} = 2.1\%$, find the spot rates s_4, s_8, and s_{10}.

Solution. (a) By formula (2.26), the forward rate $f_{2,5}$ satisfies the equation:

$$(1 + f_{2,5})^3 = \frac{(1 + s_5)^5}{(1 + s_2)^2} = \frac{(1.031)^5}{(1.023)^2} = 1.113$$

$$\Longrightarrow f_{2,5} = \sqrt[3]{1.113} - 1 \approx .0363.$$

Similarly,

$$(1 + f_{5,10})^5 = \frac{(1 + s_{10})^{10}}{(1 + s_5)^5} = \frac{(1.035)^{10}}{(1.031)^5} = 1.211$$

$$\Longrightarrow f_{5,10} = \sqrt[5]{1.211} - 1 \approx .0390.$$

(b) Formula (2.25) allows us to write:

$$(1 + f_{1,4})^3 (1 + s_1) = (1 + s_4)^4$$

$$\begin{aligned}
\Longrightarrow \quad (1 + s_4)^4 &= (1.011)^3 (1.009) = 1.043 \\
\Longrightarrow \quad s_4 &= \sqrt[4]{1.043} - 1 \approx .0106.
\end{aligned}$$

With this in hand, we can compute:

$$(1 + f_{4,8})^4 (1 + s_4)^4 = (1 + s_8)^8$$

$$\begin{aligned}
\Longrightarrow \quad (1 + s_8)^8 &= (1.016)^4 (1.0106)^4 = 1.0923 \\
\Longrightarrow \quad s_8 &= \sqrt[8]{1.0923} - 1 \approx .0111.
\end{aligned}$$

Continuing in the same fashion,

$$(1 + f_{8,10})^2 (1 + s_8)^8 = (1 + s_{10})^{10}$$

$$\begin{aligned}
\Longrightarrow \quad (1 + s_{10})^{10} &= (1.021)^2 (1.0111)^8 = 1.139 \\
\Longrightarrow \quad s_{10} &= \sqrt[10]{1.139} - 1 \approx .0131. \quad \blacksquare
\end{aligned}$$

Forward rates are current forecasts of what the yield rates are expected to be in the future. When that future comes, the actual rates may differ. The last example of this section shows how an investor might profit from this fact.

Example 6. Reconsider Example 5(a) in which spot rates for 2- and 5-year zero-coupon bonds were, respectively, $s_2 = 2.3\%$ and $s_5 = 3.1\%$. We computed that the implied forward rate was approximately $f_{2,5} = 3.63\%$. Suppose that investor Steve thinks he knows better than the bond market does, and the true forward rate $f_{2,5}$ will be lower than this. Can Steve earn a profit by buying and selling (issuing) bonds?

Solution. Steve thinks that bonds issued at time 2 will have a lower rate of return, hence a higher price, than the market thinks. He should try purchasing longer term 5-year bonds and financing the purchase by selling 2-year bonds initially. When time 2 comes and he must pay the 2-year bondholders, he can finance that by selling 3-year bonds. Specifically, suppose that Steve purchases a face value of $10,000 of 5-year bonds at time 0, which costs:

$$\frac{\$10,000}{(1.031)^5} = \$8584.34.$$

To pay for these, Steve issues 2-year bonds for a face value of F satisfying:

$$\$8584.34 = \frac{F}{(1.023)^2} \implies F = \$8584.34 \cdot (1.023)^2 = \$8983.76.$$

Time 2 arrives, and Steve needs to pay his bondholder $8983.76. To do this, he issues a 3-year bond. But suppose that, instead of a rate of 3.63%, the rate of return is actually 3.3%. Then he issues 3-year bonds with face value F satisfying:

$$\$8983.76 = \frac{F}{(1.033)^3} \implies F = \$8983.76 \cdot (1.033)^3 = \$9902.83.$$

When time 5 comes, Steve can then redeem his $10,000 5-year bonds, pay the 3-year bondholders $9902.83, and profit by $10,000 - \$9902.83 = \$97.17$. ∎

Important Terms and Concepts

Term structure of interest rates - The relationship between the maturity time and the yield rate.

Yield curve - A graph relating the yield rate of bonds to the term-to-maturity of the bond. A normal yield curve increases, so that longer terms require higher yields.

Inversion of a yield curve - The yield curve decreases over a range of maturities, signaling a potential downturn in the economy.

STRIPS - Parts of a coupon bond consisting of coupon payments only, or in the case of the maturity time, the coupon payment plus the redemption value.

Spot rate of interest - The yield rate of a zero-coupon bond for the specified maturity.

Forward rate of interest - The yield rate of a zero-coupon bond bought in the future at some time m and maturing at time n.

Exercises 2.3

1. Use the U.S. Treasury data from May 8, 2013 presented at the start of the section to compute the 2-year spot rate. Recall that the bond listed is par value, so that the coupon rate and the yield rate are the same. Also, coupons are semi-annual, but bonds of maturity 1 year or less are by nature without coupons. Interpolate linearly as necessary to approximate unlisted yield rates.

2. Suppose that a zero-coupon, 1-period bond has yield rate 4.2%, a 2-period bond with coupon rate 1% has yield rate 4.5%, and a 3-period bond with coupon rate 2% has yield rate 5%. Find the 1, 2, and 3-period spot rates. Let the face values of these bonds be equal but general, and in the process of making the computation, show that the face value does not matter.

3. In this problem we consider the effect on the spot rates of an inverted yield curve. Suppose that a \$100 1-period bond with no coupons and yield rate 3% exists, as well as a \$100 2-period par value bond with yield rate 2%, a \$100 3-period bond with coupon rate 1% and yield rate 1.5%, and a \$100 4-period bond with coupon rate 2% and yield rate 2.5%. Find the spot rates s_1, s_2, s_3, and s_4.

4. Suppose that the 1-period spot rate $s_1 = 5\%$, and that there is a \$1000 2-period bond with yield rate 4%. Compute the spot rate s_2 if: (a) the coupon rate of the 2-period bond is 2%; (b) the coupon rate is 4%; (c) the coupon rate is 6%. (Keep a large number of significant figures in the calculation to pick up on any differences.)

5. Derive a general formula for the 2-period spot rate s_2 given the 1-period spot rate s_1 and the yield rate j_2 and coupon rate r_2 of a 2-period bond of face value F.

6. Derive a general recursive relationship between the n-period spot rate s_n and the previous spot rates s_i, $i < n$, knowing the n-period yield rate j_n in the case of constant coupon rate r for all maturities. Does this relationship depend on the face value F or the particular value of r?

7. Suppose that the spot rates for 1 and 2 periods are 3% and 2%, so the spot rate curve is inverted. Find (a) the yield rate on a \$500 2-period bond with coupon rate 2%; (b) the yield rate on a \$2000 2-period bond with coupon rate 4%.

8. If it is known that $s_1 = 1.2\%$, $s_2 = 1.5\%$, $s_3 = 2.0\%$, and $s_4 = 2.2\%$, find the initial price and yield rate of a 4-period \$1000 bond with coupon rate 1%.

9. In the problem of Example 2(a), show that you get the same yield rate regardless of the face value.

10. Use the idea of expressing the initial price of a bond as the sum of discounted values of STRIPS to show that, if the term structure of spot rates is increasing, i.e., $s_1 < s_2 < \cdots < s_n$, then the yield rate j_n must be less than s_n. Also, show that, if the structure of spot rates is decreasing, i.e., $s_1 > s_2 > \cdots > s_n$, then $j_n > s_n$.

11. Show that a forward rate can be computed from two spot rates without assuming that bonds can be bought in non-integer amounts. (Hint: Suppose that the initial prices P_m and P_n of m-period and n-period zero-coupon bonds have been rounded to the nearest cent, and consider the least common multiple of the integers $100P_m$ and $100P_n$.)

12. Suppose that the yield rate on a zero-coupon, 1-period \$1000 bond is 3.2%, the yield rate on a 2% coupon, 2-period \$1000 bond is 3.4%, and the yield rate on a 3% coupon, 3-period \$1000 bond is 3.8%. Find the forward rates $f_{1,2}$, $f_{1,3}$, and $f_{2,3}$.

13. Given the table of spot rates below, find all forward rates $f_{i,j}$ that are possible to find. (Notice that the spot rate curve is inverted.)

time	1	2	5	10
spot rate	4.7%	4.0%	3.8%	4.0%

14. Redo the computation of s_{10} in Example 4 using the bond synthesis approach, and check that the solution is the same as the one obtained using the shortcut approach in the example.

15. Suppose that the following forward rates are known: $f_{1,3} = .036$, $f_{3,4} = .034$, and $f_{4,7} = .038$. Assume also that the 1-period spot rate is $s_1 = .03$. Find the spot rates s_3, s_4, and s_7.

16. Assume that the 1-period spot rate $s_1 = 2.3\%$, and that $f_{1,2} = 2.0\%$ and $f_{1,3} = 2.5\%$. Find the yield rate of a 3-period, 1% coupon bond of face value \$1000.

17. In this section we glossed over an important idea in the valuation of financial assets that we will be taking fuller advantage of later, called **arbitrage**. Roughly speaking, a transaction is an arbitrage opportunity if an investor has a positive probability of profit and no probability of loss, while investing no money at the beginning of the transaction. We have used the method of synthetic bonds to compute spot rates, arguing that, if two investments in bonds result in the same final value, then the initial values of the investments must be the same as well. In the computation of s_5 in Example 4, show that an arbitrage opportunity exists if the forward rate is $f_{1,5} = 2\%$, the 1-year spot rate is $s_1 = 1\%$, and s_5 is anything other than $\sqrt[5]{(1.01)(1.02)^4} - 1$. (Hint: issue one bond initially and use the proceeds to buy another. Consider both cases where the spot rate s_5 is lower

or higher than this value.)

18. Suppose that the current 1-year spot rate is $s_1 = .025$ and the current 3-year spot rate is $s_3 = .034$. Find the implied forward rate for time 1 to time 3. Jim believes that this forward rate is actually a low estimate and will be higher when year 1 has passed. Explain how Jim can profit by buying and selling (issuing) bonds.

19. In this chapter we have been talking about bonds as if the payoff at maturity is guaranteed. When there is a chance that some or all of the face value of the bond may not be paid, investors will demand a so-called *risk premium*, in the form of a discounted price or equivalently an increased yield rate. Consider a group of five zero-coupon corporate bonds of $10,000 and maturities 1, 2, 3, 4, and 5 years, and stated yield rates 3.6%, 3.8%, 4%, 4.2%, and 4.4% per year, respectively. But assume that there is a chance of default of half the face value at the final time, which is 1%, 2%, 3%, 4%, and 5%, respectively, for the five maturities. What will investors really pay initially for such bonds? (Hint: consider the weighted average or *expected* redemption value, using the given chances of default as the weights. Find the present values of these expected redemption amounts.) What true spot rate corresponds to the initial price of each of these bonds? Sketch a graph of the term structure of these spot rates.

Chapter 3

Discrete Probability for Finance

Very rarely in the preceding chapters have we allowed the possibility that financial quantities were not perfectly predictable. Loan interest rates remained fixed and known throughout time, annuity payments and other streams of payments that figured into present value calculations were predictable and regular, and defaults on bond payments never happened. Clearly the assumption of determinism of all inputs to a financial problem is inadequate to model the real world. Variable interest rates on loans do happen, investors may miss payments into annuities at random times, bonds can default, and, most important, some of the most interesting investment objects, like common stocks, definitely do not have a predictable evolution in value over time. To cope with these and other sources of random behavior in financial problems, we must establish a good grounding in basic probability theory. You may have seen much of the material of this chapter in other courses, especially elementary statistics, and consequently some of it will be readily reviewed and understood. But the chapter is a worthwhile read, because it adopts a financial point of view, and mostly uses examples from this area of application. After you finish, the stage will be well set for the coming chapters on portfolio optimization and valuation of derivative assets. For more information, you can check out references [7] and [8], or any of a number of fine probability books on the market.

To get an idea of the concepts and methods that we will need, consider the tree diagram in Figure 3.1, which is a simple but unexpectedly powerful model called the ***binomial branch process*** for the motion of the price of an asset as time proceeds. Each level in this tree corresponds to a time, and the nodes in the graph at a particular level give the possible prices of the asset at that time. We call the level of the root of the tree level 0, then level 1, level 2, etc., as we move from the left to the right. This particular diagram only displays possible prices up through time 2, but it can be extended to as many levels as desired. There is an intial price of S_0 at time 0 which is known, and possible prices at time 1 of

$(1+b)S_0$ or $(1+a)S_0$. The former occurs with probability q and the latter with probability $1 - q$. At each point in time, the price can go "up" by the factor $(1 + b)$ or "down" by the factor $(1 + a)$, and for this model, the probabilities of doing each remain stable with time. For example, from price $(1 + b)S_0$ at time 1, the possible next prices are $(1 + b)^2 S_0$ or $(1 + a)(1 + b)S_0$ at time 2 with probabilities q and $1 - q$. Notice that two paths in the tree lead to the same price $(1 + a)(1 + b)S_0$ at time 2. To check that you understand the model, you should fill in the third and fourth levels of the diagram.

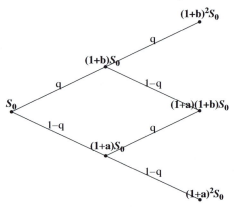

Figure 3.1: Binomial branch process model of price of asset.

Binomial branch processes are remarkably rich examples in the sense that many of the ideas and techniques from probability that are useful generally in financial problems also pertain to these processes. In this model, we would be concerned with the set of all possible paths from the left to the right, which we will call the **sample space** of the random experiment of observing the price process through time. The probabilities of paths through the diagram will certainly be important, leading to the concept of a **probability measure** on a sample space. Sample spaces and probability measures are the subjects of Section 3.1.

The price at a particular time is a random numerical quantity called a **random variable**. In Figure 3.1, the random variable X_2 could denote the price at time 2, which could take on any of the three values in level 2 of the diagram. The **probability distribution of a random variable** assigns probabilities to each of the possible values of the variable, and for a price process like this one, calculating the distribution at a future time is the best we can do to forecast the behavior of the process, since we will not know in advance what path it will follow through the tree. Random variables and distributions are discussed in Section 3.2.

As we move farther along in time, there are more and more potential values of the price variable, and a way of summarizing its average value and its variability becomes useful; thus we move to the topic of **expectation of random variables** in Section 3.3.

For time-dependent processes like this one, we may very well be able to observe part of their history and attempt to use it to forecast what could happen later. So we would like to have **conditional probabilities** of events involving the asset price in the future given that we know some information about the past and present. Section 3.4 addresses the subject of conditional probability.

The dynamics of the binomial branch process are determined by what is assumed about whether its next move depends in a probabilistic way on the branches that it has taken previously in its motion, and so Section 3.5, which focuses on **independence** and **dependence**, also has tie-ins to this model.

Finally, if such a model (extended to many short time intervals) is to be calibrated to real data, we must have straightforward ways of **estimating parameters** like a, b, and q from data. Section 3.6 covers estimation, and also looks at methods for computer simulation of processes like this one. This is important because many questions that can be asked about financial models do not lend themselves to easy analytical solutions. But simulating the model many times can give insights into the questions and allow us to compute approximate solutions.

3.1 Sample Spaces and Probability Measures

In financial situations, we frequently observe **random phenomena** or perform **random experiments**. These terms are left undefined, but we should recognize them when we see them. For example, we observe whether a bond defaults at some point prior to its maturity; we observe the progression of Friday closing prices of Caterpillar stock on the New York Stock Exchange; we simulate a binomial branch process many times and count how often it ends in a value higher than it started; we flip four coins and count the number of heads, etc. In all cases there is a phenomenon that can be observed or an experiment performed. Until the scenario has played out, we cannot know exactly what its outcome will be. Using logic, historical data, and reasonable assumptions, we may be able to say how likely each outcome will be. But once the phenomenon is observed in its entirety, we know what outcome was the one that happened.

The basic tool for modeling random phenomena is set theory. In the following definition, we give names to the most basic probabilistic objects, and the language of set theory is prominent.

Definition 1. The **sample space** Ω is the set of all possible outcomes of a random experiment or phenomenon, that is, all possible indecomposable results that could happen. Its elements $\omega \in \Omega$ are called **outcomes**. A subset $A \subset \Omega$ of the sample space, that is, a set of outcomes, is called an **event**. An event **occurs** if the final outcome of the experiment belongs to that event.

Notice that, while outcomes, like points in a geometrical space, cannot be broken down into simpler pieces, events may have more than one outcome in them. Also, by convention we call the empty set $\emptyset$, that is, the set with no outcomes, an event. The whole sample space Ω is also a set of outcomes, so it is an event as well.

Example 1. A fair coin is tossed four times in succession. Describe the sample space. How many elements does it have? List the outcomes in the event "at least three heads." Also, list the outcomes in the event "exactly two heads."

Solution. The experiment whose results we are observing involves the faces that appear on all four coins, and since the coins are flipped in succession, outcomes should be ordered lists, each entry of which is heads or tails. In set theoretic language, the sample space Ω is the set of all outcomes of the form (x_1, x_2, x_3, x_4) where each x_i can be either H or T. In this case, the sample space is small enough to list out completely:

$$\Omega = \begin{cases} (H,H,H,H), \\ (H,H,H,T), & (H,H,T,H), & (H,T,H,H), & (T,H,H,H), \\ (H,H,T,T), & (H,T,H,T), & (H,T,T,H), & (T,H,H,T), \\ & (T,H,T,H), & (T,T,H,H), \\ (T,T,T,H), & (T,T,H,T), & (T,H,T,T), & (H,T,T,T), \\ (T,T,T,T) \end{cases}$$

There are 16 possible outcomes, which we have laid out strategically so that each line contains outcomes for which there is a fixed number of heads: 4, 3, 2, 1, and 0 heads from top to bottom. The event "at least three heads" is the set of outcomes:

$$A = \{(H,H,H,H),(H,H,H,T),(H,H,T,H),(H,T,H,H),(T,H,H,H)\}.$$

This has 5 of the 16 outcomes in the sample space. If we are willing to assume that all outcomes are equally likely, then it is intuitive to assign a likelihood of 5/16 to event A. If the outcome of the experiment is (H,T,H,H), then the event A has occurred in the language of Definition 1. The event "exactly two heads" is the set of outcomes:

$$B = \{(H,H,T,T),(H,T,H,T),(H,T,T,H),(T,H,H,T),$$
$$(T,H,T,H),(T,T,H,H)\}$$

Since B has six outcomes, the equally likely outcome hypotheses would lead us to assign a likelihood of $6/16 = 3/8$ to event B. Any individual outcome would receive probability 1/16, and the total probability of an event is simply the sum of the probabilities of the outcomes in that event. ∎

Sequences of coin flips as in Example 1 may strike you as straightforward experiments in which the ideas of probability are well illustrated, albeit in a dull way. But they are more than just pedagogical tools. Look again at the binomial branch process of Figure 3.1. The path that it follows can be characterized as a sequence of two moves, where each can be either up or down. The sample space of all paths is $\Omega = \{(U,U),(U,D),(D,U),(D,D)\}$, where in a sense to be defined carefully later, one move is independent of the other. Although it is not necessarily true that up is as likely as down on a single move, this is just the same as the sample space for a sequence of two independent coin flips, with U replacing H and D replacing T. So any results that we can derive for a sequence of n coin flips will also apply to an n-stage binomial branch process.

3.1.1 Counting Rules

Computing many elementary discrete probabilities involves counting, because the probability of an event is often modeled as the number of ways in which the event can occur divided by the number of ways in which the whole phenomenon can come out.

One of the simplest, yet most powerful counting tools is the **Fundamental Counting Principle**. Suppose that a speculator in Canadian wheat believes that there are three possible prices per ton at the end of this week ($260, $250, and $240 per ton), and three possible currency exchange rates (1.20, 1.10, or 1.00 Canadian dollars per U.S. dollar) between U.S. and Canadian dollars. The wheat prices can be viewed as three possible branches on the first level of a tree such as the one in Figure 3.2, and the exchange rates three possible branches emanating from each first-level branch. It is easy to see that there are nine possible sequences of the form (wheat price, exchange rate), which is the product of the number of wheat prices times the number of exchange rates.

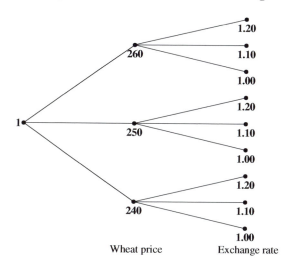

Figure 3.2: Tree display of sample space for (wheat price, exchange rate) pair.

We state this obvious fact as our first theorem.

Theorem 1 (Fundamental Counting Principle, two stages) If outcomes of a random experiment consist of two stages, and the number of stage 2 possibilities is the same for each stage 1 outcome, then the total number of outcomes of the experiment is the number of ways the first stage can occur times the number of ways that the second can occur.

The counting principle generalizes to multiple-stage experiments. For instance, in a three-stage binomial branch process, there are two outcomes U, D for each current price. Therefore there are $2 \cdot 2 \cdot 2 = 8$ outcomes for the entire sequence of moves, explicitly:

$$\{(U, U, U), (U, U, D), (U, D, U), (D, U, U), (U, D, D),$$
$$(D, U, D), (D, D, U), (D, D, D)\}$$

Theorem 2 (Fundamental Counting Principle, many stages) If a random experiment has k stages and:

(a) stage 1 has n_1 possible outcomes;
(b) for each stage 1 outcome, stage 2 has n_2 possible outcomes;
(c) for each combined stage 1, stage 2 outcome, stage 3 has n_3 possible outcomes;

etc., then the entire experiment has

$$n_1 \cdot n_2 \cdots \cdot n_k$$

possible outcomes.

For example, if a four-character identification string can be any four letters in sequence, with the exception that a letter cannot be used consecutively, and case does not matter, then there are 26 choices for the first letter, 25 for the second, 25 for the third, and also 25 for the fourth letter, since each new letter is forbidden to be the same as the one just chosen. By the Fundamental Counting Principle, there are $26 \cdot 25 \cdot 25 \cdot 25 = 406,250$ such character strings.

Another simple rule of counting elements of sets is the sum rule. If several sets have no elements in common, then it is intuitively obvious that the number of elements in the union of all the sets is the sum of the numbers of elements in the individual sets. This provides a "divide and conquer" strategy which lets you break a big problem into smaller, more manageable cases. We state this as our third theorem.

Theorem 3 (Sum Rule) If an event A can be split into disjoint subsets A_i, $i = 1, 2, ..., n$:

$$A = A_1 \cup A_2 \cup A_3 \cup \cdots \cup A_n,$$

then the number of outcomes in A is the total of the numbers of outcomes in each of the subsets.

Example 2. Consider a four-stage binomial branch process. How many U-D sequences are there in the sample space of outcomes? How many outcomes are such that U occurs on the first and third moves? How many outcomes are such that there are at least three U's?

Solution. A typical outcome is of the form:

$$(--, --, --, --)$$

where the blanks can be filled in by U or D symbols. By the Fundamental Counting Principle, there are $2^4 = 16$ possible sequences. The possible moves are displayed in Figure 3.3 for a particular situation where the initial price is \$50 and at each step the price either goes up by a factor of 1.1 or down by a factor of .9.

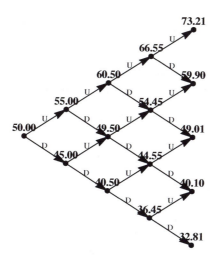

Figure 3.3: Possible transitions of a particular four-stage binomial branch process.

For the second question, the first and third positions are fixed:

$$(U, --, U, --)$$

There are two ways to fill in position 2 and for each of these there are two ways to fill in position 4, so there are 2^2 such sequences.

For the third question, use the sum rule to break the count into cases. First, all four positions could be U:

$$(U, U, U, U),$$

or else there could be exactly one D and three Us. Furthermore, in the second case the lone D is either in the first, or second, or third, or fourth position, so we get four more outcomes, as listed below, for a total of five:

$$(D, U, U, U) \ (U, D, U, U) \ (U, U, D, U) \ (U, U, U, D). \ \blacksquare$$

If all outcomes in the last example are equally likely, then the likelihood of U on both the first and third moves would be $4/16 = 1/4$, and the likelihood of at least three Us would be $5/16$. When would all outcomes be equally likely? We will have a better handle on this later, but it shouldn't be surprising that only if ups and downs are equally likely on any one move can the 16 outcomes in the sample space of paths be equally likely.

The sum rule has an easy consequence for the problem of counting the number of outcomes in the complement A^c of an event A, that is, those outcomes in the sample space that are not in A. Since A and A^c are disjoint, the total number of outcomes in A plus the total number of outcomes in A^c must equal the total number of outcomes in $A \cup A^c = \Omega$; in symbols:

$$n(A) + n\left(A^c\right) = n(\Omega) \implies n\left(A^c\right) = n(\Omega) - n(A). \tag{3.1}$$

For example, the number of paths in a five-stage binomial branch process is $n(\Omega) = 2^5 = 32$, and the number of paths with at least one U would be 32 minus the number of paths with no Us, of which there is only one, hence the number of paths with at least one U is $32 - 1 = 31$.

There is a class of counting problems that finds application in financial mathematics in which a group of objects is sampled without replacement from a larger group. A sample of this kind that is done with regard to order is called a **permutation**, and if order does not matter the sample is called a **combination**. We can count the number of possible permutations and combinations easily, as illustrated in the proofs of the next two theorems.

Theorem 4 The number of permutations, that is, ordered samples of size k from a set of size n is:

$$P_{n,k} = \frac{n!}{(n-k)!}. \tag{3.2}$$

Proof. To form a permutation is essentially to fill in a list of k blanks $(\_, \_, \_, \cdots, \_)$ with n possible objects, not reusing any that have previously been put into the list. There are n ways of filling in the first blank, then $n-1$ ways of filling in the second, $n-2$ ways of filling in the third, etc., and for the last, or k^{th} position there remain $n - (k-1) = n - k + 1$ members of the full group that can be selected. By the Fundamental Counting Principle, there are:

$$n \cdot (n-1) \cdot (n-2) \cdots (n-k+1) = \frac{n!}{(n-k)!} \tag{3.3}$$

such permutations.

Theorem 5 The number of combinations, that is, unordered samples of size k from a set of size n is:

$$C_{n,k} = \binom{n}{k} = \frac{n!}{k!(n-k)!}. \tag{3.4}$$

Proof. It is most effective to count combinations indirectly, using the result on permutations. Determining a unique permutation can be looked at as first determining an unordered subset of k objects that will belong to the list, and then ordering these k objects. Using an argument similar to the one in the proof of Theorem 3.4, there are $k! = k \cdot (k-1) \cdots \cdots 2 \cdot 1$ ways of ordering the k objects once they are selected. But stage 1 of this procedure to count permutations is just selecting a combination. So, by the counting principle,

$$\text{number of permutations} = \text{number of combinations} \cdot k!$$

$$\Longrightarrow P_{n,k} = C_{n,k} \cdot k! \Longrightarrow C_{n,k} = \frac{P_{n,k}}{k!} = \frac{n!}{k!(n-k)!}. \tag{3.5}$$

Example 3. In setting up her will, Sue is trying to decide how to divide her estate among three possible beneficiaries, her dissolute cousin Abelard, the Benevolent Order of Bird Watchers, and her cat Claudia (A, B, and C for short). She will give 50% of the estate to one of them, 30% to another, and 20% to the third. With these constraints, in how many ways can she set up her will?

Solution. The decisions that Sue must make are which of the three receives 50% of her wealth, which 30%, and which 20%. Her will can be thought of as a list of the form $(\_, \_, \_)$, where the elements of the list tell which heir gets which share. Since the proportions differ, the order in which the beneficiaries A, B, and C are listed in the will matters. So Sue is choosing a permutation of three letters from the set $\{A, B, C\}$, which can be done in $P_{3,3} = \frac{3!}{(3-0)!} = 3 \cdot 2 \cdot 1 = 6$ ways.
■

Example 4. In a four-stage binomial branch process, how many U-D sequences are there in which half of the moves are up moves? What about an eight-stage binomial branch process?

Solution. It is helpful to look back at Example 1 in which the sample space for the flip of four coins was displayed. The four-stage binomial branch process has an identical sample space, with the U symbol taking the place of H and the D symbol taking the place of T. By direct counting, there are six sequences with two up symbols. But this answer can also be deduced by using combinations. Outcomes are of the form $(\_, \_, \_, \_)$. To specify a unique sequence with exactly two ups, we must pick, without regard to order, two different list positions from the group of positions $\{1, 2, 3, 4\}$ in which the U symbols are to be placed. This is a combination of two objects drawn from four, and there are $\binom{4}{2} = \frac{4 \cdot 3 \cdot 2 \cdot 1}{(2 \cdot 1)(2 \cdot 1)} = 6$ such combinations. The advantage of this way of thinking is that, in larger experiments where it is more difficult to write out the complete sample space, the counting can nevertheless be done analogously without further difficulty. For the eight-stage branch process, outcomes look like $(\_, \_, \_, \_, \_, \_, \_, \_)$. To select four positions among the eight to receive up symbols can be done in the following number of ways:

$$\binom{8}{4} = \frac{8!}{4! \cdot 4!} = \frac{8 \cdot 7 \cdot 6 \cdot 5 \cdot 4 \cdot 3 \cdot 2 \cdot 1}{(4 \cdot 3 \cdot 2 \cdot 1)(4 \cdot 3 \cdot 2 \cdot 1)} = \frac{8 \cdot 7 \cdot 6 \cdot 5}{4 \cdot 3 \cdot 2 \cdot 1} = 70. \blacksquare$$

3.1.2   Probability Models

In some of the examples above, we suggested that events could be given a likelihood of occurring, at least if we were willing to make certain assumptions. We are ready now to be specific about how the likelihood of events is to be measured.

Definition 2. A ***probability measure*** P on the sample space of a random experiment assigns a likelihood value $P[A]$ between 0 and 1 to each event A.

The extremes of 0 and 1 represent impossibility and complete certainty, respectively. In the interest of simplicity of the definition, we have suppressed three intuitively obvious technical conditions that we will have to require in order for a function P as in the definition to be a valid probability measure, often called the ***axioms of probability***. These are:

(a) $P[\Omega] = 1$, where Ω is the sample space;

(b) For all events A, $P[A] \geq 0$;

(c) If A and B are disjoint events, then $P[A \cup B] = P[A] + P[B]$. More generally, if $A_1, A_2, ... A_n$ are events such that no two of them intersect, then

$$P[A_1 \cup A_2 \cup \cdots \cup A_n] = P[A_1] + P[A_2] + \cdots + P[A_n]. \qquad (3.6)$$

Three other simple and important properties that follow from these defining conditions are:

(d) $P[\emptyset] = 0$, where $\emptyset$ is the empty event containing no outcomes;

(e) For all events A, $P[A] \leq 1$;

(f) If A^c is the complement of event A, then:

$$P[A] + P[A^c] = 1. \qquad (3.7)$$

To see why (d) is true, notice that $\emptyset$ and Ω are two events with empty intersection, and their union is just Ω. By property (a), and then property (c), we have:

$$
\begin{aligned}
1 &= P[\Omega] \\
&= P[\Omega \cup \emptyset] \\
&= P[\Omega] + P[\emptyset] \\
&= 1 + P[\emptyset] \\
\Longrightarrow\ & P[\emptyset] = 0.
\end{aligned}
$$

You are asked for a proof of (f) in the Exercises. Property (e) follows easily from (f) and (b) (you should justify that for yourself).

In many probability problems, probability measures are constructed so that the probability of an event is the "size" of the event as a proportion of the "size" of the whole sample space. In discrete probability, "size" often means number of outcomes in the event, which is why we devoted some attention to counting methods. In what follows, we denote by $n(A)$ the number of elements in a finite set A. For example, if you roll a fair die once, since there are six possible faces that could land on top, the sample space is $\Omega = \{1, 2, 3, 4, 5, 6\}$. Since three of the faces are odd, it seems reasonable that the probability that an odd number is rolled is $n(\text{odd})/n(\Omega) = 3/6$.

In general, in an **equiprobable** or **uniform** finite sample space model in which we assume that outcomes are equally likely, the probability of an event A is

$$
P[A] = \frac{n(A)}{n(\Omega)}.
$$

Example 5. Recall the Canadian wheat example, in which a speculator in Canadian wheat believes that there are three possible prices per bushel (the first level of the tree below) at the end of this week, and three possible currency exchange rates (the second level) between U.S. and Canadian dollars, illustrated in Figure 3.2. If we suppose that the nine possible (price, exchange rate) pairs are equiprobable, then, for example, $P[\text{price is } \$250] = 3/9$, since there are 3 paths through the tree that pass through the price node denoted 250. Also, $P[\text{exchange rate is rate } \$1.10 \text{ or } \$1.00] = 6/9$, which you can see is the same as the sum $P[\text{exchange rate is rate } \$1.10] + P[\text{exchange rate is } 1.00]$. Each particular path (i.e., outcome), such as "price is \$260 and exchange rate is 1.10", has probability $1/9$. ∎

Example 6. On eight successive days, the information of whether a stock went up or down from the previous day's closing price is recorded. With what probability will there be exactly four ups among the eight price changes, if on each day the stock is just as likely to go up as to go down?

Solution. The sample space consists of 8-tuples:

$$
(--, --, --, --, --, --, --, --)
$$

each component of which can be U or D. Since each component can take on two potential values, $n(\Omega) = 2^8$.

We will suppose that the equal up-down assumption justifies that each such sequence is as likely as each other. To determine an outcome with four ups, we just need to choose four positions from the available eight in which to put the Us, which can be done in $\binom{8}{4} = 70$ ways, as we saw in Example 4. Thus,

$$P[\text{exactly 4 ups}] = 70\big/2^8. \ \blacksquare$$

3.1.3   More Properties of Probability

Two slightly more subtle rules of probability are worth knowing for the ensuing development: the general sum rule and the Law of Total Probability.

Probability rule (c) above covers the case of unions of pairwise disjoint events; what about the case where the events overlap? How does the formula for the probability of a union change? Consider events A and B in Figure 3.4. If we try to compute $P[A \cup B]$ as just the sum $P[A] + P[B]$ then we would be including outcomes in the intersection $A \cap B$ twice, thus overestimating the true probability. If the probabilities of these outcomes were subtracted from $P[A] + P[B]$, then we would get the correct value.

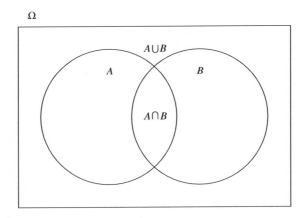

Figure 3.4: Union of two non-disjoint events.

Now, if there were three overlapping events such as A, B, and C in Figure 3.5, then we might try the same approach, but we need to be careful. As a first pass at computing $P[A \cup B \cup C]$, we compute $P[A] + P[B] + P[C]$. But we realize that outcomes in any of the intersections $A \cap B$, $A \cap C$, or $B \cap C$ are double counted. Those should be subtracted in order to correct, leaving us at our second try:

$$P[A] + P[B] + P[C] - (P[A \cap B] + P[A \cap C] + P[B \cap C]).$$

However, we have gone too far. Consider outcomes in all of A, B, and C, that is, the set $A \cap B \cap C$. Those outcomes were counted three times in the sum $P[A] + P[B] + P[C]$, but also subtracted away three times in the sum

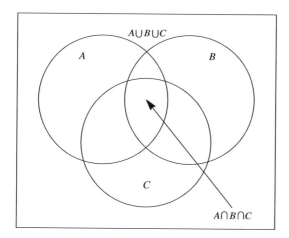

Figure 3.5: Union of three non-disjoint events.

$P[A \cap B] + P[A \cap C] + P[B \cap C]$, so they have no net contribution to the total probability. Hence, we need to add their probabilities back in again.

The last paragraph gives good intuitive motivation for the following theorem, which is sometimes referred to as the Law of Inclusion and Exclusion. Rigorous proofs of these are given in courses in probability and also in discrete mathematics.

Theorem 6 Let A and B be arbitrary events in a sample space Ω. Then

$$P[A \cup B] = P[A] + P[B] - P[A \cap B]. \tag{3.8}$$

If A, B, and C are arbitrary events, then

$$\begin{aligned}
P[A \cup B \cup C] &= P[A] + P[B] + P[C] \\
&\quad -(P[A \cap B] + P[A \cap C] + P[B \cap C]) \\
&\quad +P[A \cap B \cap C].
\end{aligned} \tag{3.9}$$

The cases of two and three events are covered explicitly in the theorem. It turns out that, as more events are included, the pattern continues. For four events you add all three-way intersection probabilities and subtract the four-way intersection probability. For five events you subtract all four-way intersection probabilities and add the five-way intersection probability, etc.

Example 7. A bank has issued three loans which may or may not default. Let A, B, and C be the events that these loans default, and suppose that A has probability .2, B has probability .1, and C also has probability .1. If we assume

that the probability that any pair defaults together is the product of the individual default probabilities, and similarly for the event that all three loans default
together, what is the probability that at least one loan will default? None of
the loans? What is the probability that either the first or the second loan will
default?

Solution. The problem statement gives us all of the following:

$$P[A] = .2, P[B] = .1, P[C] = .1$$

$$P[A \cap B] = (.2)(.1) = .02, P[A \cap C] = (.2)(.1) = .02,$$

$$P[B \cap C] = (.1)(.1) = .01$$

$$P[A \cap B \cap C] = (.2)(.1)(.1) = .002.$$

Thus, by the law of inclusion and exclusion, the probability that at least one
loan will default, which is the probability of $A \cup B \cup C$, is

$$
\begin{aligned}
P[A \cup B \cup C] &= P[A] + P[B] + P[C] - (P[A \cap B] + P[A \cap C] \\
&\quad + P[B \cap C]) + P[A \cap B \cap C] \\
&= .2 + .1 + .1 - (.02 + .02 + .01) + .002 \\
&= .4 - .05 + .002 = .352.
\end{aligned}
$$

The event that no loans default is the complement of the event that at least one
defaults, hence

$$P[\text{no loans default}] = 1 - P[\text{at least one loan defaults}] = 1 - .352 = .648.$$

The event that either the first or the second loan will default involves only A
and B, so the two set version of the law of inclusion and exclusion yields that

$$P[A \cup B] = P[A] + P[B] - P[A \cap B] = .2 + .1 - .02 = .28. \ \blacksquare$$

The last property of probability that we will look at in this section is actually the most basic among a family of theorems each of which is called the
Law of Total Probability. We will see versions again when we look at conditional probability and later when we introduce expectation of random variables.
To see the idea, consider Figure 3.6.

We have an event A in a sample space Ω, and we also have a collection of
other events $B_1, B_2, ..., B_n$ which do not overlap each other and which together
make up the whole sample space. (Actually, we only need their union to cover
A; see Exercise 17.) Such a group of sets is called a **partition** of Ω. Then event
A splits up into the union of the events $A \cap B_i$, for $i = 1, 2, ..., n$. To be more
precise, since each pair B_i and B_j do not intersect, it must be that each pair
$A \cap B_i$ and $A \cap B_j$ also do not intersect. (Try justifying this carefully.) Also,
any outcome in A is in Ω, and hence is in one of the events B_i. Such an outcome

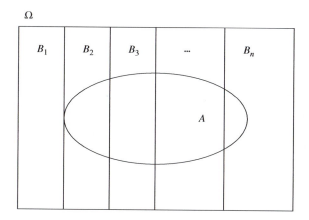

Figure 3.6: Law of Total Probability.

is therefore in $A \cap B_i$, hence the union of the events $A \cap B_i$, for $i = 1, 2, ..., n$ completely covers A and must also trivially be a subset of A. So,

$$A = \bigcup_{i=1}^{n} (A \cap B_i),$$

and the union is a disjoint union. Thus, the probability of A is the sum of the probabilities of the events $A \cap B_i$. This is the content of the following theorem.

Theorem 7 (Law of Total Probability, intersection form) Let A be an event in a sample space Ω. Then, if $B_1, B_2, ..., B_n$ is a partition of Ω, that is, a collection of pairwise disjoint events whose union is Ω, we have

$$P[A] = \sum_{i=1}^{n} P[A \cap B_i]. \tag{3.10}$$

The way to think of this theorem is that it implements a "divide and conquer" strategy for computing the probability of A. If the sample space can be split into simpler pieces using the partition (B_i), so that each $P[A \cap B_i]$ is easy to compute, then we find those and reassemble the pieces to get $P[A]$. This is frequently a useful tool in situations involving multiple time periods or multiple stages in an experiment. We used it subtly in Example 5, which refers to Figure 3.2. To compute, for example, $P[\text{exchange rate} = 1.2]$, we can break up the event into the union of disjoint events: "exchange rate $= 1.2$ and price $= 260$," "exchange rate $= 1.2$ and price $= 250$," and "exchange rate $= 1.2$ and price $= 240$." These three each have probability $1/9$, hence the total probability of the event "exchange rate $= 1.2$" is $3/9$.

Example 8. Consider a two-step binomial branch process with unequal path probabilities as listed in the following table:

path	(U, U)	(U, D)	(D, U)	(D, D)
probability	.36	.24	.24	.16

Find the probability that the second step is down.

Solution. One approach is to just add the probabilities of the outcomes with the second step down, but let us see how the Law of Total Probability fits in. Let A stand for the event that the second step is down. With the information we have, we could cope with the problem if we knew what the first step was. So let B_1 be the event that the first step was up, and let B_2 be the event that it was down. Then,

$$
\begin{aligned}
P[A] &= P\left[A \cap B_1\right] + P\left[A \cap B_2\right] \\
&= P[\text{second step down and first step up}] \\
&\quad + P[\text{second step down and first step down}] \\
&= P[(U, D)] + P[(D, D)] \\
&= .24 + .16 = .40.
\end{aligned}
$$

So in this easy situation, the LTP really does boil down to adding two outcome probabilities. We will see more useful applications later. ∎

Important Terms and Concepts

Binomial branch process - A model for the price of a risky asset in which at each time and current price, the next price can be one of two values: the current price times a constant $1 + b$ or the current price times a constant $1 + a$, with probabilities q and $1 - q$, respectively.

Outcome of an experiment - A non-decomposable result that could happen when the experiment is conducted.

Sample space of an experiment - The set of all possible individual outcomes of the experiment.

Event - A set of outcomes, that is, a subset of the sample space.

Fundamental counting principle - The number of possible outcomes of a multi-stage experiment is the product of the numbers of ways of doing each stage.

Probability measure - A function that assigns a number between 0 and 1 to each event, with the stipulations that the whole sample space has probability 1, and the total probability of the union of disjoint events is the sum of their probabilities.

Equiprobable models - Outcomes are equally likely, so that $P[E] = n(E)/n(\Omega)$.

Sum rule - The number of outcomes (or probability) of a union of disjoint events is the total of the numbers of outcomes (probabilities) of the events.

Permutations - Ordered sequences drawn without replacement from a population.

Combinations - Unordered sequences drawn without replacement from a population.

Generalized sum rule (Law of inclusion-exclusion) - If A and B are arbitrary events, then $P[A \cup B] = P[A] + P[B] - P[A \cap B]$. This can be extended to many events, subtracting paired intersections, adding three-way intersections, subtracting four-way intersections, etc.

Law of total probability - If events B_i, $i = 1, 2, ..., n$ partition the sample space Ω, then, for any event A, $P[A] = \sum_{i=1}^{n} P[A \cap B_i]$.

Exercises 3.1

1. Draw a tree for a four-stage binomial branch process in which the initial price is $S_0 = 10$, and at each transition the price either increases by 4% or decreases by 3%. Let the up and down probabilities be general. Find the sample space for the experiment of observing the final price after four moves.

2. Write the outcomes of the roll of two fair dice. How many are there? Count the number of ways of getting each possible sum 2, 3, ... , 12.

3. Write the outcomes in the sample space of the experiment of observing the path of a binomial branch process with five steps. Find the number of paths such that (a) there are exactly three up moves; (b) there are no more than two up moves; (c) there are at least three up moves.

4. (a) In how many distinct ways can the letters MATH be rearranged? (b) How about the letters MATHFINANCE?

5. On one particular day, eight stocks are to be ranked in terms of their daily rate of return. Assuming no ties, in how many ways can this random phenomenon occur?

6. A group of five bonds is to be rated in terms of their riskiness for default as high, medium, or low risk. In how many distinct ways can this be done? How many of these are such that two of the bonds are high risk, one is medium risk, and two are low risk?

7. The rates of return on an account initially containing $1000 for three time periods will be 1%, 2%, or 3% in some order that is randomly chosen. Find the sample spaces for the phenomenon of (a) observing the value of the account at the end of each of the three time periods; (b) observing the value of the account at the end of the third period. What is unusual about the sample space in case (b)?

8. Show that $\binom{n}{k} = \binom{n}{n-k}$, and interpret this result in terms of the number of ways of getting k ups in an n-stage binomial branch process.

9. Suppose that stocks A, B, and C each will either go up in price, stay the same, or go down in price.
(a) List the outcomes in the sample space of the random phenomenon that observes the price outcomes of the three together;
(b) If P is a probability measure under which all of these outcomes are equally likely, find:
  (i) $P[\text{stock } A \text{ goes up}]$;
  (ii) $P[\text{both } A \text{ and } B \text{ go up}]$;
  (iii) $P[\text{all stocks go down}]$.

10. In the context of Exercise 9, again, if P is a probability measure under which all of these outcomes are equally likely, find
(a) $P[A \text{ goes up or } C \text{ goes down}]$;
(b) $P[B \text{ goes up or } C \text{ does not go up}]$.

11. Amanda will pick a portfolio of three stocks at random from among four companies in the manufacturing sector, three in the technology sector, and three in the communications sector of the market. How many possible portfolios could she construct in this way? How many if she wants one stock in each sector? In the latter case, are all 10 companies equally likely to be selected? Why or why not?

12. For a binomial branch process with 20 steps, how many paths are there that have 12 ups? 15 ups? If all paths are equally likely, find the probabilities of 12 ups and 15 ups.

13. As a variation on Exercise 7, suppose that an account contains $1000 initially, and in each of the first three time periods, the rate of return is equally likely to be 1%, 2%, or 3% (repetitions allowed). Assume that each sequence of three rates of return is as likely as each other sequence. Find the sample space for the phenomenon of observing the final price, and describe a probability measure on that sample space.

14. Joseph has given loans to two of his brothers: Dan and Levi. The loan to Dan is in the amount of 500 shekels and Levi has borrowed 300 shekels. The probability that Levi will default on the loan is twice the probability that Dan

defaults. The probability that at least one of them defaults is .4. If Dan defaults, then 1/3 of the time Levi also defaults. Use this information to construct a probability model for the phenomenon of observing how many shekels are repaid to Joseph.

15. If A is an event and A^c is the complement of A, prove that

$$P[A] + P[A^c] = 1.$$

16. In the Canadian wheat example of Figure 3.2, assume that each of the nine possible sequences is as likely as each other. Find the probabilities that, in Canadian dollars, the price per ton is (a) more than 290; (b) less than 270.

17. Show that the law of total probability formula (3.10) remains true if you no longer assume that $\bigcup_{i=1}^{n} B_i$ is all of the sample space Ω, but if you assume instead that $A \subset \bigcup_{i=1}^{n} B_i$.

18. Three stocks can either go up or down during a particular day. Stock 1 goes up with probability .38, stock 2 goes up with probability .42, and stock 3 goes up with probability .35. The probability that both stocks 1 and 2 go up is .10, the probability that both stocks 1 and 3 go up is .13, and the probability that both stocks 2 and 3 go up is .07. The probability that all three stocks go up simultaneously is .05. Find the probability that none of the stocks goes up.

19. Consider a two-step binomial branch process with $S_0 = 40$, $b = .02$, $a = -.01$. Find each of the three possible price values s at time 2. Now assume that up moves occur with probability .4 and down moves occur with probability .6, and assume also that the probability of any path is the product of the probabilities of the branches on that path. Use the Law of Total Probability to find, for each s, $P[S_2 = s]$, where S_2 denotes the random price at time 2.

3.2 Random Variables and Distributions

The mathematical object that we introduce here allows us to deal directly with the randomness exhibited by financial quantities. Intuitively speaking, a **random variable** is a quantitative variable in a random experiment or phenomenon. Its value cannot be predicted before the experiment is performed. Framing the idea more carefully, a random variable is a function that attaches a number to each of the possible outcomes in the sample space of a random experiment. While outcomes in a sample space, like paths of ups and downs $(U, D, U, ...)$, may not be numerical in form, in many situations we are most interested in some numerical quality of outcomes, like the number of ups. So random variables can make the necessary conversion from non-numerical outcomes to numerical values of interest.

Example 1. Exercise 2 of Section 3.1 sets the stage for us. Suppose that two fair dice are rolled, one red and one white, and their sum is recorded. Call that sum X. What are the outcomes in the sample space of this experiment? What value does X have at each of these outcomes? How likely is it that X will equal each of its possible values 2, 3, 4, ... , 12?

Solution. The sample space is the set of pairs (x_1, x_2) which give the faces of die 1 (red) and die 2 (white). In set theoretic notation, it can be written as:

$$\Omega = \{(x_1, x_2) \,|\, x_1, x_2 \in \{1, 2, ..., 6\}\}.$$

The fairness assumption indicates that an appropriate probability model is to give each of the 36 outcomes probability equal to 1/36. The random variable X works on outcomes by addition:

$$X(x_1, x_2) = x_1 + x_2.$$

A helpful display of the 36 equally likely outcomes in the sample space, grouped by common sum, is below.

outcomes	$\longrightarrow$	value of X
$(1,1)$	$\longrightarrow$	2
$(1,2),(2,1)$	$\longrightarrow$	3
$(1,3),(2,2),(3,1)$	$\longrightarrow$	4
$(1,4),(2,3),(3,2),(4,1)$	$\longrightarrow$	5
$(1,5),(2,4),(3,3),(4,2),(5,1)$	$\longrightarrow$	6
$(1,6),(2,5),(3,4),(4,3),(5,2),(6,1)$	$\longrightarrow$	7
$(2,6),(3,5),(4,4),(5,3),(6,2)$	$\longrightarrow$	8
$(3,6),(4,5),(5,4),(6,3)$	$\longrightarrow$	9
$(4,6),(5,5),(6,4)$	$\longrightarrow$	10
$(5,6),(6,5)$	$\longrightarrow$	11
$(6,6)$	$\longrightarrow$	12

This table has a visual impact that tells an important story: not all possible sums are created equal. A sum of 7 is the most likely, because 6 of the 36 equally likely outcomes in the sample space are taken by the function X to 7. Sum values of 2 and 12 are the least likely, since only one outcome maps to these values of X. The lengths of the lines in the left column give a visual measure of the relative likelihoods of the values of X. Computing probabilities by counting numbers of outcomes and dividing by 36, we get

value of X:	2	3	4	5	6	7	8	9	10	11	12
probability:	$\frac{1}{36}$	$\frac{2}{36}$	$\frac{3}{36}$	$\frac{4}{36}$	$\frac{5}{36}$	$\frac{6}{36}$	$\frac{5}{36}$	$\frac{4}{36}$	$\frac{3}{36}$	$\frac{2}{36}$	$\frac{1}{36}$

. ■   (3.11)

Example 2. Consider the flip of three fair coins in succession. Let X be the number of heads observed. What value does X have at each outcome? What is the probability that X will equal each of its possible values?

Solution. This time we draw a picture of the sample space and the set of values that X can take on, showing which outcomes are taken to which possible X values. The set of possible X values, i.e., the range of the function X, is called its **state space**, and the values are called **states**. Assuming that the eight outcomes are equally likely, Figure 3.7 shows that:

$$P[X = 0] = \frac{1}{8}, P[X = 1] = \frac{3}{8}, P[X = 2] = \frac{3}{8}, P[X = 3] = \frac{1}{8}. \ \blacksquare \qquad (3.12)$$

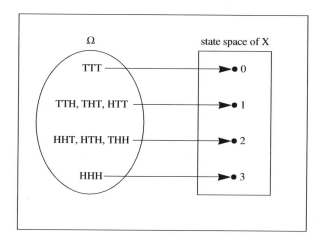

Figure 3.7: Action of X = number of heads in three coin flips.

Here are the formal definitions that we will need. The previous examples should have already given you good intuition about their meaning.

Definition 1. A **discrete random variable** X is a function whose domain is the sample space Ω of a random phenomenon and whose range is a finite or countable set E, which is called its **state space**. A point $x = X(\omega)$ is called a possible **state** of the random variable X.

Definition 2. The **probability mass function** (p.m.f.) $f(x)$ of a discrete random variable X is a function taking the state space of X to the interval $[0, 1]$ such that

$$f(x) = P[X = x], \text{ for each } x \in E. \qquad (3.13)$$

We need to discuss a few important concepts before returning to examples.

The probability mass function quantifies how probability is distributed among possible states, and so it is called by some people the **probability distribution** of the random variable. Since there are other ways (see the discussion of the **cumulative distribution function** below) of characterizing this distribution of probability, the term probability distribution is more generic than probability mass function. Notice that in Examples 1 and 2 we found probability mass functions, which are listed in displays (3.11) and (3.12).

Since at least one of the possible states of a random variable X must occur when the phenomenon is observed, and since each likelihood of occurrence must be non-negative and less than or equal to 1, we have the following regularity conditions that must be satisfied in order for a function f to qualify as a valid p.m.f.:

$$f(x) \geq 0 \text{ for all } x \in E, \text{ and } \sum_{x \in E} f(x) = 1. \tag{3.14}$$

For example, the function $f(1) = 1/3$, $f(2) = 1/6$, $f(3) = 1/4$, $f(4) = 1/3$ is not a valid probability mass function, since the total of the probability masses is

$$\frac{1}{3} + \frac{1}{6} + \frac{1}{4} + \frac{1}{3} = \frac{4}{12} + \frac{2}{12} + \frac{3}{12} + \frac{4}{12} = \frac{13}{12}.$$

Reduction of any of the probability masses by $1/12$, for instance, redefining $f(2) = 1/12$, would result in a valid p.m.f.

We can display a distribution visually using the following graphical tool. A **probability histogram** is a bar chart whose bars sit atop a horizontal axis, located at the possible states x of a random variable, and whose bar heights are the probability masses $f(x)$. For Example 2, the probability histogram for $X =$ number of heads in three coin flips is shown in Figure 3.8.

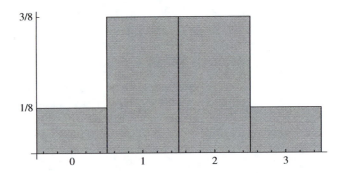

Figure 3.8: Probability histogram for $X =$ number of heads in three coin flips.

The notation $P[X = x]$ hides a detail that is worth paying attention to. Probability is a function on the set of events in a sample space, so "$X = x$" must be viewed as an event in order for P to apply to it. It is indeed an event, that is, the set of outcomes ω that X maps onto x, i.e.,

$$\{X = x\} = \{\omega \in \Omega | X(\omega) = x\}. \tag{3.15}$$

So the probability measure P on Ω can apply to this event, which implies that X is really inheriting its probability distribution from the measure P. Frequently though, we will keep the sample space of the random phenomenon in the background and take the probability distribution as our point of origin.

In the absence of a defined sample space with a probability measure on it, probability distributions can also be estimated using historical data. For example, if a stock has behaved in such a way that in the last 100 trading days it has gone up by \$.50 20 times, up by \$.25 30 times, down by \$.25 30 times, and down by \$.50 20 times, then we would estimate that the probability mass function of the random variable X giving the daily change in value is:

$$P[X = .5] = \frac{20}{100}; P[X = .25] = \frac{30}{100};$$

$$P[X = -.25] = \frac{30}{100}; P[X = -.5] = \frac{20}{100}.$$

For sets B inside the state space of random variable X, we can calculate the probability that X takes a value in B by summing the probability masses of all states in B. The formula is

$$P[X \in B] = \sum_{x \in B} f(x) = \sum_{x \in B} P[X = x]. \tag{3.16}$$

This is justified by the additivity axiom of probability, since, for different x values, the sets of outcomes $\{\omega \in \Omega | X(\omega) = x\}$ are disjoint. The set of outcomes $X \in B = \{\omega \in \Omega | X(\omega) \in B\}$ is the disjoint union of the sets $\{\omega \in \Omega | X(\omega) = x\}$ taken over all states $x \in B$. The probability of the event $X \in B$ is therefore the sum of the probabilities of the events $\{X = x\}$. For example, for the two dice experiment in Example 1, the probability of rolling at least a 10 is:

$$P[X \geq 10] = f(10) + f(11) + f(12) = \frac{3}{36} + \frac{2}{36} + \frac{1}{36} = \frac{6}{36}.$$

Example 3. On 8 successive days, the information of whether a stock went up or down from the previous day's closing price is recorded. Suppose that ups are twice as likely as downs, and that the probability of any path is the product of the probabilities of the steps along that path. What is the probability mass function of the random variable $X =$ number of ups among the 8 days? Use it to calculate the probability of at least 6 up moves.

Solution. First, the random variable X can take on the states $0, 1, 2, ..., 8$ depending on how many days resulted in an up observation. Next, let's try to get a better handle on the probabilistic structure of this phenomenon. It is given that, in any one step, $P[\text{up}] = 2P[\text{down}]$. Together with the regularity requirement that $P[\text{up}] + P[\text{down}] = 1$, it is easy to see that $P[\text{up}] = p = 2/3$ and $P[\text{down}] = 1 - p = 1/3$. The assumption about path probabilities implies, for example, that an outcome like (U, D, D, U, U, D, U, U) should be given probability

$$P[(U,D,D,U,U,D,U,U)] = \frac{2}{3} \cdot \frac{1}{3} \cdot \frac{1}{3} \cdot \frac{2}{3} \cdot \frac{2}{3} \cdot \frac{1}{3} \cdot \frac{2}{3} \cdot \frac{2}{3} = \left(\frac{2}{3}\right)^5 \left(\frac{1}{3}\right)^3,$$

which is the product of factors formed as the one-step up probability raised to the number of up moves in the path times the one-step down probability raised to the number of down moves. In general, for any particular sequence with exactly k Us, and hence $8 - k$ Ds, the probability of that sequence is $(2/3)^k (1/3)^{8-k}$. The total probability that the stock goes up exactly k times is:

$$g(k) = P[X = k] = \binom{8}{k}\left(\frac{2}{3}\right)^k \left(\frac{1}{3}\right)^{8-k}, k = 0, 1, 2, ..., 8 \qquad (3.17)$$

because there are are $\binom{8}{k}$ sequences with k U symbols. Figure 3.9 is a histogram of this probability mass function.

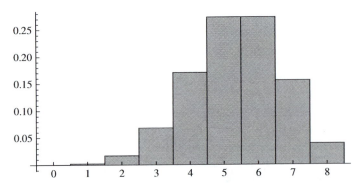

Figure 3.9: Probability histogram for $X =$ number of up moves, $n = 8$, $p = 2/3$.

To finish the example, we need to find the probability of at least 6 up moves. This is:

$$
\begin{aligned}
P[X \geq 6] &= P[X = 6] + P[X = 7] + P[X = 8] \\
&= \binom{8}{6}\left(\frac{2}{3}\right)^6 \left(\frac{1}{3}\right)^2 + \binom{8}{7}\left(\frac{2}{3}\right)^7 \left(\frac{1}{3}\right)^1 + \binom{8}{8}\left(\frac{2}{3}\right)^8 \left(\frac{1}{3}\right)^0 \\
&= .468221.
\end{aligned}
$$

Remark. The p.m.f. in the previous example is a special case of an important distribution called the **binomial distribution**, whose p.m.f. has the general form:

$$g(k; n, p) = \binom{n}{k}p^k(1 - p)^{n-k}, k = 0, 1, 2, ..., n. \qquad (3.18)$$

There are two parameters, n and p, that determine the shape of the distribution of probability. We will be very interested in this probability distribution later (see also Exercises 13 and 24 of this section).

Example 4. The interest rate given by a bank to mortgage applicants is 5% this month. Next month it will either go up by .5% or down by .5% with probabilities .6 and .4, respectively. The month after, it will go up by .2% with probability .8 or up by .3% with probability .2. Describe the sample space of this random phenomenon. Form a probability measure on the sample space assuming that the probability of a sequence of two interest rate changes is the product of the individual rate change probabilities. If X is the interest rate after these two months, what is the p.m.f. of X?

Solution. We may characterize the sample space as a set of four pairs, ranging through all of the possible interest rate changes over the two months.

$$\Omega = \{(.5\%, .2\%), (-.5\%, .2\%), (.5\%, .3\%), (-.5\%, .3\%)\}.$$

By the given probabilities and the multiplication rule, the probabilities of these four outcomes are:

$$P[(.5\%, .2\%)] = (.6)(.8) = .48;$$
$$P[(-.5\%, .2\%)] = (.4)(.8) = .32;$$
$$P[(.5\%, .3\%)] = (.6)(.2) = .12;$$
$$P[(-.5\%, .3\%)] = (.4)(.2) = .08.$$

The final interest rate X may by any of the following values:

$$X(.5\%, .2\%) = 5\% + .5\% + .2\% = 5.7\%;$$
$$X(-.5\%, .2\%) = 5\% - .5\% + .2\% = 4.7\%;$$
$$X(.5\%, .3\%) = 5\% + .5\% + .3\% = 5.8\%;$$
$$X(-.5\%, .3\%) = 5\% - .5\% + .3\% = 4.8\%.$$

Hence X is a 1-1 mapping of outcomes to states, and it follows that the p.m.f. of X is

$$f(x) = \begin{cases} .48 & \text{if } x = 5.7\%; \\ .32 & \text{if } x = 4.7\%; \\ .12 & \text{if } x = 5.8\%; \\ .08 & \text{if } x = 4.8\%. \end{cases}$$

A probability histogram of this mass function is given in Figure 3.10, which highlights that there are two widely separated parts of the state space of X, due to the crucial decision point in the first month of whether the interest rate will go up or down. ∎

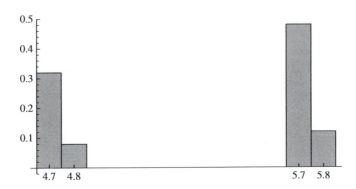

Figure 3.10: Probability histogram for X = interest rate (%) after 2 months.

To take a glimpse ahead at the idea of **expectation** that will be covered later, what might we mean by an "average" or "expected" interest rate in Example 4? There are four possible rates, 4.7, 4.8, 5.7, and 5.8, of which we could simply take an arithmetical average. But that ignores the fact that rates 4.7 and 5.7 are much likelier than the other two rates. They should contribute more to a probabilistic notion of average rate. What we will do is to define the expected interest rate, denoted by $E[X]$ or by μ_x, as a weighted average of states using the probability masses as weights:

$$\mu_x = E[X] = (.32)(4.7) + (.08)(4.8) + (.48)(5.7) + (.12)(5.8) = 5.32.$$

You might recognize such a combination from calculus as a dot product of vectors $(.32, .08, .48, .12) \cdot (4.7, 4.8, 5.7, 5.8)$, the first being the vector of probability masses and the second being the vector of states of X.

Section 3.3 will explore expectations of random variables and functions of random variables much more deeply.

3.2.1   Cumulative Distribution Functions

There is an alternative to the probability mass function that also completely characterizes the distribution of probability among states of a random variable, as in the definition below.

Definition 3. The **cumulative distribution function** (c.d.f.) $F(x)$ of a discrete random variable X is the function

$$F(x) = P[X \leq x], x \in \mathbb{R}. \tag{3.19}$$

The idea is that, as the argument x increases from $-\infty$ to ∞, more and more probability weight is accumulated as x sweeps past the states of X. In Example 2, where X = number of heads in three coin flips, we found that the probability mass function is:

$$f(x) = \begin{cases} 1/8 & \text{if } x = 0; \\ 3/8 & \text{if } x = 1; \\ 3/8 & \text{if } x = 2; \\ 1/8 & \text{if } x = 3. \end{cases}$$

For any $x < 0$, $P[X \leq x] = 0$, since no probability is accumulated until we hit the lowest state 0. For values of x between 0 and 1, including 0 and excluding 1, state 0 contributes its probability of $1/8$ to the total. When x passes by 1, the total probability to the left of x is $1/8 + 3/8 = 1/2$ because state 1 now contributes to the total. Similarly, when x passes by 2, the total probability increases to $1/8 + 3/8 + 3/8 = 7/8$, and when x passes by three, all states have been included, so the cumulative probability is 1. Between states, the c.d.f. remains flat. A piecewise-defined formula for the c.d.f. is therefore:

$$F(x) = \begin{cases} 0 & \text{if } x < 0; \\ 1/8 & \text{if } 0 \leq x < 1; \\ 4/8 & \text{if } 1 \leq x < 2; \\ 7/8 & \text{if } 2 \leq x < 3; \\ 1 & \text{if } x \geq 3. \end{cases}$$

A graph of the c.d.f. is shown in Figure 3.11, which is consistent with the formula.

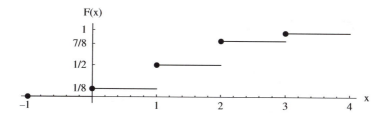

Figure 3.11: Cumulative distribution function of X = number of heads in three coin flips.

The other thing to notice about the c.d.f. in Figure 3.11 is that it has jumps precisely at the states of the random variable, and the sizes of the jumps are the probability masses attributed to those states by the p.m.f. Thus, if we know the p.m.f., we can graph the c.d.f. and write its formula, and if we know the c.d.f., we can look at where the jumps are and what their magnitude is and infer the p.m.f. In this sense, the probability mass function and the cumulative distribution function are equivalent ways of specifying the probability distribution of the random variable.

You can practice with c.d.f.'s in a sequence of exercises: 14–17. We will not need them heavily for awhile, but when we move to the world of continuous probability, they turn out to be natural tools to use in order to calculate probabilities of events involving random variables.

3.2.2   Random Vectors and Joint Distributions

In many financial problems more than one random variable is involved. For instance, an investor may be considering allocating wealth among six risky assets with rates of return $R_1, R_2, R_3, ..., R_6$ that are random, and these rates might depend on each other in some way. The goal would be to compute the probability distribution of the rate of return on the entire investment, or to compute some aspects of the rate of return, such as the expected value. As another example, variable interest rates on a credit card may change occasionally to new levels. Monthly interest rates $M_1, M_2, ..., M_{12}$ may be in force during a year, and we would be interested in computing the probability distribution of the final balance. In yet another example, a bank may have 10 loans outstanding to local small businesses, and each business has some chance of defaulting on their loan. Say $I_j = 1$ if business j defaults and it equals 0 otherwise, $j = 1, 2, ..., 10$. Defaults may be correlated in some way between businesses, and the bank would be interested in such things as the probability distribution of the number of defaults, or of the amount lost.

To be able to model such situations, we need to introduce the concept of the joint distribution of several random variables.

Definition 4. The ***joint probability mass function*** $f(x_1, x_2, ..., x_n)$ of n discrete random variables $X_1, X_2, ..., X_n$ is a function taking the Cartesian product of the state spaces of the X_i to the interval $[0, 1]$ such that:

$$f(x_1, x_2, \ldots, x_n) = \begin{array}{l} P[X_1 = x_1, X_2 = x_2, \ldots, X_n = x_n] \\ \text{for each } x_1 \in E_1, x_2 \in E_2, \ldots, x_n \in E_n. \end{array} \tag{3.20}$$

Note that, in the notation of formula (3.20), the commas mean "and," so that we are referring to the probability that simultaneously all of the random variables X_i take values equal to their corresponding states x_i.

It can be useful to think of the random variables in the definition in a group, and also to assemble the joint states in a group. So we will sometimes use the language ***random vector*** to refer to the list of random variables $\boldsymbol{X} = (X_1, X_2, ..., X_n)$. A ***state*** of this random vector is an element $\boldsymbol{x} = (x_1, x_2, ..., x_n)$ of n-dimensional space. The equation in (3.19) defining the joint p.m.f. can be rewritten:

$$f(\boldsymbol{x}) = P[\boldsymbol{X} = \boldsymbol{x}], \boldsymbol{x} \in E_1 \times E_2 \times \cdots \times E_n. \tag{3.21}$$

We see that we haven't really defined an entirely new concept; our random variable just has n components taking values in an n-dimensional state space. Although we have written the set of states of the random vector $\boldsymbol{X}$ as the Cartesian product of all of the individual state spaces, it may well be that certain combinations of states $(x_1, x_2, ..., x_n)$ are not allowed, or have probability 0.

Example 5. Suppose that two assets have five possible pairs of rates of return (R_1, R_2), as indicated in the table below, with probabilities as listed in the table.

Find (a) the probability that asset 1 has a rate of return greater than or equal to .04 and (b) the probability that asset 2 has a rate of return less than .03. If an investor starts with $1000 and places half of that in the first and half in the second asset, find (c) the probability that the final wealth will be at least $1020.

value of (R_1, R_2)	$(-.01, 0)$	$(.01, .03)$	$(.02, .02)$	$(.04, .05)$	$(.05, .04)$
probability	.3	.1	.2	.2	.2

Solution. (a) In this question, the joint probability mass function of the random vector $R = (R_1, R_2)$ has been given in tabular form rather than as a formula. But we still can find probabilities of events involving the pair of rates of return by adding up state probabilities for those states that satisfy the event. There are two states in which the rate of return on asset 1 is at least .04: $(.04, .05)$, and $(.05, .04)$. The total probability of these states is $.2 + .2 = .4$.

(b) Similarly, there are two states in which the rate of return on asset 2 is less than .03: $(-.01, 0)$ and $(.02, .02)$. Their total probability is $.3 + .2 = .5$.

(c) The investor will put $500 into asset 1 and $500 into asset 2. The final wealth will be

$$\$500 \, (1 + R_1) + \$500 \, (1 + R_2) = \$1000 + \$500 \, (R_1 + R_2).$$

Thus, the final wealth is expressible in an easy way in terms of the sum of the two rates of return. For this final wealth to be at least $1020, we must have:

$$\$1000 + \$500 \, (R_1 + R_2) \geq \$1020 \Longrightarrow R_1 + R_2 \geq \frac{1020 - 1000}{500} = .04.$$

The only state for which this inequality does not hold is the first one, $(-.01, 0)$, whose probability is .3. Hence,

$$P[\text{final wealth} \geq \$1020] = 1 - .3 = .7. \blacksquare$$

Example 6. The interest rate on an account is 2% at the start of the year. At the beginning of each quarter it may change. Assume that the rate will be the same as the previous quarter with probability .6, or it will increase by .2% with probability .4. Assume also that the probability of a sequence of changes is the product of the probabilities of each individual change. Find the joint distribution of the rates R_2, R_3, R_4 at the beginnings of the second, third, and fourth quarters, and use that to find the distribution of the final value of an initial deposit of $10,000 into this account.

Solution. Since the interest rate R_1 is deterministic, the states of the random vector $R = (R_2, R_3, R_4)$ correspond to paths of length 2 starting from either of the two states 2% and 2.2% in level 1 of the tree, as shown in Figure 3.12.

These paths are itemized in the table below, and the probabilities of each state are computed using the given assumptions about moves from one quarter to the next.

state	probability
(2%, 2%, 2%)	(.6)(.6)(.6) = .216
(2%, 2%, 2.2%)	(.6)(.6)(.4) = .144
(2%, 2.2%, 2.2%)	(.6)(.4)(.6) = .144
(2%, 2.2%, 2.4%)	(.6)(.4)(.4) = .096
(2.2%, 2.2%, 2.2%)	(.4)(.6)(.6) = .144
(2.2%, 2.2%, 2.4%)	(.4)(.6)(.4) = .096
(2.2%, 2.4%, 2.4%)	(.4)(.4)(.6) = .096
(2.2%, 2.4%, 2.6%)	(.4)(.4)(.4) = .064

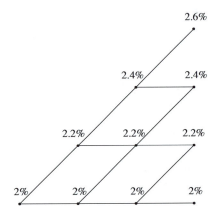

Figure 3.12: Interest rate tree.

With this joint distribution in hand, we can find the distribution of the final value, which symbolically is:

$$\$10,000(1.02)\left(1 + R_2\right)\left(1 + R_3\right)\left(1 + R_4\right) = \$10,200\left(1 + R_2\right)\left(1 + R_3\right)\left(1 + R_4\right).$$

For instance, for the first two states, rounded to the nearest dime, the final values are:

$$\$10,200(1.02)(1.02)(1.02) = \$10,824.30,$$

$$\$10,200(1.02)(1.02)(1.022) = \$10,845.50.$$

Continuing the computation, we get the final values in the expanded table below.

state	final value	probability
(2%, 2%, 2%)	10824.30	.216
(2%, 2%, 2.2%)	10845.50	.144
(2%, 2.2%, 2.2%)	10866.80	.144
(2%, 2.2%, 2.4%)	10888.10	.096
(2.2%, 2.2%, 2.2%)	10888.10	.144
(2.2%, 2.2%, 2.4%)	10909.40	.096
(2.2%, 2.4%, 2.4%)	10930.80	.096
(2.2%, 2.4%, 2.6%)	10952.10	.064

The distribution of the final value is essentially described by the second and third columns of this table, except that, with rounding, there are only distinct states, with the value \$10,888.10 being common to both the fourth and fifth rows, so that its total probability is $.096 + .144 = .240$. ∎

Example 7. The National Bank of Desperationville has issued four loans, each for \$10,000. Each of these loans has a chance of defaulting, and the default of one loan can affect the probability of default for another. Define indicator random variables I_1, I_2, I_3, I_4 to indicate defaults of the four loans by:

$$I_j = \begin{cases} 1 & \text{if loan } j \text{ defaults;} \\ 0 & \text{otherwise.} \end{cases}$$

Given the joint p.m.f. $f(i_1, i_2, i_3, i_4)$ tabulated below, find the p.m.f. of the total amount lost by the bank due to defaults.

state	probability
(0, 0, 0, 0)	.5
(0, 0, 0, 1)	.05
(0, 0, 1, 0)	.05
(0, 1, 0, 0)	.05
(1, 0, 0, 0)	.05
(0, 0, 1, 1)	.01
(0, 1, 0, 1)	.01
(1, 0, 0, 1)	.01
(0, 1, 1, 0)	.01
(1, 0, 1, 0)	.01
(1, 1, 0, 0)	.01
(0, 1, 1, 1)	.05
(1, 0, 1, 1)	.05
(1, 1, 0, 1)	.05
(1, 1, 1, 0)	.05
(1, 1, 1, 1)	.04

Solution. The states have been laid out in the table conveniently, with the state having no defaults first, the four states having one default next, the six states with two defaults after that, then the four states with three defaults, and finally the state in which all loans default. If X denotes the p.m.f. of the total amount lost, then X can equal \$0 if there are no defaults, \$10,000 if there is one, \$20,000 if there are two, \$30,000 if there are three, or \$40,000 if there are four defaults. We can simply total the probabilities in the table of joint states that lead to particular numbers of defaults; for instance, in the case of two defaults,

$$P[X = 20000] = .01 + .01 + .01 + .01 + .01 + .01 = .06.$$

You should check the other cases that yield the p.m.f. $g(x)$ for X below:

$$g(x) = \begin{cases} .5 & \text{if } x = 0; \\ .2 & \text{if } x = 10000; \\ .06 & \text{if } x = 20000; \\ .2 & \text{if } x = 30000; \\ .04 & \text{if } x = 40000. \ \blacksquare \end{cases}$$

Important Terms and Concepts

Random variable - A numerical valued function on a sample space.

State space - The set of values that a random variable can take on.

Probability mass function - The function $f(x) = P[X = x]$ describing the distribution of probability weight among possible states of a random variable X.

Cumulative distribution function - The function $F(x) = P[X \leq x]$ describing how probability weight accumulates as a cutoff point x increases.

Probability histogram - A bar chart in which the probability masses $f(x)$ are plotted as bar heights above the states x.

Joint probability mass function - The function $f(x_1, x_2, ..., x_n)$ describing the likelihood that simulataneously n random variables equal given values in their state spaces.

Exercises 3.2

1. Make a table as in Example 1 of the outcomes in the sample space for a four-step binomial branch process, displaying the action of the random variable X that counts the number of up moves. With what probability does X take on each of its possible values if paths are equally likely?.

2. Draw a diagram similar to Figure 3.7 displaying the sample space of possible outcomes for the experiment of flipping four fair coins in succession, and the

action of the random variable Y that counts the number of tails. Find the state space and the probability distribution of Y.

3. Argue that a valid probability mass function f determines a valid probability model on the state space of its associated random variable X, that is, it satisfies axioms (a), (b), and (c) of Section 3.1.

4. A simple slot machine has three spinning wheels, each with 16 positions on it. Nine of the positions are blank, one is marked with a cherry icon, three with a single bar icon, two with a double bar, and one with a triple bar. A quarter lets you spin the wheel. Suppose that the only winning combination is triple cherries, paying $1000.

   (a) Find the probability distribution of the net amount you win on one spin. Formulate a notion of the "expected winnings" on one spin, and compute it.

   (b) Make reasonable assumptions and find the probability distribution of the net winnings on two spins.

5. Each of two containers has four slips of paper numbered 1, 2, 3, and 4. A slip is drawn from each container at random. Find the probability mass function of the total of the two numbers drawn.

6. Verify that each of the following defines a valid probability mass function.

(a) $f(x) = \begin{cases} 1/8 & \text{if } x \in \{0, 1, 2, 3\}, \\ 1/4 & \text{if } x \in \{4, 5\}; \\ 0 & \text{otherwise.} \end{cases}$

(b) $g(i) = \begin{cases} \dfrac{2i}{n(n+1)} & i \in \{1, 2, ..., n\}; \\ 0 & \text{otherwise.} \end{cases}$

7. Verify that each of the following defines a valid probability mass function.

(a) $h(k) = \dfrac{1}{4}\left(\dfrac{3}{4}\right)^k, k = 0, 1, 2, ..., \infty$

(b) $p(n) = e^{-2} \cdot \dfrac{2^n}{n!}, n = 0, 1, 2, ..., \infty$

8. In the sequential flip of five fair coins, what is the probability of flipping (a) fewer than four heads; (b) two or more heads? Be sure to use random variable modeling in solving the problem.

9. In a group of four stocks whose prices change "independently" of each other (by which we mean that the probability of a joint event involving the stocks is the product of the individual probabilities), such that each goes up this week

with probability .3, find the probability mass function of the number that go up, and use it to compute the probability that at most two of them go up.

10. A stock is currently priced at $50. Each week it will either go up or down by $1 with equal probability. Find the probability mass function of the price of the stock after 3 weeks.

11. A sum of $1000 is deposited in an account earning compound interest at nominal rate 3% compounded monthly. At the beginning of the second quarter, the rate may change: it may be 2.8% with probability .4, or stay at 3% with probability .6. Similarly, at the beginning of the third quarter it may go down by .2%, or stay the same, this time with equal probability. And at the beginning of the fourth quarter, the nominal rate may go up by .2% with probability .3, or stay the same with probability .7. The rate stays constant from there until the end of the year. Find the probability mass function of the final value in the account at the end of the year. Assume that the probability of a sequence of rates is the product of the probabilities of the individual rates.

12. Egbert is applying for a mortgage in the amount of $100,000 on a term of 30 years with beginning of month payments and interest compounded monthly. His nominal yearly rate is not fixed until he submits the application, and he feels that the yearly rate will be 4.8% with probability 1/8, 5.0% with probability 1/4, 5.2% with probability 1/4, 5.4% with probability 1/4, and 5.6% with probability 1/8. Find the probability distribution of his monthly payments. Using the idea suggested in the section, compute his expected monthly payment.

13. An important discrete probability distribution that has come up several times is called the **binomial distribution**. Its p.m.f. is below. Show that this defines a valid p.m.f.

$$f(k) = P[X = k] = \binom{n}{k} p^k (1 - p)^{n-k} \, , k = 0, 1, 2, ..., n.$$

14. Give a piecewise-defined formula for the cumulative distribution function corresponding to the probability mass function below. (Here c is a constant; solve for it.)

$$f(k) = c \cdot k, k = 1, 2, 3, 4, 5.$$

15. For a random variable X with the cumulative distribution function below, find the probability mass function and compute $P[1 < X \le 5]$.

$$F(x) = \begin{cases} 0 & \text{if } x < 0; \\ .2 & \text{if } 0 \le x < 1; \\ .3 & \text{if } 1 \le x < 3; \\ .7 & \text{if } 3 \le x < 5; \\ 1 & \text{if } x \ge 5. \end{cases}$$

16. Show that the cumulative distribution function of a discrete random variable must be a non-decreasing function. (Hint: first use the axioms of probability to prove that if one event A is contained in another event B, then $P[A] \le P[B]$.)

17. Show that, if X is a discrete random variable with c.d.f. $F(x)$ and p.m.f. $f(x)$, then

(a) $P[a < X \le b] = F(b) - F(a);$

(b) $P[a < X < b] = F(b) - F(a) - f(b);$

(c) $P[a \le X < b] = F(b) - F(a) - f(b) + f(a).$

18. Two fair dice are rolled, one after the other. Find (a) the joint distribution of the first die and the total of the two dice; (b) the probability that the first die is at least four and the total is at least nine.

19. In the situation described in Example 4, where the interest rate begins at 5% and undergoes two changes, suppose that, in place of the product assumption, the joint distribution of the changes is as in the table below. Recalculate the p.m.f. of the final interest rate and the expected value of the final interest rate. Define appropriate random variables.

$(1^{st}\text{change}, 2^{nd}\text{change})$	$(.5\%, .2\%)$	$(-.5\%, .2\%)$	$(.5\%, .3\%)$	$(-.5\%, .3\%)$
probability	.2	.4	.3	.1

20. A two-step binomial branch process with initial state $S_0 = 100$ is such that an up move occurs with probability .5 and multiplies the current value by a factor of 1.1, while a down move multiplies it by .9. Assume that the probability of any path is the product of the single-step probabilties along that path. Find the joint probability mass function of S_1 and S_2.

21. In the setting of Example 7, instead of the given joint distribution of the default random variables $I_j, j = 1, 2, 3, 4$, suppose that defaults are **mutually independent**, in the sense that the probability that all loans in any subcollection default is the product of the probabilities that the individual loans default. Assume that the probability that any individual loan defaults is .02. Find the joint distribution of the random variables I_j, and use it to find the p.m.f. of the amount X lost by the bank.

22. In general, random variables $X_1, X_2, ..., X_n$ are called **mutually independent** if, for any subsets of their respective state spaces $B_1 \subset E_1, B_2 \subset E_2, ..., B_n \subset E_n$,

$$P[X_1 \in B_1, X_2 \in B_2, ..., X_n \in B_n] = P[X_1 \in B_1] \cdot P[X_2 \in B_2] \cdots P[X_n \in B_n].$$

Show that the X_is are mutually independent if and only if their joint p.m.f. factors into the product of their individual p.m.f.s.

23. Suppose that two assets have random rates of return R_1 and R_2 that follow the joint distribution described in the table below. (a) Find, for each possible r_1 and r_2, both $P[R_1 = r_1]$ and $P[R_2 = r_2]$. (b) A portfolio consisting of \$500 in asset 1 and \$1000 in asset 2 is constructed. Find the probability mass function of the final value X of this portfolio.

(r_1, r_2)	$(-.01, -.02)$	$(0, .01)$	$(0, .02)$	$(.02, .03)$	$(.04, .04)$	$(.05, .02)$
probability	.1	.2	.3	.2	.1	.1

24. Consider a general binomial branch process with initial state s, up probability p, and up and down ratios $u = 1 + b$ and $d = 1 + a$. Find the p.m.f. of the price S_n after n steps. Also, find the p.m.f. of the logged price $\log(S_n)$. (Assume independence of steps, that is, the probability of a path is the product of the probabilities of the steps along that path.)

3.3   Discrete Expectation

We now turn to numerical measures that summarize aspects of a probability distribution, highlighted by the **mean**, a measure of center, and the **variance**, a measure of spread of a distribution. Variance and its related measure, the **standard deviation** of a distribution, help us understand and quantify the idea of risk. In so doing, these quantities turn out to be crucial in many problems in financial mathematics, including the optimization of portfolios of risky assets to be studied in Chapter 4 and the valuation of options that is introduced in Chapter 5.

3.3.1   Mean

In the last section when we defined the probability mass function of a random variable, we illustrated with an example of the probability mass function of a random variable X giving the daily change in value of a stock, specifically:

$$P[X = .5] = .2; P[X = .25] = .3;$$
$$P[X = -.25] = .3; P[X = -.5] = .2.$$

It is natural to think of characterizing the average change in value. The idea was well foreshadowed in the last section: compute a weighted average of the states of X, using the probabilities as weights:

$$\text{average value of } X = (.5)(.2) + (.25)(.3) + (-.25)(.3) + (-.5)(.2) = 0.$$

The average value of a random variable is not necessarily a possible state, but is rather a center of mass of the collection of states, allowing for different state probability masses. The probability histogram in Figure 3.13 illustrates this idea; 0 is a center of symmetry of the graph, but is not itself a state.

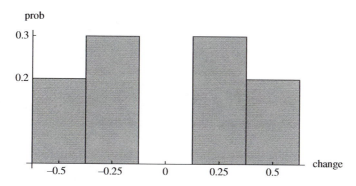

Figure 3.13: Histogram of stock price change.

When a probability distribution on a finite set of states is symmetric about a vertical line, then the mean will be the point of symmetry. For another random variable Y with non-symmetric mass function:

$$g(0) = .3; g(.25) = .3; g(.5) = .2; g(.75) = .1; g(1) = .1,$$

the average value would be:

$$(0)(.3) + (.25)(.3) + (.5)(.2) + (.75)(.1) + (1)(.1) = .35.$$

The probability histogram for this distribution is in Figure 3.14, which shows the center of mass at .35. Roughly speaking, if you were to put a fulcrum under the x-axis like a seesaw at this point, the axis would perfectly balance (see Exercise 1).

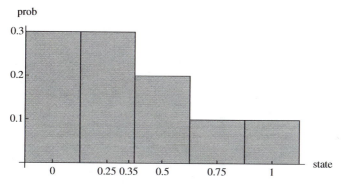

Figure 3.14: Histogram of stock price change.

With these ideas in mind, we give the formal definition.

Definition 1. If X is a discrete r.v. with p.m.f. $p(x) = P[X = x]$, then the **mean** or **expected value of** X (or, of the distribution of X) is:

$$\mu = E[X] = \sum_x xp(x) = \sum_x xP[X = x]. \qquad (3.22)$$

The sum is taken over all states x in the state space of X.

Remark. Similarly, we can define the expected value of a function g of X as:

$$E[g(X)] = \sum_x g(x)p(x). \qquad (3.23)$$

This can be useful in cases where g represents a reward or cost that is earned, which is dependent on the value of a random variable X. We will also need this extension of the definition in order to define the variance of a distribution, which we will do shortly.

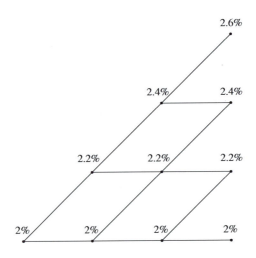

Figure 3.15: Interest rate tree.

Example 1. In Example 6 of the preceding section, we had a model in which the interest rate on an account was 2% at the start of the year, and in subsequent quarters the rate was unchanged with probability .6, or increased by .2% with probability .4. A tree describing the possible sequences of rates is reproduced in Figure 3.15. In that example, we were able to find the p.m.f. of the final value of an initial deposit of $10,000, displayed in the table below.

final value	probability
$10,824.30	.216
$10,845.50	.144
$10,866.80	.144
$10,888.10	.240
$10,909.40	.096
$10,930.80	.096
$10,952.10	.064

Then the expected year-end value of the investment is:

$$
\begin{aligned}
\mu \;=\;& (\$10,824.30)(.216) + (\$10,845.50)(.144) \\
&+ (\$10,866.80)(.144) + (\$10,888.10)(.240) \\
&+ (\$10,909.40)(.096) + (\$10,930.80)(.096) \\
&+ (\$10,952.10)(.064) \\
=\;& \$10,875.40. \ \blacksquare
\end{aligned}
$$

Example 2. If a random variable X has probability mass function:

$$
P[X = 0] = 1 - p, \, P[X = 1] = p, \tag{3.24}
$$

then the mean of the distribution is:

$$
E[X] = 0 \cdot (1 - p) + 1 \cdot p = p. \tag{3.25}
$$

We could also compute the expected value of the function $g(X) = X^2$, which is:

$$
E\left[X^2\right] = 0^2 \cdot (1 - p) + 1^2 \cdot p = p. \tag{3.26}
$$

This will have implications later to the computation of the variance of this X, and, more important, the variance of a binomial random variable with probability parameter p. $\blacksquare$

Example 3. In Example 7 of the previous section we derived the p.m.f. of the amount of money X lost by a bank due to defaults on four of its loans, each for $10,000. This was:

$$
g(x) = \begin{cases}
.5 & \text{if } x = \$0; \\
.2 & \text{if } x = \$10,000; \\
.06 & \text{if } x = \$20,000; \\
.2 & \text{if } x = \$30,000; \\
.04 & \text{if } x = \$40,000.
\end{cases}
$$

The expected amount lost is therefore

$$
\begin{aligned}
E[X] \;=\;& (.5)(\$0) + (.2)(\$10,000) + (.06)(\$20,000) \\
&+ (.2)(\$30,000) + (.04)(\$40,000) \\
=\;& \$10,800.
\end{aligned}
$$

Another way of modeling the problem would be to define a random variable N to be the number of loans that default, which means that the function $g(N) = \$10,000N$ is the one whose expectation is to be taken. The probability weights are the same for N as they were for X, but the states are $0, 1, 2, 3, 4$. (See Exercise 9.) ■

Example 4. In a bull market, each week a certain "penny stock" stays the same in value with probability 3/4, goes up by \$.05 with probability 1/8, or goes up by \$.10 with probability 1/8. The diagram below displays the possibilities. Assume that the probability of sequence of changes is equal to the product of the probabilities of the individual changes along the path. Find the expected value of the stock after 2 weeks if the initial price of the stock is \$.50.

initial	week 1	week 2
		.70
		.65
	.60	.60
	.55	.55
.50	.50	.50

Solution. Let us derive the p.m.f. of the price X at week 2. For final price .50, there is only one path, in which the price stays the same in both weeks. By the assumed product rule,

$$P[\text{stock price is .50 at week 2}] = (3/4)(3/4) = 9/16.$$

For price .55, there are two paths, $(.50, .50, .55)$ and $(.50, .55, .55)$, whose total probability is:

$$P[\text{stock price is .55 at week 2}] = (3/4)(1/8) + (1/8)(3/4) = 6/32.$$

To reach a final price of .60, the price could stay at .50 at week 1 and then jump by .10, or go to .55 at week 1 then jump by .05, or go to .60 at week 1 and stay there at week 2. Therefore,

$$\begin{aligned} P[\text{stock price is .60 at week 2}] &= (3/4)(1/8) + (1/8)(1/8) + (1/8)(3/4) \\ &= 13/64. \end{aligned}$$

To reach .65, the price could go up to .55 the first week and then to .65, or up to .60 the first week and then up to .65. Thus,

$$P[\text{stock price is .65 at week 2}] = (1/8)(1/8) + (1/8)(1/8) = 2/64.$$

Finally, for price .70, the only path goes to .60 the first week and then to .70 the second week.

$$P[\text{stock price is .70 at week 2}] = (1/8)(1/8) = 1/64.$$

You can check that this list of probabilities forms a valid p.m.f. Then the expected price at week 2 is:

$$(.5)\left(\frac{9}{16}\right) + (.55)\left(\frac{6}{32}\right) + (.60)\left(\frac{13}{64}\right) + (.65)\left(\frac{2}{64}\right) + (.70)\left(\frac{1}{64}\right) = .5375.$$

As we have mentioned before, this is the same as the mathematical dot product of the vector of states with the vector of probabilities:

$$(.5, .55, .60, .65, .70) \cdot \left(\frac{9}{16}, \frac{6}{32}, \frac{13}{64}, \frac{2}{64}, \frac{1}{64}\right) = .5375. \quad \blacksquare$$

The next theorem, called **linearity of expectation**, is very useful, both theoretically and in the quite common financial situations in which payments occur at several time points. We will now be able to deal with the case where such payments are random, and compute, for example, expected net present values.

Theorem 1. If X and Y are random variables, and a and b are constants, then

$$E[aX + bY] = aE[X] + bE[Y]. \tag{3.27}$$

(This generalizes to many random variables as well.)

We will be content to motivate the theorem without giving all of the details. The left side of formula (3.27) is a sum of possible values of $aX + bY$ weighted by their joint probabilities $P[X = x, Y = y] = f(x, y)$. Then, by linearity of summation,

$$
\begin{aligned}
E[aX + bY] &= \sum_x \sum_y (ax + by) P[X = x, Y = y] \\
&= a \cdot \sum_x \sum_y x P[X = x, Y = y] + b \sum_x \sum_y y P[X = x, Y = y].
\end{aligned}
$$

It is not hard to argue that the first double sum reduces to $E[X]$ and the second to $E[Y]$.

Example 5. In the previous section, we introduced the very important binomial distribution with parameters n and p, which, among other things, models the distribution of the number of up moves in n steps, in which the probability of each up move is p. The p.m.f. is:

$$f(k; n, p) = \binom{n}{k} p^k (1 - p)^{n-k}, k = 0, 1, 2, ..., n.$$

Exercise 10 asks you to simplify the sum below, which would give the mean value of this distribution:

$$\mu = E[X] = \sum_{k=0}^{n} k \cdot \binom{n}{k} p^k (1-p)^{n-k}. \qquad (3.28)$$

Here we will compute μ in a more subtle but far less computationally intensive way, using linearity of expectation.

Let X_i be 1 or 0 according to whether the ith jump is up or down, respectively, for $i = 1, 2, ..., n$. Then the X_is all have the same probability distribution, which is exactly the one in Example 2. We found there that $E[X_i] = p$. Now the total number X of up jumps among n is the sum of the X_i. Thus,

$$E[X] = E\left[\sum_{i=1}^{n} X_i\right] = \sum_{i=1}^{n} E[X_i] = \sum_{i=1}^{n} p = np. \qquad (3.29)$$

So we have derived with very little labor the result that the mean of the binomial distribution with parameters n and p is np. Note that this result applies to any phenomenon in which a random variable counts the number of times among n trials of a simple binary experiment that a particular event occurs. (We will need to make an assumption of independence of trials, which will be fleshed out in a later section.) ∎

Example 6. Suppose that a hotel room which rents for $100/night is either occupied, with probability 2/3, or unoccupied, with probability 1/3. If the value of money discounts by .0005 per night, find the expected present value of the proceeds from the room for the next year.

Solution. To model the situtation, let X_i be 1 or 0 according to whether the room is occupied or not on day i. Then the revenue on day i is $100 X_i$. The present value of that revenue is $.9995^i (100X_i)$. By linearity, the expected total present value over the 365 days is:

$$
\begin{aligned}
E\left[\sum_{i=1}^{365} .9995^i (\$100X_i)\right] &= (\$100) \sum_{i=1}^{365} .9995^i E[X_i] \\
&= (\$100) \sum_{i=1}^{365} (2/3 \cdot 1 + 1/3 \cdot 0).9995^i \\
&= (\$100) \cdot (2/3) \sum_{i=1}^{365} .9995^i \\
&= (\$100) \cdot (2/3) \left(\frac{1 - .9995^{366}}{1 - .9995} - 1\right) \\
&= \$22,236. \ \blacksquare
\end{aligned}
$$

Example 7. Most of the time in this book we consider random variables with finite state space. But occasionally we encounter countably infinite state spaces, and the definition of expected value is the same; however, it is possible that the sum, which is an infinite series, does not converge, which leaves the mean undefined. Let us compute, if we can, the mean of the random variable counting the number of down moves prior to the first up move in a binomial branch process with up probability p. (The distribution that results is called the **geometric distribution** with parameter p.)

Solution. Let X be the random variable that counts the number of down moves prior to the first up move. There is no limit to how large X could be, although, intuitively, large values of X will have low probability. So the state space of X is $E = \{0, 1, 2, ...\}$.

We must find the p.m.f. of X before embarking on the computation of the expected value. Assuming as always that the probability of a path is the product of the probabilities of the steps within the path, we observe that the event that X equals a value k is the same as the event that the first k moves are down and the last move is up, which takes place with probability $(1 - p)(1 - p) \cdots (1 - p)p$. In this product there are k factors of the down probability $(1 - p)$. Therefore, the p.m.f. of X is

$$P[X = k] = p(1 - p)^k, k = 0, 1, 2, ...$$

Next, an expression for the expected value of X is:

$$E[X] = \sum_{k=0}^{\infty} k \cdot p(1 - p)^k = p(1 - p) \sum_{k=0}^{\infty} k \cdot (1 - p)^{k-1}.$$

This sum may or may not turn out to have a finite value. The strategy of factoring out p and $1 - p$ has left us with a series of the form:

$$\sum_{k=0}^{\infty} k \cdot x^{k-1},$$

where $x = 1 - p$. But this is the formal derivative of the geometric series $\sum_{k=0}^{\infty} x^k$, which does converge absolutely if $|x| < 1$. As long as $1 - p < 1$, that is, $p > 0$, we can differentiate the geometric series to get:

$$
\begin{aligned}
\sum_{k=0}^{\infty} k \cdot x^{k-1} &= \frac{d}{dx} \left(\sum_{k=0}^{\infty} x^k \right) \\
&= \frac{d}{dx} \left(\frac{1}{1 - x} \right) \\
&= \frac{d}{dx} (1 - x)^{-1} \\
&= (-1)(1 - x)^{-2}(-1) \\
&= \frac{1}{(1 - x)^2}.
\end{aligned}
$$

Plugging in $x = 1 - p$ in the last formula yields:

$$\sum_{k=0}^{\infty} k \cdot (1 - p)^{k-1} = \frac{1}{(1 - (1 - p))^2} = \frac{1}{p^2}.$$

Finally, we can combine these results and write:

$$E[X] = p(1-p) \sum_{k=0}^{\infty} k \cdot (1-p)^{k-1} = p(1-p) \cdot \frac{1}{p^2} = \frac{1-p}{p}.$$

This formula for the mean gives fairly intuitive results in specific cases; for instance, when the up probability is $\frac{1}{2}$, the expected number of downs until the first up is $(1-1/2)/(1/2) = 1$. When the up probability is $\frac{1}{4}$, the expectation of the number of downs is $(1-1/4)/(1/4) = 3$. ■

3.3.2   Variance

We have looked at the average or expected value of a random variable, which measures central tendency of the distribution of the random variable in the spirit of the center of mass in physics. But what makes random models distinct from deterministic ones is the fact that probability mass spreads out in some manner away from that average. To understand riskiness of financial objects, we must understand that variability.

Example 8. Which of the three discrete distributions in the histograms of Figure 3.16 seems to have the largest "variability"? The smallest? In each case the state space is $E = \{0, 1, 2, 3, 4\}$ and the mean is the center of symmetry $\mu = 2$, but in (a) there is mass $1/4$ on each of states 0, 2, and 4 and mass $1/8$ on the other states; in (b) there is equal mass of $1/5$ on all states; and in (c) there is a heavy mass of $5/8$ on state 2, $1/8$ on each of states 1 and 3, and $1/16$ on each of states 0 and 4.

Solution. One way of thinking about the question is to imagine that the states are possible rates of return on an investment, in units of %. The least variable rate of return is certainly the one in histogram (c), since you are very sure that you will receive a rate of return near to the mean of 2%. But which of histograms (a) or (b) represents the "most variable" distribution isn't entirely clear. In (b) there isn't much information to go on because all of the five possible rates or return are equally likely. One might say that it is the least predictable. But should we say that it is the most "risky" as a consequence? The distribution in (a) has the most probability weight on the extreme states of 0 and 4. In a sense, then, perhaps (a) is most variable, or "risky." Clearly we need a way of quantifying risk before we can make a decision. ■

The most common measure of risk is defined as follows.

Definition 2. If X is a discrete r.v. with p.m.f. $p(x) = P[X = x]$, then the **variance** of X (or, of the distribution of X) is:

$$\sigma^2 = \text{Var}(X) = E\left[(X - \mu)^2\right] = \sum_x (x-\mu)^2 P[X = x] = \sum_x (x-\mu)^2 p(x). \quad (3.30)$$

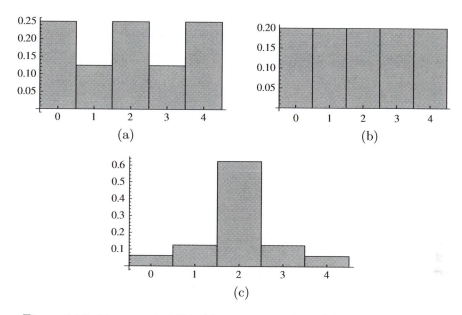

Figure 3.16: Three probability histograms, each with mean value $\mu = 2$.

The sum is taken over all states x in the state space of X. The square root of the variance $\sigma = \sqrt{\sigma^2}$ is called the **standard deviation** of X.

The expectation that defines variance will be large in cases where heavy probability mass is given to states x such that $(x - \mu)^2$ is large. So variance and standard deviation measure the degree to which the probability weight for a distribution spreads out away from the mean.

A frequently useful computational formula for variance is:

$$\text{Var}(X) = E\left[X^2\right] - \mu^2. \tag{3.31}$$

This is easily derived using linearity of expectation by expanding the right side of formula (3.30) as:

$$
\begin{aligned}
\text{Var}(X) &= \sum_x (x - \mu)^2 P[X = x] \\
&= \sum_x \left(x^2 - 2\mu x + \mu^2\right) p(x) \\
&= \sum_x x^2 p(x) - 2\mu \sum_x x p(x) + \mu^2 \sum_x p(x) \\
&= E\left[X^2\right] - 2\mu E[X] + \mu^2 \cdot 1 \\
&= E\left[X^2\right] - 2\mu^2 + \mu^2 \\
&= E\left[X^2\right] - \mu^2.
\end{aligned}
$$

Another very useful property of variance is listed in Exercise 13: $\text{Var}(aX + b) = a^2\text{Var}(X)$. One of the important implications of this is that changes in scale affect variance severely, but translations by constants like b do not affect the variance at all. For example, if X represents a rate of return expressed as a decimal and a dollars are invested, then aX is the growth in value of the investment, and its variance is a^2 times the variance of the rate of return itself. The standard deviation, on the other hand, would grow proportionately with a, because

$$\sigma_{aX} = \sqrt{\text{Var}(aX)} = \sqrt{a^2 \text{Var}(X)} = |a| \cdot \sigma_X.$$

Example 9. Find the variance for all of the three discrete distributions in Example 8.

Solution. Recall that the mean for all of these distributions is 2. For the distribution in (a), using the definition of variance:

$$\begin{aligned} \text{Var}(X) &= \frac{1}{4}(0-2)^2 + \frac{1}{8}(1-2)^2 + \frac{1}{4}(2-2)^2 \\ &\quad + \frac{1}{8}(3-2)^2 + \frac{1}{4}(4-2)^2 \\ &= \frac{9}{4}. \end{aligned}$$

Hence the standard deviation is $\sqrt{9/4} = 3/2$.

Moving to the distribution in (b) that gives equal weight of $1/5$ to each state, we can easily find the variance by the computational formula (3.31):

$$\begin{aligned} \text{Var}(X) &= E\left[X^2\right] - \mu^2 \\ &= \left(\frac{1}{5} \cdot 0^2 + \frac{1}{5} \cdot 1^2 + \frac{1}{5} \cdot 2^2 + \frac{1}{5} \cdot 3^2 + \frac{1}{5} \cdot 4^2 \right) - 2^2 \\ &= 2 \end{aligned}$$

The standard deviation of this distribution would be $\sqrt{2}$. We see that this distribution is less variable than the first by the measure of variance, which answers the open question in Example 7. Finally, using the defining formula again for the third distribution,

$$\begin{aligned} \text{Var}(X) &= \frac{1}{16}(0-2)^2 + \frac{1}{8}(1-2)^2 + \frac{5}{8}(2-2)^2 \\ &\quad + \frac{1}{8}(3-2)^2 + \frac{1}{16}(4-2)^2 \\ &= \frac{3}{4}. \end{aligned}$$

So the standard deviation would be $\sigma = \sqrt{3}/2$, which we see is much less than the standard deviations for the other two distributions. A lot of probability mass is given to the middle term $(2-2)^2 = 0$ and not much to the outer terms $(0-2)^2$ and $(4-2)^2$, which reduces the variance and standard deviation compared to the first two distributions. ∎

Example 10. In Example 2 we found that the mean of the two-point distribution taking value 1 with probability p and 0 with probability $1 - p$ was exactly p. The variance of this distribution, by the computational formula, would be:

$$\sigma^2 = \text{Var}(X) = E[X^2] - \mu^2$$
$$= (p(1^2) + (1-p)(0)^2) - p^2$$
$$= p - p^2$$
$$= p(1-p).$$

Going a little further, recall that we computed the mean of a binomial random variable X with parameters n and p by letting X_i be 1 or 0 according to whether the ith jump is up or down, respectively, for $i = 1, 2, ..., n$ and observing that $E[X] = E[\sum_{i=1}^{n} X_i]$. Although we don't quite have the tools yet, when we study independence of random variables later, we will see that:

$$\text{Var}[X] = \text{Var}\left[\sum_{i=1}^{n} X_i\right] = \sum_{i=1}^{n} \text{Var}(X_i). \tag{3.32}$$

Since each X_i has variance equal to $p(1-p)$, the variance of the binomial distribution becomes $np(1-p)$. ∎

Example 11. Find the variance and standard deviation of the week 2 price of the penny stock in Example 4.

Solution. From that example, if X is the week 2 price, then the distribution of X was characterized by:

$$P[X = .50] = 9/16; P[X = .55] = 6/32; P[X = .60] = 13/64;$$

$$P[X = .65] = 2/64; P[X = .70] = 1/64,$$

and the mean of X was $\mu = .5375$. Then the variance and standard deviation are:

$$\sigma^2 = (.5)^2\left(\frac{9}{16}\right) + (.55)^2\left(\frac{6}{32}\right) + (.60)^2\left(\frac{13}{64}\right)$$
$$+ (.65)^2\left(\frac{2}{64}\right) + (.70)^2\left(\frac{1}{64}\right) - .5375^2$$
$$= .00242188;$$

$$\sigma = \sqrt{\sigma^2} = .04921. ∎$$

3.3.3 Chebyshev's Inequality

There is a famous result in probability called **Chebyshev's inequality** that helps to give meaning to the variance and standard deviation. The import of it is that any random variable is quite likely to take a value close to its mean, where closeness is measured in units of standard deviations. The precise statement of the theorem is next, and we leave it without proof. (It is usually proved in a course in advanced undergraduate probability.)

Theorem 2. If X is a random variable with mean μ and standard deviation σ, then, for all $k > 0$,

$$P[|X - \mu| \le k \cdot \sigma] \ge 1 - \frac{1}{k^2}. \tag{3.33}$$

Equivalently,

$$P[|X - \mu| > k \cdot \sigma] \le \frac{1}{k^2}. \tag{3.34}$$

Note that formula (3.34) follows immediately from formula (3.33), and conversely, by complementation. Looking at particular cases, for $k = 2$,

$$P[|X - \mu| \le 2 \cdot \sigma] \ge 1 - \frac{1}{2^2} = .75.$$

For $k = 3$,

$$P[|X - \mu| \le 3 \cdot \sigma] \ge 1 - \frac{1}{3^2} = \frac{8}{9} \approx .889.$$

Hence Chebyshev's inequality lets us make powerful statements, applicable to all random variables, such as: "It is at least 75% likely for a random variable to take a value within 2 standard deviations of its mean," and "It is at least 88.8% likely for a random variable to take a value within 3 standard deviations of its mean," and "It is no more than 25% likely for X to differ from its mean by more than 2 standard deviations." But the power involved in being able to make such claims for all distributions entails sacrifice in precision. For most distributions that one encounters, the Chebyshev bound $1 - \frac{1}{k^2}$ is conservative; the actual probability is higher. We see this in our final example.

Example 12. Find the exact probability that X takes a value within 2 and within 3 standard deviations of its mean if: (a) X has the binomial distribution with parameters $n = 5$ and $p = .5$; (b) X has a distribution that gives equal mass to the eight states $1, 2, 3, 4, 5, 6, 7, 8$.

Solution. (a) We have seen that the mean of the binomial distribution is $np = 5(.5) = 2.5$, and the variance is equal to $np(1 - p) = 5(.5)(.5) = 1.25$. The standard deviation is then $\sigma = \sqrt{1.25} \approx 1.12$. Since X takes integer values, we can write:

$$\begin{aligned}
P[|X - \mu| \le 2 \cdot \sigma] &= P[\mu - 2\sigma \le X \le \mu + 2\sigma] \\
&= P[2.5 - 2(1.12) \le X \le 2.5 + 2(1.12)] \\
&= P[.26 \le X \le 4.74] \\
&= P[1 \le X \le 4].
\end{aligned}$$

For this distribution, the corresponding probability is:

$$\binom{5}{1}(.5)^1(.5)^4 + \binom{5}{2}(.5)^2(.5)^3 + \binom{5}{3}(.5)^2(.5)^3 + \binom{5}{4}(.5)^4(.5)^1 = .9375.$$

This probability substantially exceeds the 75% that is guaranteed by Chebyshev's inequality. Considering the 3-standard deviation interval, notice that $\mu - 3\sigma = 2.5 - 3(1.12) < 0$ and $\mu + 3\sigma = 2.5 + 3(1.12) > 5$, hence

$$P[|X - \mu| \le 3 \cdot \sigma] = P[\mu - 3\sigma \le X \le \mu + 3\sigma] = P[0 \le X \le 5] = 1.$$

So it is certain that X will take a value within 3 standard deviations of its mean in this case. The Chebyshev bound of 88.9% is again conservative.
(b) Each of the eight states is given equal probability, which must be $\frac{1}{8}$ in order for the masses to total to 1. Hence the mean of the distribution is:

$$\mu = \frac{1}{8}(1 + 2 + 3 + 4 + 5 + 6 + 7 + 8) = \frac{36}{8} = \frac{9}{2} = 4.5.$$

The variance is:

$$\begin{aligned}
\sigma^2 &= E\left[X^2\right] - \mu^2 \\
&= \frac{1}{8}\left(1^2 + 2^2 + 3^2 + 4^2 + 5^2 + 6^2 + 7^2 + 8^2\right) - \left(\frac{9}{2}\right)^2 \\
&= \frac{204}{8} - \frac{81}{4} \\
&= \frac{21}{4} = 4.25.
\end{aligned}$$

You can compute that the standard deviation is $\sigma = \sqrt{\sigma^2} \approx 2.06$. Thus,

$$\begin{aligned}
P[|X - \mu| \le 2 \cdot \sigma] &= P[\mu - 2\sigma \le X \le \mu + 2\sigma] \\
&= P[4.5 - 2(2.06) \le X \le 4.5 + 2(2.06)] \\
&= P[.38 \le X \le 8.62] \\
&= P[1 \le X \le 8] = 1.
\end{aligned}$$

So not only is it true that at least 75% of the probability mass of this distribution lies within plus or minus 2 standard deviations from the mean, but in fact all of the mass does so. Of course this is also true for a 3-standard deviation interval.
■

Important Terms and Concepts

Mean of a distribution (random variable) - The weighted average of states using the probabilities as weights: $\mu = E[X] = \sum_x x P[X = x]$.

Expected value of a function of a random variable - The weighted average of the functional values $g(x)$ of the states: $E[g(X)] = \sum_x g(x)p(x)$.

Variance of a distribution (random variable) - The weighted average of squared deviations of the states from the mean: $\sigma^2 = \text{Var}(X) = E\left[(X - \mu)^2\right]$.

Standard deviation of a distribution (random variable) - The square root of the variance.

Linearity of expectation - The expected value of a linear combination of random variables is the linear combination of expected values: $E[aX + bY] = aE[X] + bE[Y]$.

Computational formula for variance - $\text{Var}(X) = E\left[X^2\right] - \mu^2$.

Chebyshev's inequality - A random variable X has probability of at least $1 - \frac{1}{k^2}$ of taking a value within k standard deviations of its mean.

Exercises 3.3

1. In physics, the ***torque*** on an object about a point is the force being exerted times the distance between the point at which the force is applied and the object. For a lever like a seesaw as in the diagram, a torque of $F \cdot d$ is being applied on the right of the fulcrum of the lever. For the non-symmetric distribution shown in Figure 3.14, show that, if the axis is the lever, the fulcrum is at the mean, and the force is the probability mass, then the total torque to the right of the mean is the same as the total torque to the left of the mean. Therefore the mean is located at what we might call the balance point of the distribution.

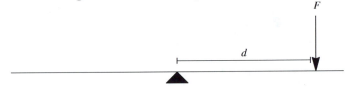

Exercise 1

2. The interest rate on a 5-year car loan for \$20,000 will be 7.5%, 8%, 8.5%, or 9% with equal probability. (These are nominal yearly rates, which will be compunded monthly.) Find the expected value of the monthly payment on the loan, assuming beginning of the month payments.

3. Find the mean, variance, and standard deviation of the random variable X in Exercise 15 of Section 3.2, whose cumulative distribution function is below.

$$F(x) = \begin{cases} 0 & \text{if } x < 0; \\ .2 & \text{if } 0 \le x < 1; \\ .3 & \text{if } 1 \le x < 3; \\ .7 & \text{if } 3 \le x < 5; \\ 1 & \text{if } x \ge 5. \end{cases}$$

4. In Example 5 of Section 3.2, there were two assets such that the joint distribution of their rates of return was as in the table below. If an investor invests $500 in the first asset and $500 in the second asset, find the mean, variance, and standard deviation of the final value of the investment.

(r_1, r_2)	$(-.01, 0)$	$(.01, .03)$	$(.02, .02)$	$(.04, .05)$	$(.05, .04)$
probability	.3	.1	.2	.2	.2

5. A discrete distribution places equal weight on states 1, 2, and 3, and each of states 4, 5, and 6 is twice as likely as each of 1, 2, and 3. Find the mean and variance of this distribution.

6. A two-state distribution with states a and b is such that the probability that b occurs is $p = .4$. Find values of a and b, if any, that make the mean of the distribution equal to .2 and the variance equal to 2.16.

7. The **discrete uniform distribution** on state space $E = \{a, a + 1, ...b - 1, b\}$ puts equal probability mass on all states. Find the mean of this distribution.

8. The **Poisson distribution with parameter** μ has countable state space $E = \{0, 1, 2, ...\}$ and probability mass function $f(k) = \frac{e^{-\mu} \cdot \mu^k}{k!}$. If the number of dividends received on a company's stock in 1 year has the Poisson distribution with parameter 2, find the expected value of the number of dividends. For general parameter μ, guess and derive a formula for the mean of the Poisson distribution.

9. In Example 3, write out the computation of the expected value of $g(N) = \$10,000N$ and check that the result is the same as the expected value of X. What property of expectation is illustrated?

10. Redo the computation of the binomial mean using only the definition of expected value. (Hint: after writing the appropriate sum, cancel what can be canceled and multiply and divide by appropriate constants to manipulate the sum into a form to which you can apply the binomial theorem $(a + b)^n = \sum_{k=0}^{n} \binom{n}{k} a^k b^{n-k}$.)

11. Referring to the penny stock in Example 11, suppose that another such stock had a mean week 2 price of .54 and a variance of .0022. Is it clear which

is preferable? Why? What if the variance of the second stock was .003?

12. Compute the variance and standard deviation of the amount lost by the bank in Example 3.

13. Prove that if a and b are constants, then $\text{Var}(aX + b) = a^2\text{Var}(X)$.

14. Suppose that an oil company is deciding whether to invest $1 million to explore a drilling site. The chance that oil will be found at the site is 30%. If oil is found, the site will yield $5 million with probability p, or $1.5 million with probability $1 - p$. How large must p be in order that the oil company expects to at least break even?

15. Find the mean and the variance of the number of up moves in a three-step binomial branch process with up probability $p = \frac{1}{2}$. Also find the mean and variance of the number of down moves.

16. Suppose that an investor will split $10,000 evenly between a riskless savings account earning 3% per year compounded monthly and a mutual fund currently selling for $50 share and selling for a random price X at the end of a year, where X has a distribution putting equal weight on each possible price in the state space $\{50, 51, 52, 53, 54\}$. Find the mean and variance of the yearly rate of return on this portfolio.

17. For a random variable X having the distribution in Figure 3.16(a), find the probability that X is more than 2 standard deviations from its mean.

18. Let X have the probability mass function given below. Find (a) the probability that X will be within 2 standard deviations of its mean; (b) the probability that X will differ by more than 3 standard deviations from its mean.

$$f(x) = \begin{cases} \dfrac{1}{12}x & x = 1, 2, 3; \\ \dfrac{7}{12} - \dfrac{1}{12}x & x = 4, 5, 6. \end{cases}$$

19. Referring to Exercise 1, show in general that the mean μ is the unique point on the horizontal axis such that the torque applied by states to the right of μ exactly equals the torque applied by states to the left of μ.

20. Show that mean μ minimizes the quantity $E\left[(X - a)^2\right]$ with respect to a.

3.4 Conditional Probability

In many problems, especially in studies of random economic events that take place over time, we are interested in computing probabilities of events using partial knowledge that some other event has already occurred. These kinds of probabilities are called **conditional probabilities**. One need look no further than the binomial branch process to see that it is an interesting problem to find the probability that the state at a future time n is a particular possible state given the state at the current time m.

3.4.1 Fundamental Ideas

A simple and intuitive way to begin is with the following example.

Example 1. A basket contains eight apples and six oranges. A piece of fruit is selected at random. The randomness assumption implies that the probability that it is an apple is $8/14$. If two pieces are selected and the first is not known, the probability that the second is an apple is still $8/14$. But if two pieces are selected and it is observed that the first is an apple, the probability that the second is an apple is $7/13$, because the population from which the second fruit was drawn contains 7 apples and 13 pieces of fruit in total. ∎

Financial mathematics has little to do with apples and oranges, however (unless, of course, an investor is interested in apple and orange futures). More relevant to our story is the next example.

Example 2. In Figure 3.17(a) is a tree carried over from an earlier example for a four-step binomial branch process modeling an asset price. The initial price is 50, and each up step multiplies the price by 1.1 while each down step multiplies the price by .9. If it is observed that the first two moves are up, then the possible price paths are just those emanating from node 60.50 on the graph, shown in part (b) of the figure. Supposing for simplicity that up moves are as likely as down moves, and a path probability is the product of the probabilities of the steps on that path, then we would say that the conditional probability that the price at time 4 is 73.21 given that the first two moves are up is $\frac{1}{2} \cdot \frac{1}{2} = \frac{1}{4}$. If X_k denotes the random price at time k, then the last sentence translates as $P[X_4 = 73.21 \,|\, X_2 = 60.50] = \frac{1}{4}$, where the vertical bar "|" is read as "given that." ∎

The main idea is that, if an event A is known to have happened, the sample space reduces to those outcomes in A, and every other event B reduces to its part on A, that is, $A \cap B$. This motivates the following definition.

Definition 1. If A is an event with $P[A] > 0$, and B is an event, then the **conditional probability** of B given A is:

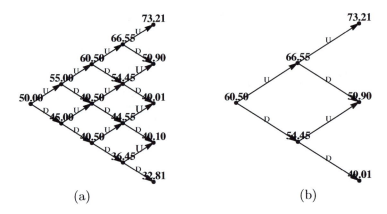

Figure 3.17: (a) Stock price tree. (b) Conditional stock price subtree given 2 up moves.

$$P[B|A] = \frac{P[A \cap B]}{P[A]}. \tag{3.35}$$

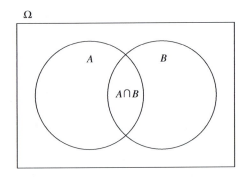

Figure 3.18: Conditional probability $P[B|A]$ reduces sample space to A and event B to $A \cap B$.

The Venn diagram in Figure 3.18 illustrates the idea. Once it is known that event A occurred, any outcomes in the whole sample space Ω that are not contained in A are irrelevant to finding the probability of other events. In essence, event B is reduced to its part on A, namely, $A \cap B$. To recompute the probability of B given that A has occurred, you need to compare the probabilities of the outcomes in $A \cap B$ to those in the new, reduced sample space A, just as to find the conditional probability that $X_4 = 73.21$ in Example 2, you must trim the original tree down to the one rooted at node 60.50.

We can also talk about the **conditional probability measure** $P[\bullet|A]$ defined by formula (3.35), since, for fixed A of non-zero probability, it is easy to show (see Exercise 11) that $P[B|A]$ as a function of B is a valid probability measure on sample space A.

Example 3. Find (a) the conditional probability that, in two dice rolls, the sum is greater than 7 given that it is greater than 5; (b) the conditional probability that both of two moves in a binomial branch process are Us given that at least one is a U, assuming that up on a single step has probability .7 and the probability of a two-step path is the product of the individual step probabilities.

Solution. (a) If it is known that the event $A =$ "sum greater than 5" occurred, then the outcomes

$$(1,1), (1,2), (1,3), (1,4), (2,1), (2,2), (2,3), (3,1), (3,2), (4,1)$$

are cut out of the usual sample space of 36 outcomes, leaving the outcomes:

$$\begin{aligned} A \;=\; &\{(1,5), (1,6), (2,4), (2,5), (2,6), (3,3), (3,4),\\ &(3,5), (3,6), (4,2), (4,3), (4,4), (4,5), (4,6),\\ &(5,1), (5,2), (5,3), (5,4), (5,5), (5,6), (6,2),\\ &(6,3), (6,4), (6,5), (6,6)\}. \end{aligned}$$

Within this reduced sample space, the set of outcomes B for which the sum is greater than 7 is:

$$\begin{aligned} B \;=\; &\{(2,6), (3,5), (3,6), (4,4), (4,5), (4,6),\\ &(5,3), (5,4), (5,5), (5,6), (6,2),\\ &(6,3), (6,4), (6,5), (6,6)\}. \end{aligned}$$

Then A has $36 - 10 = 26$ equally likely outcomes, of which 15 are in B. Thus, $P[B|A] = \frac{15}{26}$. Using the definition directly, we would get the same result:

$$\begin{aligned} P[B|A] \;&=\; \frac{P[A\cap B]}{P[A]}\\ &=\; \frac{P[\text{sum} > 5 \text{ and sum} > 7]}{P[\text{sum} > 5]}\\ &=\; \frac{P[\text{sum} > 5 \text{ and sum} > 7]}{P[\text{sum} > 5]}\\ &=\; \frac{15/36}{26/36} = \frac{15}{26}. \end{aligned}$$

(b) Outcomes are not equally likely, so counting outcomes will not help here. Given the assumptions in the problem statement, the probability measure on the sample space would be characterized by the table below:

outcome	(U,U)	(U,D)	(D,U)	(D,D)
probability	$.7^2 = .49$	$(.7)(.3) = .21$	$(.3)(.7) = .21$	$.3^2 = .09$

Then, if B is the event that both moves are up and A is the event that at least one move is up, we have:

$$
\begin{aligned}
P[B|A] &= \frac{P[A \cap B]}{P[A]} \\
&= \frac{P[(U,U)]}{P[\{(U,U),(U,D),(D,U)\}]} \\
&= \frac{.49}{.49 + .21 + .21} \\
&= \frac{.49}{.91} = .538462. \quad \blacksquare
\end{aligned}
$$

The definition of conditional probability can be rewritten as:

$$
P[A \cap B] = P[A] \cdot P[B|A]. \tag{3.36}
$$

We call this equation the ***multiplication rule*** for conditional probabilities, whose purpose is to enable the computation of intersection probabilities using assumptions about conditional probabilities. In formula (3.36), think of a chain of events in which the occurrence of A will be observed first. The probability both A and B occur is the probability that A occurs times the probability that B occurs given A.

This is actually quite an intuitive idea. Suppose that a hat contains slips labeled A, B, C, D, and E, and we draw out two at random, in succession, and without replacement. The probability that we draw B and then D is the probability that B is drawn first, namely, $1/5$, times $1/4$, which is the probability that D is drawn from the reduced group $\{A, C, D, E\}$ given that B was drawn first. The result of $\frac{1}{5} \cdot \frac{1}{4} = \frac{1}{20}$ coincides with the easy counting argument that observes there are $5 \cdot 4 = 20$ equally likely outcomes in this experiment, of which one is the desired outcome (B, D).

We now have the ability to talk more carefully about what we are really assuming about transitions in the binomial branch model. To say that, whatever value X_n the process currently has at time n, it will go up by a factor of $(1 + b)$ with probability p or down by a factor of $(1 + a)$ with probability $(1 - p)$ is really to say that:

$$
\begin{aligned}
P\left[X_{n+1} = (1 + b)x \,|\, X_n = x\right] &= p; \\
P\left[X_{n+1} = (1 + a)x \,|\, X_n = x\right] &= 1 - p.
\end{aligned} \tag{3.37}
$$

We also are assuming that the past history of process values before time n, $X_0, X_1, ..., X_{n-1}$, has no bearing on these conditional probabilities. We will have more to say on that issue below when we introduce memoryless processes called ***Markov processes***. The multiplication rule would allow us to say that, for two-step transitions,

$$
P\left[X_{n+2} = (1 + b)^2 x \,|\, X_n = x\right]
$$

$$
\begin{aligned}
&= P\left[X_{n+2} = (1+b)^2x, X_{n+1} = (1+b)x \,|\, X_n = x\right] \\
&= P\left[X_{n+1} = (1+b)x \,|\, X_n = x\right] \\
&\quad \cdot P\left[X_{n+2} = (1+b)^2x | X_{n+1} = (1+b)x, X_n = x\right] \\
&= P\left[X_{n+1} = (1+b)x \,|\, X_n = x\right] \\
&\quad \cdot P\left[X_{n+2} = (1+b)^2x | X_{n+1} = (1+b)x\right] \\
&= p \cdot p = p^2.
\end{aligned}
\tag{3.38}
$$

In the first line, given $X_n = x$, there is only one path that leads to $(1+b)^2x$ in two steps, namely, the one that first moves to $(1+b)x$. The second line is the multiplication rule as applied to the conditional probability measure $P\left[\bullet \,|\, X_n = x\right]$. The change to the second factor in the third line follows from the binomial branch assumption that the past before time $n+1$ does not influence the state at time $n+2$, given the state at time $n+1$. This computation is a more rigorous way of verifying that, for two moves of a binomial branch process with up probability p,

$$
P[(U,U)] = P[U \text{ on 1st}]P[U \text{ on 2nd}|U \text{ on 1st}] = p \cdot p = p^2.
$$

The idea of the two computations is exactly the same.

Return for a moment to the example above of the hat with slips labeled A, B, C, D, and E. Suppose that, instead of two slips, we draw three slips in succession and without replacement. It should not take you long to understand why the following is true:

$$
P\left[B \text{ on } 1^{\text{st}}, D \text{ on } 2^{\text{nd}}, A \text{ on } 3^{\text{rd}}\right] = \frac{1}{5} \cdot \frac{1}{4} \cdot \frac{1}{3}.
$$

The chance that B is drawn first is once again $\frac{1}{5}$, then the chance that D comes next given that B was first is $\frac{1}{4}$. But now the reduced space from which the third slip is sampled contains $\{A, C, E\}$, and the conditional probability that A is drawn next given both of the earlier draws is $\frac{1}{3}$. This example suggests that the multiplication rule can be generalized to sequences of three or more events, as follows.

Theorem 1. (Multiplication Rule) Let $A_1, A_2, ..., A_n$ be events such that each of the following conditional probabilities is defined. Then:

$$
\begin{aligned}
P\left[A_1 \cap A_2 \cap \cdots \cap A_n\right] &= P\left[A_1\right] \cdot P\left[A_2|A_1\right] \cdot P\left[A_3|A_1 \cap A_2\right] \\
&\quad \cdots P\left[A_n|A_1 \cap \cdots \cap A_{n-1}\right].
\end{aligned}
\tag{3.39}
$$

You will be asked to prove the case where $n = 3$ in Exercise 13.

Remark. Observe that, in the general multiplication rule, factors like

$$
P\left[A_3|A_1 \cap A_2\right], P\left[A_4|A_1 \cap A_2 \cap A_3\right],
$$

etc., occur in the formula, which are not necessarily the same as $P\left[A_3|A_2\right]$, $P\left[A_4|A_3\right]$, ... In general you must condition on the results of all past steps, not just the most recent one. In the special case of Markov processes, as we will see, only the most recent event matters.

Example 4. Suppose that a bank is looking at three of its home loans on properties in a particular neighborhood, one for $200,000, one for $150,000, and a third for $300,000. They believe that the first will default with probability .05. If the first does default, the bank estimates that the second will default with probability .1, but if the first does not default, the second will default with probability .05 as well. Similarly, if neither of the first two loans defaults, then the third defaults with probability .05. But in either case in which exactly one of the first two loans defaults, the third loan defaults with probability .1. If both of the first two loans default, the third will default with probability .15. Use this information and the multiplication rule to define a probability measure on the sample space of all triples (x_1, x_2, x_3), where x_i is either 1 or 0 according to whether loan i defaulted or not. Also, find the unconditional probabilities that loan 2 defaults and that loan 3 defaults.

Solution. Consider, for example, the outcome $(1,1,1)$ in which all three loans default. To shorten notation, let L_i be the event that loan i defaults, for $i = 1, 2, 3$. By the given information,

$$
\begin{aligned}
P\left[L_1 \cap L_2 \cap L_3\right] &= P\left[L_1\right] P\left[L_2|L_1\right] P\left[L_3|L_1 \cap L_2\right] \\
&= (.05)(.10)(.15) = .00075.
\end{aligned}
$$

As another example, consider the outcome $(1,0,1)$, in which loan 1 defaults, loan 2 does not (the complementary event to L_2), and loan 3 defaults. Then:

$$
\begin{aligned}
P\left[L_1 \cap L_2{}^c \cap L_3\right] &= P\left[L_1\right] P\left[L_2^c|L_1\right] P\left[L_3|L_1 \cap L_2^c\right] \\
&= (.05)(1 - .10)(.1) = .0045.
\end{aligned}
$$

The third conditional probability in this product is .1 because it is known that exactly one of the first two loans defaulted. Continuing in this way we can compute all of the probabilities in the table below. (You should verify these, and also verify that the probabilities sum to 1.)

outcome	probability
$(1,1,1)$	$(.05)(.10)(.15) = .00075$
$(1,1,0)$	$(.05)(.10)(1 - .15) = .00425$
$(1,0,1)$	$(.05)(1 - .10)(.1) = .0045.$
$(0,1,1)$	$(1 - .05)(.05)(.1) = .00475$
$(1,0,0)$	$(.05)(1 - .10)(1 - .1) = .0405$
$(0,1,0)$	$(1 - .05)(.05)(1 - .1) = .04275$
$(0,0,1)$	$(1 - .05)(1 - .05)(.05) = .045125$
$(0,0,0)$	$(1 - .05)(1 - .05)(1 - .05) = .857375$

Summing up the probabilities of the four cases in which loan 2 defaults (lines 1, 2, 4, and 6 of the table), we obtain:

$$P[L_2] = .00075 + .00425 + .00475 + .04275 = .0525.$$

The probability that loan 3 defaults is found by adding the probabilities in lines 1, 3, 4, and 7 of the table:

$$P[L_3] = .00075 + .0045 + .00475 + .045125 = .055125. \blacksquare$$

Example 5. The interest rate on an investment changes in a random way in each quarter of a year. The beginning value is known to be .01 (effective quarterly). The only other possible values for the other quarters are .015 and .02. The joint distribution of the interest rates in the second, third, and fourth quarters is given in the table below. Notice that this distribution implies that it is not possible for the rate to stay the same from one quarter to the next. Denote by Q_2, Q_3, and Q_4 the rates in the second, third, and fourth quarters, respectively. Find all of the following probabilities:

(a) $P[Q_2 = .015]$;
(b) $P[Q_2 = .015, Q_3 = .02]$;
(c) $P[Q_3 = .02 | Q_2 = .015]$;
(d) $P[Q_4 = .01 | Q_2 = .015, Q_3 = .02]$;
(e) $P[Q_4 = .01 | Q_3 = .02]$.

outcome	$(.02, .01, .02)$	$(.02, .01, .015)$	$(.02, .015, .02)$	$(.02, .015, .01)$
probability	.075	.075	.21	.14

outcome	$(.015, .02, .01)$	$(.015, .02, .015)$	$(.015, .01, .02)$	$(.015, .01, .015)$
probability	.09	.21	.1	.1

Solution. (a) To find $P[Q_2 = .015]$ we simply add the probabilities of the four cases in the second portion of the table that correspond to the event that the second quarter rate is .015:

$$P[Q_2 = .015] = .09 + .21 + .1 + .1 = .5.$$

(b) The joint probability $P[Q_2 = .015, Q_3 = .02]$ would be the total of the probabilities of the first two outcomes in the second portion of the table, which is:

$$P[Q_2 = .015, Q_3 = .02] = .09 + .21 = .3.$$

(c) By parts (a) and (b) and the definition of conditional probability,

$$P[Q_3 = .02 | Q_2 = .015] = \frac{P[Q_2 = .015, Q_3 = .02]}{P[Q_2 = .015]} = \frac{.3}{.5} = .6.$$

(d) To find $P[Q_4 = .01 | Q_2 = .015, Q_3 = .02]$, we should first find

$$P[Q_2 = .015, Q_3 = .02, Q_4 = .01].$$

This is just a given outcome probability, .09 in the second portion of the table. Then by part (b), we have:

$$P\left[Q_4 = .01 \,|\, Q_2 = .015, Q_3 = .02\right] = \frac{P\left[Q_2 = .015, Q_3 = .02, Q_4 = .01\right]}{P\left[Q_2 = .015, Q_3 = .02\right]}$$

$$= \frac{.09}{.3} = .3.$$

(e) To use the definition of conditional probability, we will need both $P\left[Q_3 = .02\right]$ and $P\left[Q_3 = .02, Q_4 = .01\right]$. These may be found by picking out outcomes:

$$P\left[Q_3 = .02, Q_4 = .01\right] = P[\{(.015, .02, .01)\}] = .09;$$

$$P\left[Q_3 = .02\right] = P[\{(.015, .02, .01), (.015, .02, .015)\}] = .09 + .21 = .3.$$

Then:

$$P\left[Q_4 = .01 \,|\, Q_3 = .02\right] = \frac{P\left[Q_3 = .02, Q_4 = .01\right]}{P\left[Q_3 = .02\right]} = \frac{.09}{.3} = .3.$$

Notice that we happened to get the same answer for $P\left[Q_4 = .01 \,|\, Q_3 = .02\right]$ as we did for $P\left[Q_4 = .01 \,|\, Q_2 = .015, Q_3 = .02\right]$. In this case the observed value for Q_2 had no effect on the conditional probability that $Q_4 = .01$, given that $Q_3 = .02$. (See also Exercise 18.) ■

Example 6. In a general binomial branch process with up and down ratios $1 + b$ and $1 + a$ and up probability p, compute:

$$P\left[X_{n+2} = (1 + a)(1 + b)x \,|\, X_n = x\right].$$

Solution. There are two disjoint cases that lead the price process in two steps from a value of x to a value of $(1+a)(1+b)x$: (U, D), and (D, U). The values of X_{n+1} are $(1+b)x$ and $(1+a)x$, respectively, in these two cases. By the sum rule, the multiplication rule, and the assumed independence of future states from past states given the present state, we can compute:

$$
\begin{aligned}
P\,[X_{n+2} &= (1 + a)(1 + b)x \,|\, X_n = x] \\
&= \; P\,[X_{n+2} = (1 + a)(1 + b)x, X_{n+1} = (1 + b)x \,|\, X_n = x] \\
&\quad + P\,[X_{n+2} = (1 + a)(1 + b)x, X_{n+1} = (1 + a)x \,|\, X_n = x] \\
&= \; P\,[X_{n+1} = (1 + b)x \,|\, X_n = x] \\
&\quad \cdot P\,[X_{n+2} = (1 + a)(1 + b)x \,|\, X_{n+1} = (1 + b)x] \\
&\quad + P\,[X_{n+1} = (1 + a)x \,|\, X_n = x] \\
&\quad \cdot P\,[X_{n+2} = (1 + a)(1 + b)x \,|\, X_{n+1} = (1 + a)x] \\
&= \; p \cdot (1 - p) + (1 - p) \cdot p = 2p(1 - p). \; ■
\end{aligned}
$$

This example illustrates a problem-solving technique that is important in probabilistic financial problems. We started with $\{X_{n+2} = (1+a)(1+b)x\}$, an event involving the state of a process two time units into the future, and broke it into cases depending on what might happen at the intervening time period $n+1$. The multiplication rule then allows us to factor the probability of each case into the product of two simpler one-step conditional probabilities. This motivates a modification of an earlier favorite theorem.

Theorem 2. (Law of Total Probability, conditional version) If $A_1, A_2, ..., A_n$ is a partition of the sample space, that is, the As are pairwise disjoint and their union is Ω, then for any event B,

$$P[B] = \sum_{i=1}^{n} P[B|A_i] P[A_i] \tag{3.40}$$

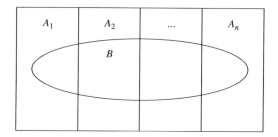

Figure 3.19: The Law of Total Probability.

Proof. The events $A_i \cap B$ are pairwise disjoint, and their union is A (see Figure 3.19). Thus, by the additivity property of probability and the multiplication rule,

$$P[B] = \sum_{i=1}^{n} P[A_i \cap B] = \sum_{i=1}^{n} P[B|A_i] P[A_i].$$

The thought process involved in using the Law of Total Probability is that the event B is difficult to assign a probability value to directly, but if it is known that A_i occurs, the conditional probability of B given A_i is more tractable. Some people call this technique "conditioning and unconditioning," where the word "unconditioning" refers to taking a weighted average of these conditional probabilities with the probabilities $P[A_i]$ as weights.

Here is another example of the application of this useful theorem.

Example 7. Aunt Mathilda tells her stockbroker nephew Colin that she is interested in selecting a portfolio of two among three possible stocks. She will divide her money equally between the two stocks in the portfolio she selects. The possible rates of return of the stocks and their probabilities of occurrence are in the tables below. Although Mathilda tells Colin that she will study the three companies before she tells him which portfolio she will purchase, her actual intention is to pick a portfolio at random. Find the joint distribution of the two rates of return on the stocks she selects for the portfolio, and find the distribution of the overall rate of return on her portfolio. Assume that the joint probability of a pair of rates is the product of the individual probabilities in the pair.

	rate of return	.03	.04	.05
stock 1:	probability	.4	.4	.2
stock 2:	rate of return	.04	.05	
	probability	.5	.5	
stock 3:	rate of return	.03	.04	
	probability	.6	.4	

Solution. Let (R_1, R_2) be the rates of return on the lower numbered and higher numbered stocks that are selected. The possible stock combinations that could be chosen are (stock 1, stock 2), (stock 1, stock 3), and (stock 2, stock 3). The probability is $1/3$ that Aunt Mathilda selects each of these. If we knew which portfolio she selected, we could list and find the probabilities of the possible pairs of rates of return, which suggests that we should condition and uncondition the events $(R_1, R_2) = (r_1, r_2)$ whose probabilities we need on the partition formed by the events that portfolio 1, 2, or 3 was selected.

There are nine possible pairs of rates of return (r_1, r_2): $(.03, .03)$, $(.03, .04)$, $(.03, .05)$, $(.04, .03)$, $(.04, .04)$, $(.04, .05)$, $(.05, .03)$, $(.05, .04)$, $(.05, .05)$. For the first pair,

$$
\begin{aligned}
P\left[(R_1, R_2) = (.03, .03)\right] &= P\left[(R_1, R_2) = (.03, .03)|\text{stock 1, stock 2}\right] \\
&\quad \cdot P[\text{stock 1, stock 2}] \\
&\quad + P\left[(R_1, R_2) = (.03, .03)|\text{stock 1, stock 3}\right] \\
&\quad \cdot P[\text{stock 1, stock 3}] \\
&\quad + P\left[(R_1, R_2) = (.03, .03)|\text{stock 2, stock 3}\right] \\
&\quad \cdot P[\text{stock 2, stock 3}] \\
&= (.4)(0)\tfrac{1}{3} + (.4)(.6)\tfrac{1}{3} + (0)(.6)\tfrac{1}{3} \\
&= .08.
\end{aligned}
$$

Notice that only the stock 1, stock 3 portfolio can give rise to a rate of return of .03 for both stocks in the portfolio. For the pair $(.03, .04)$,

$$
\begin{aligned}
P\left[(R_1, R_2) = (.03, .04)\right] &= P\left[(R_1, R_2) = (.03, .04)|\text{stock 1, stock 2}\right] \\
&\quad \cdot P[\text{stock 1, stock 2}] \\
&\quad + P\left[(R_1, R_2) = (.03, .04)|\text{stock 1, stock 3}\right] \\
&\quad \cdot P[\text{stock 1, stock 3}] \\
&\quad + P\left[(R_1, R_2) = (.03, .04)|\text{stock 2, stock 3}\right] \\
&\quad \cdot P[\text{stock 2, stock 3}] \\
&= (.4)(.5)\tfrac{1}{3} + (.4)(.4)\tfrac{1}{3} + (0)(.4)\tfrac{1}{3} \\
&= .12.
\end{aligned}
$$

For the pair $(.03, .05)$,

$$
\begin{aligned}
P\left[(R_1, R_2) = (.03, .05)\right] &= P\left[(R_1, R_2) = (.03, .05)|\text{stock 1}, \text{stock 2}\right] \\
&\quad \cdot P[\text{stock 1}, \text{stock 2}] \\
&\quad + P\left[(R_1, R_2) = (.03, .05)|\text{stock 1}, \text{stock 3}\right] \\
&\quad \cdot P[\text{stock 1}, \text{stock 3}] \\
&\quad + P\left[(R_1, R_2) = (.03, .05)|\text{stock 2}, \text{stock 3}\right] \\
&\quad \cdot P[\text{stock 2}, \text{stock 3}] \\
&= (.4)(.5)\tfrac{1}{3} + (.4)(0)\tfrac{1}{3} + (0)(0)\tfrac{1}{3} \\
&= .067.
\end{aligned}
$$

We will leave the other six computations for the exercises. The results are as follows:

value of (R_1, R_2)	$(.03, .03)$	$(.03, .04)$	$(.03, .05)$	$(.04, .03)$	$(.04, .04)$
probability	.08	.12	.067	.18	.1867

value of (R_1, R_2)	$(.04, .05)$	$(.05, .03)$	$(.05, .04)$	$(.05, .05)$
probability	.067	.14	.1267	.033

Since Aunt Mathilda's investment is to be split evenly between the two stocks in her portfolio, the rate of return on the whole portfolio will be the average of the rates on the two stocks in the portfolio (see Exercise 17). We can compute the averages of each of the pairs above, totaling the probabilities when cases correspond to the same average. With $R = (R_1 + R_2)/2$ as the overall portfolio rate, we get:

$$
P[R = .03] = P\left[(R_1, R_2) = (.03, .03)\right] = .08;
$$

$$
\begin{aligned}
P[R = .035] &= P\left[(R_1, R_2) = (.03, .04)\right] + P\left[(R_1, R_2) = (.04, .03)\right] \\
&= .12 + .18 \\
&= .3;
\end{aligned}
$$

$$
\begin{aligned}
P[R = .04] &= P\left[(R_1, R_2) = (.03, .05)\right] + P\left[(R_1, R_2) = (.04, .04)\right] \\
&\quad + P\left[(R_1, R_2) = (.05, .03)\right] \\
&= .067 + .1867 + .14 \\
&= .3933;
\end{aligned}
$$

$$
\begin{aligned}
P[R = .045] &= P\left[(R_1, R_2) = (.04, .05)\right] + P\left[(R_1, R_2) = (.05, .04)\right] \\
&= .067 + .1267 \\
&= .1933;
\end{aligned}
$$

$$
P[R = .05] = P\left[(R_1, R_2) = (.05, .05)\right] = .033. \blacksquare
$$

3.4.2   Conditional Distributions of Random Variables

Now that we have a proper definition in hand of the conditional probability of an event given another event, it is a short step to talk about the conditional distribution of a random variable given the value of another random variable. In fact, we have already seen the idea. Refer back to Example 2 and Figure 3.18 for a moment, in which there was a particular four-step binomial branch process with initial state 50, up ratio $1 + b = 1.1$, and down ratio $1 + a = .9$. If up and down moves were assumed to be equally likely, then we used the reduced sample space idea to reason that $P[X_4 = 73.21 \,|\, X_2 = 60.50] = \frac{1}{4}$. This would also follow directly from the definition of conditional probability, since

$$
\begin{aligned}
P[X_4 = 73.21 \,|\, X_2 = 60.50] &= \frac{P[X_2 = 60.50, X_4 = 73.21]}{P[X_2 = 60.50]} \\
&= \frac{P[\text{all four moves are } U]}{P[\text{first two moves are } U]} \\
&= \frac{(1/2)^4}{(1/2)^2} = \frac{1}{4}.
\end{aligned}
$$

Given that $X_2 = 60.50$, there are three possible states for X_4, namely, 73.21, 59.90, and 49.01, whose probabilities of occurrence are easy to determine from Figure 3.18(b):

$$
P[X_4 = 73.21 \,|\, X_2 = 60.50] = \frac{1}{4};
$$

$$
P[X_4 = 59.90 \,|\, X_2 = 60.50] = \frac{1}{2};
$$

$$
P[X_4 = 49.01 \,|\, X_2 = 60.50] = \frac{1}{4}.
$$

This list of probability masses describes what is called the **conditional distribution** of X_4, given $X_2 = 60.50$. The general definition of the conditional probability mass function of one random variable given another is next.

Definition 2. Let X and Y be two random variables, and let x be a possible state of X for which $P[X = x] > 0$. Then the **conditional probability mass function** of Y given $X = x$ is:

$$
q(y|x) = P[Y = y | X = x] = \frac{P[X = x, Y = y]}{P[X = x]}, \tag{3.41}
$$

where y ranges over all possible values of Y.

We will illustrate by reconsidering a couple of earlier examples. In the process, we will see how useful that direct appeals to the defining formula (3.41), as well as the Law of Total Probability, can be.

Example 8. In the home loan default example (Example 4), let X_i be the indicator random variables of loan defaults, where X_i is either 1 or 0 according to whether loan i defaulted or not, $i = 1, 2, 3$. Determine the conditional distribution of X_2 given each possible value of X_1, and of X_3 given each possible value of X_2.

Solution. The problem statement immediately gives the conditional probability mass function of X_2 given each of $X_1 = 1$ and $X_1 = 0$:

$$
\begin{aligned}
q(1|1) &= P[X_2 = 1 | X_1 = 1] = .1; \\
q(0|1) &= P[X_2 = 0 | X_1 = 1] = 1 - .1 = .9; \\
q(1|0) &= P[X_2 = 1 | X_1 = 0] = .05; \\
q(0|0) &= P[X_2 = 0 | X_1 = 0] = 1 - .05 = .95.
\end{aligned}
$$

The distribution of X_3 given X_2 is not as clear from the problem information, since, as described, the conditional probabilities depend on both the values of X_2 and X_1. But we can work with the table of outcome probabilities, repeated below, and the definition of the conditional p.m.f.

outcome	probability
$(1, 1, 1)$	$(.05)(.10)(.15) = .00075$
$(1, 1, 0)$	$(.05)(.10)(1 - .15) = .00425$
$(1, 0, 1)$	$(.05)(1 - .10)(.1) = .0045.$
$(0, 1, 1)$	$(1 - .05)(.05)(.1) = .00475$
$(1, 0, 0)$	$(.05)(1 - .10)(1 - .1) = .0405$
$(0, 1, 0)$	$(1 - .05)(.05)(1 - .1) = .04275$
$(0, 0, 1)$	$(1 - .05)(1 - .05)(.05) = .045125$
$(0, 0, 0)$	$(1 - .05)(1 - .05)(1 - .05) = .857375$

For example, in the case that loan 2 defaults, that is, $X_2 = 1$, we have:

$$
\begin{aligned}
q(1|1) = P[X_3 = 1 | X_2 = 1] &= \frac{P[X_2 = 1, X_3 = 1]}{P[X_2 = 1]} \\
&= \frac{P[\{(1, 1, 1), (0, 1, 1)\}]}{P[\{(1, 1, 1), (1, 1, 0), (0, 1, 1), (0, 1, 0)\}]} \\
&= \frac{.00075 + .00475}{.00075 + .00425 + .00475 + .04275} \\
&= .104762,
\end{aligned}
$$

hence,

$$
\begin{aligned}
q(0|1) &= P[X_3 = 0 | X_2 = 1] \\
&= 1 - P[X_3 = 1 | X_2 = 1] \\
&= 1 - .104762 = .895238.
\end{aligned}
$$

For the conditional mass function of X_3 given $X_2 = 0$, we can make similar computations:

$$
\begin{aligned}
q(1|0) &= P[X_3 = 1 \,|\, X_2 = 0] \\
&= \frac{P[X_2 = 0, X_3 = 1]}{P[X_2 = 0]} \\
&= \frac{P[\{(1,0,1),(0,0,1)\}]}{P[\{(1,0,1),(1,0,0),(0,0,1),(0,0,0)\}]} \\
&= \frac{.0045 + .045125}{.0045 + .0405 + .045125 + .857375} \\
&= .0523747,
\end{aligned}
$$

and therefore,

$$
\begin{aligned}
q(0|0) &= P[X_3 = 0 \,|\, X_2 = 0] \\
&= 1 - P[X_3 = 1 \,|\, X_2 = 0] \\
&= 1 - .0523747 = .947625. \ \blacksquare
\end{aligned}
$$

We now look back at another relevant problem, Example 4 from Section 3.3, in the context of conditional distributions.

Example 9. In a bull market, each week a certain "penny stock" stays the same in value with probability 3/4, goes up by \$.05 with probability 1/8, or goes up by \$.10 with probability 1/8. Find the possible values of the stock after 1, 2, and 3 weeks and their probabilities, if the initial price of the stock is \$.50. Find the conditional distribution of the week 3 value given that the week 1 value is \$.55.

Solution. We can extend the earlier price diagram to a third week; since the largest possible price at the end of week 2 is \$.70, possible prices of \$.75 and \$.80 are added to the states for week 2:

initial	week 1	week 2	week 3
			.80
			.75
		.70	.70
		.65	.65
	.60	.60	.60
	.55	.55	.55
.50	.50	.50	.50

If X_i represents the price at the end of week i, then, by the given information,

$$
P[X_1 = .50] = \frac{3}{4}; \ P[X_1 = .55] = \frac{1}{8}; \ P[X_1 = .60] = \frac{1}{8}.
$$

In the earlier example we were able to compute the probability mass function of X_2:

$$P[X_2 = .50] = \frac{9}{16}; P[X_2 = .55] = \frac{6}{32}; P[X_2 = .60] = \frac{13}{64};$$

$$P[X_2 = .65] = \frac{2}{64}; P[X_2 = .70] = \frac{1}{64}.$$

(If you look back at the computation in the earlier example, you will see that we were actually using the conditional probability version of the Law of Total Probability without labeling it as such.) We can compute the distribution of X_3 similarly by conditioning and unconditioning on X_2. Here we do the first few computations; you are asked to finish in Exercise 19.

$$P[X_3 = .50] = P[X_2 = .50] \cdot P[X_3 = .50 | X_2 = .50] = \frac{9}{16} \cdot \frac{3}{4} = \frac{27}{64};$$

$$
\begin{aligned}
P[X_3 = .55] \;&=\; P[X_2 = .50] \cdot P[X_3 = .55 | X_2 = .50] \\
&\quad + P[X_2 = .55] \cdot P[X_3 = .55 | X_2 = .55] \\
&=\; \frac{9}{16} \cdot \frac{1}{8} + \frac{6}{32} \cdot \frac{3}{4} \\
&=\; 27/128;
\end{aligned}
$$

$$
\begin{aligned}
P[X_3 = .60] \;&=\; P[X_2 = .50] \cdot P[X_3 = .60 | X_2 = .50] \\
&\quad + P[X_2 = .55] \cdot P[X_3 = .60 | X_2 = .55] \\
&\quad + P[X_2 = .60] \cdot P[X_3 = .60 | X_2 = .60] \\
&=\; \frac{9}{16} \cdot \frac{1}{8} + \frac{6}{32} \cdot \frac{1}{8} + \frac{13}{64} \cdot \frac{3}{4} \\
&=\; 63/256.
\end{aligned}
$$

Summarizing, the distribution of X_3 is characterized in the following table. You should check to be sure that the probability masses sum to 1.

value of X_3	.50	.55	.60	.65	.70	.75	.80
probability	$\frac{27}{64}$	$\frac{27}{128}$	$\frac{63}{256}$	$\frac{37}{512}$	$\frac{21}{512}$	$\frac{3}{512}$	$\frac{1}{512}$

We are also asked to find the conditional distribution of X_3 given that $X_1 = .55$. Knowledge of this X_1 value excises part of the original diagram, leaving the following possibilities:

week 1	week 2	week 3
		.75
		.70
	.65	.65
	.60	.60
.55	.55	.55

We can compute the desired probabilites by conditioning and unconditioning on the value of X_2. We do this using the conditional probability measure $P[\bullet | X_1 = .55]$.

$$
\begin{aligned}
P[X_3 = .55 | X_1 = .55] &= P[X_2 = .55 | X_1 = .55] \\
&\quad \cdot P[X_3 = .55 | X_2 = .55, X_1 = .55] \\[4pt]
&= \frac{3}{4} \cdot \frac{3}{4} = \frac{9}{16};
\end{aligned}
$$

$$
\begin{aligned}
P[X_3 = .60 | X_1 = .55] &= P[X_2 = .55 | X_1 = .55] \\
&\quad \cdot P[X_3 = .60 | X_2 = .55, X_1 = .55] \\[4pt]
&\quad + P[X_2 = .60 | X_1 = .55] \\
&\quad \cdot P[X_3 = .60 | X_2 = .60, X_1 = .55] \\[4pt]
&= \frac{3}{4} \cdot \frac{1}{8} + \frac{1}{8} \cdot \frac{3}{4} \\[4pt]
&= 6/32;
\end{aligned}
$$

$$
\begin{aligned}
P[X_3 = .65 | X_1 = .55] &= P[X_2 = .55 | X_1 = .55] \\
&\quad \cdot P[X_3 = .65 | X_2 = .55, X_1 = .55] \\[4pt]
&\quad + P[X_2 = .60 | X_1 = .55] \\
&\quad \cdot P[X_3 = .65 | X_2 = .60, X_1 = .55] \\[4pt]
&\quad + P[X_2 = .65 | X_1 = .55] \\
&\quad \cdot P[X_3 = .65 | X_2 = .65, X_1 = .55] \\[4pt]
&= \frac{3}{4} \cdot \frac{1}{8} + \frac{1}{8} \cdot \frac{1}{8} + \frac{1}{8} \cdot \frac{3}{4} \\[4pt]
&= 13/64;
\end{aligned}
$$

$$
\begin{aligned}
P[X_3 = .70 | X_1 = .55] &= P[X_2 = .60 | X_1 = .55] \\
&\quad \cdot P[X_3 = .70 | X_2 = .60, X_1 = .55] \\[4pt]
&\quad + P[X_2 = .65 | X_1 = .55] \\
&\quad \cdot P[X_3 = .70 | X_2 = .65, X_1 = .55] \\[4pt]
&= \frac{1}{8} \cdot \frac{1}{8} + \frac{1}{8} \cdot \frac{1}{8} \\[4pt]
&= 2/64;
\end{aligned}
$$

$$
\begin{aligned}
P[X_3 = .75 | X_1 = .55] &= P[X_2 = .65 | X_1 = .55] \\
&\quad \cdot P[X_3 = .75 | X_2 = .65, X_1 = .55] \\[4pt]
&= \frac{1}{8} \cdot \frac{1}{8} = \frac{1}{64}.
\end{aligned}
$$

Notice the similarity between these two-step conditional probabilities and the unconditional distribution of X_2 above. ∎

3.4.3   Introduction to Markov Processes

In examples such as the penny stock and the binomial branch model, we often assume that the future behavior of the price process is not influenced by the past, given the present state. This property is known as the **Markov property**, and we would like to give a very brief introduction to Markov processes here. A few important properties are left to you as Exercises 24–25, and you can learn much more in references such as Hastings [6], Çinlar [4], and Ross [13].

We consider **discrete-time stochastic processes**, that is, sequences of random variables $X_0, X_1, X_2, X_3, ...$ indexed by a non-negative integer subscript. The interpretation, for us, is that X_i is the random value at time i of an economic quantity in which we are interested, such as the price of an asset, a monetary exchange rate, or the variable interest rate on a credit account. There are such things as **continuous-time stochastic processes** $(X_t)_{t \geq 0}$ where the time subscript is real valued, and they are very important in financial mathematics, but we do not discuss them in this book.

Definition 3. A discrete-time stochastic process $(X_n)_{n \geq 0}$ is a **Markov process** if:

$$P[X_{n+1} = j \,|\, X_0 = i_0, X_1 = i_1, ..., X_n = i] = P[X_{n+1} = j \,|\, X_n = i] \quad (3.42)$$

for all times n, all possible values j of X_{n+1}, and all possible path histories $i_0, i_1, i_2, ..., i_{n-1}, i$. In this case we say that the process value X_{n+1} at time $n+1$ is **conditionally independent** of the past random variables $X_1, X_2, ..., X_{n-1}$ given the present value X_n. If, furthermore, the conditional probability on the right side of the equation, $P[X_{n+1} = j \,|\, X_n = i]$, does not depend on the time n, then we call the Markov process **time-homogeneous**.

The sequence of observed values of a binomial branch process is clearly a Markov process, because regardless of what path occurred prior to time n, the value X_{n+1} must either be $(1+b)X_n$ or $(1+a)X_n$ with probabilities p and $1-p$, respectively. More formally, if the value $X_n = i$ is known, then the conditional distribution of the next state is:

$$P[X_{n+1} = (1+b)i \,|\, X_0 = i_0, X_1 = i_1, ..., X_n = i]$$
$$= P[X_{n+1} = (1+b)i \,|\, X_n = i] = p;$$

$$P[X_{n+1} = (1+a)i \,|\, X_0 = i_0, X_1 = i_1, ..., X_n = i]$$
$$= P[X_{n+1} = (1+a)i \,|\, X_n = i] = 1 - p.$$

We observe also that this is a time-homogeneous process.

For the penny stock, we had assumed that, whatever the current price is, the next price was either the current price (with probability 3/4), the current price + \$.05 (with probability 1/8), or the current price + \$.10 (with probability 1/8). This is the same as saying that the price process is a Markov process with one-step transition probabilities:

$$P\left[X_{n+1} = i \,|\, X_n = i\right] = \frac{3}{4}; \, P\left[X_{n+1} = i + .05 \,|\, X_n = i\right] = \frac{1}{8};$$

$$P\left[X_{n+1} = i + .10 \,|\, X_n = i\right] = \frac{1}{8}.$$

Both of these process are somewhat unusual as Markov processes go, because the sets of possible values of the random variables X_n change as time n changes.

In the time-homogeneous case, the probabilistic behavior of the process can be described easily either by a **transition diagram** or a **transition matrix**. The transition diagram of a Markov process is a graph with the possible states as nodes, annotated along edges by the probabilities of transitioning in one time step from each state to each other state. The associated transition matrix puts these one-step transition probabilities into a table, with a row for each current state and a column for each next state. For instance, suppose a variable interest rate on a credit card forms a Markov process with possible values 12%, 13%, 14%, and 15%, and possible transitions as in Figure 3.20. The diagram indicates, for example, that

$$P\left[X_{n+1} = 14 \,|\, X_n = 13\right] = .6, \text{ and } P\left[X_{n+1} = 13 \,|\, X_n = 12\right] = 1.$$

We observe that the interest rate is certain to bounce off the two extreme values of 12% and 15%, and that elsewhere it rises by 1% with probability .6 and falls by 1% with probability .4.

Figure 3.20: Transition diagram of a Markov process of interest rates.

The transition matrix for this Markov process would be:

	next:	12	13	14	15
	12	0	1	0	0
current:	13	.4	0	.6	0
	14	0	.4	0	.6
	15	0	0	1	0

In general, the transition matrix T of a homogeneous Markov process is a two-dimensional table $T = T(i,j)$ whose row i and column j entry is:

$$T(i,j) = P\left[X_{n+1} = j \,|\, X_n = i\right]. \tag{3.43}$$

In examples like Examples 6 and 9 we have already shown techniques for working with Markov processes. Here is another example using the process in Figure 3.20.

Example 10. Consider the interest rate process whose transition diagram and matrix are shown above, where the time steps are quarter-years. Consider a loan of $10,000 with initial nominal annual rate 12% and nominal rates in subsequent quarters following this Markov process. Assume for simplicity that interest is simple interest on the initial balance, calculated each quarter, and due at the end of the year. Find the probability mass function of the total interest that will be owed.

Solution. The first two quarters will be guaranteed, since in the initial quarter the rate is specified at 12% (or 3% quarterly rate), and by the transition structure the second quarter rate must be 13% (3.25% quarterly). In quarter 3, the rate either goes to 12% with probability .4, or to 14% (3.5% quarterly) with probability .6. If the rate went down to 12% in the third quarter, then it must be 13% in the fourth, whereas if the rate was 14% in quarter 3, then it will either be 13% in quarter 4 with probability .4, or 15% (3.75% quarterly) with probability .6. The possible interest rate paths and their probabilities are listed in the table below.

interest rates	probability	total interest
.03, .0325, .03, .0325	$1 \cdot (.4) \cdot 1 = .4$	$(.03 + .0325 + .03 + .0325)\$10,000$ $= \$1250$
.03, .0325, .035, .0325	$1 \cdot (.6) \cdot (.4) = .24$	$(.03 + .0325 + .035 + .0325)\$10,000$ $= \$1300$
.03, .0325, .035, .0375	$1 \cdot (.6) \cdot (.6) = .36$	$(.03 + .0325 + .035 + .0375)\$10,000$ $= \$1350$

Since the values of total interest in the third column are unique to the paths, the second and third columns define the p.m.f. of the random variable X, which is the total interest:

$$P[X = \$1250] = .4; \quad P[X = \$1300] = .24; \quad P[X = \$1350] = .36. \quad \blacksquare$$

Example 11. Exercise 24 presents an extremely useful computational result pertaining to the transition matrix of a Markov chain. It is worthwhile to foreshadow that here. Consider the three-state Markov chain (X_n) with the transition matrix T below. Compute: (a) $P[X_2 = 1 | X_0 = 2]$; (b) $P[X_2 = 3 | X_0 = 1]$. Do you notice anything about the relationship between the computations and the transition matrix?

$$T = \begin{pmatrix} .5 & 0 & .5 \\ .3 & .4 & .3 \\ 0 & .6 & .4 \end{pmatrix}$$

Solution. (a) By the Markov property and the Law of Total Probability, conditioning and unconditioning on the value of X_1, we find:

$$
\begin{aligned}
P[X_2 = 1 | X_0 = 2] \;=\;& P[X_1 = 1 | X_0 = 2] \cdot P[X_2 = 1 | X_1 = 1, X_0 = 2] \\
& + P[X_1 = 2 | X_0 = 2] \cdot P[X_2 = 1 | X_1 = 2, X_0 = 2] \\
& + P[X_1 = 3 | X_0 = 2] \cdot P[X_2 = 1 | X_1 = 3, X_0 = 2] \\
\;=\;& P[X_1 = 1 | X_0 = 2] \cdot P[X_2 = 1 | X_1 = 1] \\
& + P[X_1 = 2 | X_0 = 2] \cdot P[X_2 = 1 | X_1 = 2] \\
& + P[X_1 = 3 | X_0 = 2] \cdot P[X_2 = 1 | X_1 = 3] \\
\;=\;& T(2,1) \cdot T(1,1) + T(2,2) \cdot T(2,1) + T(2,3) \cdot T(3,1) \\
\;=\;& (.3)(.5) + (.4)(.3) + (.3)(0) = .27.
\end{aligned}
$$

(b) The same strategy works for the second conditional probability:

$$
\begin{aligned}
P[X_2 = 3 | X_0 = 1] \;=\;& P[X_1 = 1 | X_0 = 1] \cdot P[X_2 = 3 | X_1 = 1] \\
& + P[X_1 = 2 | X_0 = 1] \cdot P[X_2 = 3 | X_1 = 2] \\
& + P[X_1 = 3 | X_0 = 1] \cdot P[X_2 = 3 | X_1 = 3] \\
\;=\;& T(1,1) \cdot T(1,3) + T(1,2) \cdot T(2,3) + T(1,3) \cdot T(3,3) \\
\;=\;& (.5)(.5) + (0)(.3) + (.5)(.4) = .45.
\end{aligned}
$$

Look at the third and fourth lines on the right of the computation of the conditional probability $P[X_2 = 1 | X_0 = 2]$ in part (a). The quantity

$$
T(2,1) \cdot T(1,1) + T(2,2) \cdot T(2,1) + T(2,3) \cdot T(3,1) = (.3)(.5) + (.4)(.3) + (.3)(0)
$$

is the dot product of the second row of T with the first column of T:

$$
(T(2,1)\ T(2,2)\ T(2,3)) \cdot \begin{pmatrix} T(1,1) \\ T(2,1) \\ T(3,1) \end{pmatrix}.
$$

You can check also that, in part (b), $P[X_2 = 3 | X_0 = 1]$ is the dot product of the first row of T with the third column of T. There is nothing particularly special about the fact that there are three states, nor about the particular transition probabilities in the matrix. We hypothesize that:

$$
P[X_2 = j | X_0 = i] = \text{dot product of row } i \text{ of } T \text{ with column } j \text{ of } T. \quad (3.44)
$$

In fact, by time homogeneity, we would conclude that this is true for any two-step transition, not just those starting at time 0. If you are familiar with how matrices multiply, that is, the $i-j$ entry of the product is row i dotted with column j, you see from formula (3.44) that $T^2 = T \cdot T$ gives all two-step transition probabilities. In fact, more is true. It turns out that $T^3 = T^2 \cdot T$ gives all three-step transition probabilities, $T^4 = T^3 \cdot T$ gives all four-step transition probabilities, etc. ∎

3.4.4 Conditional Expectation

We noted earlier that $P[B|A]$, looked at as a function applied to events B, determines a valid probability measure. We might denote this measure by P_A to emphasize that fact. But this means that we can consider expectations of random variables Y relative to this measure, which would be defined by:

$$E_A[Y] = E[Y|A] = \sum_y y \cdot P_A[Y = y] = \sum_y y \cdot P[Y = y|A]. \tag{3.45}$$

The sums in formula (3.45) are taken over all possible values y in the state space of Y. Some of those could have conditional probability zero, given A. The most common situation in financial mathematics occurs when the event A being conditioned on is the event that another random variable assumes one of its values, that is, $A = \{X = x\}$. Then, the **conditional expectation of Y given $X = x$** would be:

$$\mu_{y|x} = E[Y|X = x] = \sum_y y \cdot P[Y = y|X = x], \tag{3.46}$$

which is a function of x rather than a constant. Once x is fixed, a numerical value results.

For example, consider again the four-step binomial branch process in Example 2. We argued earlier that:

$$P[X_4 = 73.21 \,|X_2 = 60.50] = \frac{1}{4}; \; P[X_4 = 59.90 \,|X_2 = 60.50] = \frac{1}{2};$$

$$P[X_4 = 49.01 \,|X_2 = 60.50] = \frac{1}{4}.$$

The conditional expectation of X_4 given $X_2 = 60.50$ is nothing more than the expectation of X_4 using this conditional distribution, which is:

$$E[X_4|X_2 = 60.50] = \frac{1}{4} \cdot 73.21 + \frac{1}{2} \cdot 59.90 + \frac{1}{4} \cdot 49.01 = 60.505.$$

Conditional expectations of functions of random variables are defined similarly; when conditioning on general events,

$$E_A[g(Y)] = E[g(Y)|A] = \sum_y g(y) \cdot P_A[Y = y] = \sum_y g(y) \cdot P[Y = y|A], \tag{3.47}$$

and when conditioning on particular values of another random variable,

$$E[g(Y)|X = x] = \sum_y g(y) \cdot P[Y = y|X = x]. \tag{3.48}$$

Then we could define the **conditional variance of Y** as simply the variance under the conditional probability measure, and a computational formula would be:

$$\text{Var}(Y|X = x) = E_{\{X=x\}}\left[Y^2\right] - \left(E_{\{X=x\}}[Y]\right)^2 \tag{3.49}$$
$$= E\left[Y^2 \big| X = x\right] - (E[Y|X = x])^2.$$

Conditional expectation also inherits all of the usual properties of expectation, particularly linearity:

$$E[cY + dZ|A] = cE[Y|A] + dE[Z|A]. \tag{3.50}$$

As an example, let us compute the conditional variance of X_4 given $X_2 = 60.50$ for the four-step binomial branch process. We already have that

$$E\left[X_4|X_2 = 60.50\right] = 60.505.$$

Also,

$$E\left[X_4^2|X_2 = 60.50\right] = \frac{1}{4} \cdot (73.21)^2 + \frac{1}{2} \cdot (59.90)^2 + \frac{1}{4} \cdot (49.01)^2 = 3734.43.$$

Then,

$$\text{Var}\left(X_4|X_2 = 60.50\right) = E\left[X_4^2|X_2 = 60.50\right] - (E\left[X_4|X_2 = 60.50\right])^2$$
$$= 3734.43 - 60.505^2 = 73.575.$$

Example 12. Consider again the penny stock of Example 9. Find the conditional mean and variance of the price X_3 at time 3, given that the price at time 1 was $.60.

Solution. We must find the conditional distribution of X_3 given that $X_1 = .60$ first. Because X_1 is known to be .60, the possible paths are:

week 1	week 2	week 3
		.80
		.75
	.70	.70
	.65	.65
.60	.60	.60

Similar computations to the ones done before, conditioning and unconditioning on the value of X_2, give us the desired probabilities. Here are the first two:

$$P\left[X_3 = .60\,|X_1 = .60\right] = P\left[X_2 = .60\,|X_1 = .60\right]$$
$$\cdot P\left[X_3 = .60\,|X_2 = .60, X_1 = .60\right]$$
$$= \frac{3}{4} \cdot \frac{3}{4} = \frac{9}{16};$$

$$P[X_3 = .65 \,|X_1 = .60] = P[X_2 = .60 \,|X_1 = .60]$$
$$\cdot P[X_3 = .65 \,|X_2 = .60, X_1 = .60]$$
$$+ P[X_2 = .65 \,|X_1 = .60]$$
$$\cdot P[X_3 = .65 \,|X_2 = .65, X_1 = .60]$$
$$= \frac{3}{4} \cdot \frac{1}{8} + \frac{1}{8} \cdot \frac{3}{4}$$
$$= 6/32.$$

In fact, you can check that the computations are not just similar, but identical to those carried out in Example 9, except that all prices increase by 5 cents. Thus, the complete conditional distribution of X_3 is:

$$P[X_3 = .60 \,|X_1 = .60] = \frac{9}{16}; P[X_3 = .65 \,|X_1 = .60] = \frac{6}{32};$$

$$P[X_3 = .70 \,|X_1 = .60] = \frac{13}{64}; P[X_3 = .75 \,|X_1 = .60] = \frac{2}{64};$$

$$P[X_3 = .80 \,|X_1 = .60] = \frac{1}{64}.$$

With the conditional p.m.f. of X_3 given $X_1 = .60$ in hand, we can compute the conditional mean:

$$E[X_3|X_1 = .60] = \frac{9}{16} \cdot (.60) + \frac{6}{32} \cdot (.65) + \frac{13}{64} \cdot (.70) + \frac{2}{64} \cdot (.75) + \frac{1}{64} \cdot (.80) = .6375.$$

For the conditional variance, we use the computational formula again:

$$\text{Var}(X_3|X_1 = .60) = E\left[X_3^2|X_1 = .60\right] - (E[X_3|X_1 = .60])^2$$
$$= \frac{9}{16} \cdot (.60)^2 + \frac{6}{32} \cdot (.65)^2 + \frac{13}{64} \cdot (.70)^2$$
$$+ \frac{2}{64} \cdot (.75)^2 + \frac{1}{64} \cdot (.80)^2 - (.6375)^2$$
$$= .00242. \blacksquare$$

Example 13. The first two of Aunt Mathilda's stocks in Example 7 had individual probability mass functions as below, and recall that we were assuming that the joint probability of a pair of rates is the product of the individual probabilities in the pair. Find the conditional expectation of the rate of return on the second stock given that the first experienced a rate of return of either .03 or .04.

	rate of return	.03	.04	.05
stock 1:	probability	.4	.4	.2

$$\text{stock 2:} \quad \begin{array}{lcc} \text{rate of return} & .04 & .05 \\ \text{probability} & .5 & .5 \end{array}$$

Solution. Let R_1 and R_2 be the two random variables modeling the rates of return on stocks 1 and 2. There are six joint states for the random vector (R_1, R_2), and by assumption their probabilities are as in the table below.

(r_1, r_2)	$(.03, .04)$	$(.03, .05)$	$(.04, .04)$	$(.04, .05)$	$(.05, .04)$	$(.05, .05)$
prob	$.4(.5) = .2$	$.2$	$.2$	$.2$	$.2(.5) = .1$	$.1$

We are asked for $E[R_2 | R_1 \in \{.03, .04\}]$. Since the event being conditioned on is not a single state event $R_1 = r_1$, we will have to resort to the original defining formula (3.45) for the conditional expectation:

$$\begin{aligned} E[R_2 | R_1 \in \{.03, .04\}] &= \sum_{r_2} r_2 \cdot P[R_2 = r_2 | R_1 \in \{.03, .04\}] \\ &= \sum_{y \in \{.04, .05\}} r_2 \cdot \frac{P[R_2 = r_2, R_1 \in \{.03, .04\}]}{P[R_1 \in \{.03, .04\}]}. \end{aligned}$$

The denominator is $P[R_1 \in \{.03, .04\}] = .4 + .4 = .8$. For the case $R_2 = .04$, summing the probabilities in the previous table gives:

$$P[R_2 = .04, R_1 \in \{.03, .04\}] = .2 + .2 = .4,$$

and for the case $R_2 = .05$, we also get

$$P[R_2 = .05, R_1 \in \{.03, .04\}] = .2 + .2 = .4.$$

Therefore,

$$E[R_2 | R_1 \in \{.03, .04\}] = .04 \cdot \frac{.4}{.8} + .05 \cdot \frac{.4}{.8} = .045.$$

Interestingly, it is easy to compute from the given distribution of R_2 that the unconditional expectation $E[R_2]$ is also .045. Occurrence of the event $R_1 \in \{.03, .04\}$ does not alter the mean of R_2. This is an example of a consequence of *independence*, a subject to be studied in the next section. ∎

Our last task for this section is to show yet another form of the Law of Total Probability, which can be very useful in computing expectations. To set this up, note again that expression (3.48) for the conditional expectation of a function of Y given x is a function h of real variable x, the value taken on by random variable X. Understanding this, we could denote:

$$E[g(Y)|X] = h(X), \quad \text{where } h(x) = E[g(Y)|X = x]. \tag{3.51}$$

A very famous result follows, which says that we can compute an expectation of a function of Y by conditioning and unconditioning on all possible values of another random variable X. It can be called the expectation version of the Law of Total Probability, but usually goes by the shorter name of the **Tower Law**.

Theorem 3. (Tower Law, or Law of Total Probability, expectation version) Let X and Y be discrete random variables. Then:

$$E[g(Y)] = \sum_x E[g(Y)|X = x]P[X = x] = E[E[g(Y)|X]]. \qquad (3.52)$$

Proof. Let us begin with the expression on the right side, $E[E[g(Y)|X]]$. Its meaning is $E[h(X)]$, where $h(x) = E[g(Y)|X = x]$, hence:

$$\begin{aligned} E[E[g(Y)|X]] &= E[h(X)] \\ &= \sum_x h(x)P[X = x] \\ &= \sum_x E[g(Y)|X = x]P[X = x]. \end{aligned}$$

So the middle expression in equation (3.52) is indeed equal to the expression on the right. Now, from (3.47),

$$E[g(Y)|X = x] = \sum_y g(y) \cdot P[Y = y|X = x].$$

Substituting into the previous equation yields:

$$\begin{aligned} E[E[g(Y)|X]] &= \sum_x E[g(Y)|X = x]P[X = x] \\ &= \sum_x \sum_y g(y) \cdot P[Y = y|X = x]P[X = x] \\ &= \sum_x \sum_y g(y) \cdot P[X = x, Y = y] \\ &= \sum_y g(y) \sum_x P[X = x, Y = y] \\ &= \sum_y g(y)P[Y = y] \\ &= E[g(Y)]. \end{aligned}$$

This completes the proof.

In the spirit of our other versions of the Law of Total Probability, this Tower Law suggests a "divide-and-conquer" approach to calculating a cumbersome expectation that may depend on events a few time steps in the future. For a convenient random variable X in the intervening time, find the simpler conditional expectations of $g(Y)$ given all possible X values, then recombine into a weighted average with the masses $P[X = x]$ as weights. We illustrate the technique with a favorite example.

Example 14. For the penny stock, use the Tower Law to compute $E[X_2]$, where X_2 represents the week 2 price. The relevant part of the price diagram is below, and recall that the price stays the same with probability 3/4 or goes up by .05 or by .10 each with probability 1/8.

	initial	week 1	week 2
			.70
			.65
		.60	.60
		.55	.55
	.50	.50	.50

Solution. We will condition and uncondition on X_1, the price at week 1. There are three possible values of X_1, and its distribution follows directly from assumptions:

$$P[X_1 = .50] = \frac{3}{4}; \ P[X_1 = .55] = \frac{1}{8}; \ P[X_1 = .60] = \frac{1}{8}.$$

Given $X_1 = .50$, there are three values of X_2 that are possible: .50, .55, and .60 with probabilities $\frac{3}{4}, \frac{1}{8}$, and $\frac{1}{8}$, respectively. Thus,

$$E[X_2|X_1 = .50] = (.50) \cdot \frac{3}{4} + (.55) \cdot \frac{1}{8} + (.60) \cdot \frac{1}{8} = .51875.$$

Similarly,

$$E[X_2|X_1 = .55] = (.55) \cdot \frac{3}{4} + (.60) \cdot \frac{1}{8} + (.65) \cdot \frac{1}{8} = .56875.$$

And the third conditional expectation that we need is:

$$E[X_2|X_1 = .60] = (.60) \cdot \frac{3}{4} + (.65) \cdot \frac{1}{8} + (.70) \cdot \frac{1}{8} = .61875.$$

Finally,

$$
\begin{aligned}
E[X_2] &= \sum_{x_1} E[X_2|X_1 = x_1] P[X = x_1] \\
&= E[X_2|X_1 = .50] P[X = .50] + E[X_2|X_1 = .55] P[X = .55] \\
&\quad + E[X_2|X_1 = .60] P[X = .60] \\
&= (.51875) \cdot \frac{3}{4} + (.56875) \cdot \frac{1}{8} + (.61875) \cdot \frac{1}{8} \\
&= .5375.
\end{aligned}
$$

Computing the simple three-state, one-step conditional expectations has allowed us to sidestep the problem of finding the distribution of X_2. ∎

Important Terms and Concepts

Definition of conditional probability - $P[B|A] = \frac{P[A \cap B]}{P[A]}$, provided $P[A] > 0$.

Multiplication rule - If $A_1, A_2, \ldots, A_n$ are events, then:

$$P[A_1 \cap A_2 \cap \cdots \cap A_n] = P[A_1] \cdot P[A_2|A_1] \cdot P[A_3|A_1 \cap A_2]$$
$$\cdots P[A_n|A_1 \cap \cdots \cap A_{n-1}].$$

Law of Total Probability, conditional probability version - If events A_i partition the sample space, then $P[B] = \sum_{i=1}^{n} P[B|A_i] P[A_i]$.

Conditional distribution - The collection $P[Y = y|X = x]$ of conditional probabilities ranging over y in the state space of Y for a particular value x of the random variable X.

Markov process - A sequence of random variables $X_0, X_1, X_2, \ldots$ such that each X_{n+1} has a conditional distribution given all of its predecessors that depends only on X_n and not on $X_0, \ldots, X_{n-1}$.

Transition diagram of a Markov process - A graph with states of the process as nodes, which are connected by edges showing the possible one-step transitions. It is also annotated on the edges with the transition probabilities.

Transition matrix of a Markov process - A table with a row for each current state and a column for each next state containing the one-step transition probabilities. The nth power of the transition matrix gives the n-step transition probabilities.

Conditional expectation - The expectation of a random variable using the conditional probability measure given an event of interest.

Tower Law - $E[E[g(Y)|X]] = E[g(Y)]$.

Exercises 3.4

1. Find the conditional probability that, in four successive flips of a fair coin, there are at least three heads given that there are at least two heads.

2. If three cards are drawn in succession and without replacement from an ordinary deck, what is the probability that all are aces? What is the probability that at least two of them are aces?

3. Two assets have a joint rate of return distribution as described by the following table. If R_1 and R_2 denote the two random rates, find: (a) $P[R_1 = .04]$; (b) $P[R_2 = .05|R_1 = .04]$; (c) $P[R_2 = .05]$; (d) $P[R_1 = .04|R_2 = .05]$.

	R_2				
	.03	.04	.05	.06	.07
.03	.14	.14	.03	.01	.01
R_1 .04	.14	.14	.04	.01	.01
.05	.03	.04	.05	.02	.02
.06	.01	.01	.02	.02	.02
.07	.01	.01	.02	.02	.03

		loan 2	
		1	0
loan 1	1	.02	.04
	0	.06	.88

Exercise 3   Exercise 4

4. The table above gives the joint distribution of indicator random variables I_1 and I_2 on two loans, where $I_i = 1$ if the loan defaults and $I_i = 0$ otherwise. Find all of the conditional probabilities $P[I_2 = 1 | I_1 = 1]$, $P[I_2 = 0 | I_1 = 1]$, $P[I_2 = 1 | I_1 = 0]$, and $P[I_2 = 0 | I_1 = 0]$. Also find all of the conditional probabilities below:

$$P[I_1 = 1 | I_2 = 1], \ P[I_1 = 0 | I_2 = 1], \ P[I_1 = 1 | I_2 = 0], \ P[I_1 = 0 | I_2 = 0].$$

5. In the setting of Exercise 4, suppose that, instead of the given values, the tabled probabilities are left general: p_{11} and p_{12} for the first row, and p_{21} and p_{22} for the second. But assume that we know that $P[I_1 = 1] = .05$ and $P[I_2 = 1] = .06$. Find all possible values of these entries that guarantee the condition $P[I_2 = 1 | I_1 = 1] = P[I_2 = 1]$. For any probabilities that you find, show that, in addition, $P[I_1 = 1 | I_2 = 1] = P[I_1 = 1]$.

6. A survey is taken regarding people's gender and age category and their reaction to a statement: "The federal government is doing all it can to improve the economy." Subjects can choose the responses "Agree," "Neutral," "Disagree," or "Don't Know." Suppose that 100 surveys are returned and the number of subjects in each category is tabulated, as below. If a subject is selected at random from this group, find: (a) P[Agree | male 45 or older]; (b) P[Disagree | female 45 or older]; (c) P[Agree | male]; (d) P[Disagree | female].

	Agree	Neutral	Disagree	Don't Know
male 45 or older	9	4	12	2
male under 45	12	5	7	3
female 45 or older	10	3	8	2
female under 45	15	5	2	1

7. A random variable X maps outcomes in a sample space Ω to a state space E, as shown in the diagram. Suppose that each of $\omega_1, \omega_2, \omega_3$ has the same probability, each of $\omega_4, \omega_5, \omega_6$ has the same probability, and ω_1 is twice as likely as ω_4. Find $P[X > 2 | X > 1]$.

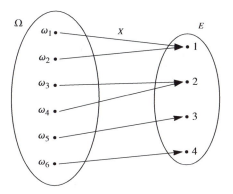

Exercise 7

8. A number is sampled at random from the list of integers $\{1, 2, 3, 4, ..., 20\}$. Find:

(a) $P[$the number is even | the number is greater than 10$]$;

(b) $P[$the number is greater than 8| the number is odd$]$.

9. Under what condition on $A \cap B$ will $P[B|A] = P[B]$? Try to interpret the meaning of this equation.

10. For a general binomial branch process with up probability p and up and down ratios $(1 + b)$ and $(1 + a)$, respectively, find with careful justification:

$$P\left[X_{10} = (1 + b)^4(1 + a)^6 x | X_8 = (1 + b)^4(1 + a)^4 x\right].$$

11. Show that $P[\bullet|A]$ defined by formula (3.35) satisfies axioms (a)–(c) of probability in Section 3.1.

12. A group of students in a Financial Mathematics class each has one primary bank credit card out of a possible 20 cards. The professor asks one of them at a time which card they have, and records whether any student has the same card as a student previously asked. Find the probabililty that at least two students in the class will have matching cards if the class size is 4, 5, 6, 7, or 8. (Hint: consider the complementary event of no matches. This is a reimagining of the famous **birthday problem** in probability theory, in which the probability is surprisingly high of finding matching birthdays in rooms containing just a few people.)

13. Prove the multiplication rule (3.39) in the case of three events using the definition of conditional probability.

14. A landlord has a rather unreliable tenant. If the tenant pays his rent this month, the chance is 90% that he will pay it next month, and if he doesn't pay

the rent this month the chance is 25% that he will also not pay next month's rent. Suppose that the tenant has paid in January. Find: (a) the probability that he pays in all of the next four months February through May; (b) the probability that he pays in none of the next four months; (c) the probability that he pays in exactly three of the next four months.

15. In a three-step binomial branch process compute:

$$P\left[X_{n+3} = (1+a)(1+b)^2 x \mid X_n = x\right].$$

16.   Finish the computations of $P\left[(R_1, R_2) = (r_1, r_2)\right]$ for the last six pairs $(.04, .03)$, $(.04, .04)$, $(.04, .05)$, $(.05, .03)$, $(.05, .04)$, $(.05, .05)$, in Example 7.

17. In Example 7 on Aunt Mathilda, a claim was made that the rate of return on an equally balanced portfolio of assets is the average of the asset rates of return. Prove this. Specifically, assume that an initial amount of wealth W is divided in half between two risky assets with random rates of return R_1 and R_2. Then show that the rate of return on the whole portfolio is $(R_1 + R_2)/2$.

18. In Example 5 compute: (a) the conditional distribution of Q_3 given Q_2 for each of the possible Q_2 values; (b) the conditional distribution of Q_4 given Q_3 for each of the possible values of Q_3.

19. Check the results on the distribution of X_3 in Example 9.

20. An investment has an interest rate that can flip back and forth between two values: 6% effective per year and 8% effective per year. The rate stays the same with probability 3/4 and changes with probability 1/4. Model this situation as a Markov process, writing the transition matrix and transition diagram. If interest is compounded annually, the rate in the first year is 6%, and the initial investment is $2000, find the probability mass function of the value of the investment after 3 years.

21. Which of the following could be transition matrices of a Markov process? Explain.

$$
\text{(a)} \begin{pmatrix} .2 & .4 & .4 \\ .3 & .5 & .2 \end{pmatrix} \quad
\text{(b)} \begin{pmatrix} .2 & .3 & .2 & .4 \\ 0 & 0 & 1 & 0 \\ .5 & .2 & .2 & .1 \\ 0 & .4 & .3 & .1 \end{pmatrix} \quad
\text{(c)} \begin{pmatrix} 0 & 1 & 0 \\ 1/2 & 0 & 1/2 \\ 1/3 & 1/3 & 1/3 \end{pmatrix}
$$

22. Let $(X_n)_{n\geq 0}$ be a Markov process with the transition diagram below. Find: (a) $P\left[X_2 = 2 \mid X_0 = 2\right]$; (b) $P\left[X_3 = 1 \mid X_1 = 3\right]$.

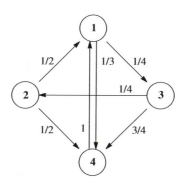

Exercise 22

23. For the Markov process with the transition matrix in Exercise 21(c), find:
(a) $P[X_2 = 3 | X_0 = 1]$; (b) $P[X_3 = 2 | X_0 = 2]$.

24. If $A = (a_{ij})$ is a matrix with m rows and p columns, and $B = (b_{ij})$ is a matrix with p rows and n columns, then the product matrix $A \cdot B$ is defined as the m row, n column matrix in which the row i column j entry is the dot product of the ith row of A with the jth column of B, that is:

$$(A \cdot B)_{ij} = a_{i1}b_{1j} + a_{i2}b_{2j} + \cdots + a_{ip}b_{pj}.$$

Using this definition and the law of total probability, prove that, if T is the transition matrix of a Markov process, then the $i - j$ entry of the matrix $T^2 = T \cdot T$ is equal to $P[X_2 = j | X_0 = i]$. Then show, by conditioning and unconditioning on the value of X_2, that the $i - j$ entry of the matrix $T^3 = T^2 \cdot T$ is equal to $P[X_3 = j | X_0 = i]$.

25. Suppose that a Markov process has an **initial distribution** p, that is, p is a row vector giving the p.m.f. values of X_0; its kth element is $p_k = P[X_0 = k]$. If T is the transition matrix of the process, then the product $p \cdot T$ as in Exercise 24 is a well-defined row vector. Use the Law of Total Probability to prove that:

$$(p \cdot T)_j = P[X_1 = j],$$

where $(p \cdot T)_j$ is the element in the jth position in $p \cdot T$. (It turns out also that $(p \cdot T^n)_j = P[X_n = j]$, but you need not show this in general here.) In Exercise 22, if initially a state is chosen at random, find $P[X_1 = 3]$.

26. A stock follows a Markov process with possible states $E = \{50, 52, 54, 56\}$ and the transition matrix below. If X_n denotes the stock price at time n, then find: (a) $P[X_2 = 54 | X_0 = 50]$; (b) $P[X_3 = 54 | X_0 = 50]$; (c) the probability that the process has not reached 54 by time 3, given that it started at 50.

$$
\begin{array}{c@{\quad}c}
 & \begin{array}{cccc} 50 & 52 & 54 & 56 \end{array} \\
\begin{array}{c} 50 \\ 52 \\ 54 \\ 56 \end{array} &
\left(\begin{array}{cccc}
0 & 1 & 0 & 0 \\
.4 & .4 & .2 & 0 \\
0 & .4 & .4 & .2 \\
0 & 0 & 1 & 0
\end{array} \right)
\end{array}
$$

27. For the stock price process in Exercise 26, find: (a) $E[X_2|X_0 = 52]$; (b) the conditional variance of X_2 given $X_0 = 52$.

28. Suppose that (X_n) is a non-time-homogeneous Markov process such that its transition probabilities from time 0 to time 1 are given by the matrix T_1 below, and the transition probabilities from time 1 to time 2 are given by T_2. Compute $P[X_2 = 1|X_0 = 2]$.

$$
T_1 = \left(\begin{array}{ccc}
1/2 & 1/2 & 0 \\
0 & 1/2 & 1/2 \\
1/3 & 1/3 & 1/3
\end{array} \right)
\quad
T_2 = \left(\begin{array}{ccc}
0 & 1 & 0 \\
1/2 & 0 & 1/2 \\
3/4 & 1/4 & 0
\end{array} \right)
$$

29. In Example 10, find the p.m.f. and the expected value of the total interest charged if the initial nominal yearly rate in the first quarter is 14%.

30. For the rates of return R_1 and R_2 in Exercise 3, find: (a) $E[R_2|R_1 = .03]$; (b) $\operatorname{Var}(R_2|R_1 = .03)$; (c) $E[R_1|R_2 = .05]$; (d) $\operatorname{Var}(R_1|R_2 = .05)$.

31. Consider random variables X and Y with the joint p.m.f.

$$
f(x,y) = c \cdot x \cdot y, \text{ if } x,y \in \{1,2,3\}; f(x,y) = 0 \text{ otherwise.}
$$

First, find the constant c that makes this a valid p.m.f. Then, compute

$$
E[Y|X = x]
$$

for each of the three x values. Interpret the results you are getting.

32. Let $Y = I_B$ be the indicator random variable of an event B; that is, it has the value 1 on B and 0 otherwise. If A is another event, show that $E[I_B|A] = P[B|A] = \frac{E[I_A \cdot I_B]}{P[A]}$. Show also that, if a random variable Y is a linear combination of indicators, $Y = \sum_{k=1}^{n} c_k \cdot I_{B_k}$, then

$$
E[Y|A] = \sum_{k=1}^{n} c_k \cdot \frac{E[I_A \cdot I_{B_k}]}{P[A]}.
$$

33. For the random variable described in Exercise 7, compute $E[X|X > 2]$.

34. For a binomial branch process with up probability .6, initial state 10, up ratio $(1+b) = 1.05$, and down ratio $(1+a) = .98$, compute the conditional mean

and variance of X_3 given $X_1 = 10.50$.

35. In Example 10, compute the conditional expectation and variance of the total interest accrued in the fourth quarter, given that the interest rate started at 15% nominal yearly.

36. If X is a Markov process show that

$$E[X_2|X_1 = x_1, X_0 = x_0] = E[X_2|X_1 = x_1].$$

37. Given a general binomial branch process, find the value of the up probability p such that $E[X_1|X_0 = x] = x$. For this special probability, called the **martingale probability**, compute $E[X_2|X_0 = x]$.

3.5   Independence and Dependence

Several times in the preceding sections we have made assumptions along the lines of: suppose that a sequence of two events has a probability equal to the product of the probabilities of the events. This was computationally convenient, but it has not been clear where such an assumption comes from or why it might be true. In this section we will answer these questions by exploring the concept of **independence**, both of events and of random variables. And as with most concepts, its negation, namely, **dependence**, is also interesting and worthy of study. Particularly, the dependence of the price behavior of different financial assets affects our ability to assess risks. This problem lies at the heart of much of financial mathematics.

3.5.1   Independent Events

Roughly speaking, events are independent if the occurrence of one of them does not change the probability that the other occurs. Notice that, despite the fact that the English definitions of the words "independent" and "disjoint" are slightly similar, their probabilistic meanings are very different. When events are disjoint, they have no outcomes in common, so that, if one is known to have occurred, then the other cannot occur. This would change the probability of the second drastically; in fact, if we know that the first event occurred, the second event then has probability zero. So disjoint events cannot be independent in the sense we are using.

The most useful property that independence has does not involve conditional probability as the intuition suggests, but instead involves the occurrence of sequences of independent events. Intuition and common sense support you well here; if two coins are flipped "independently," then it is very natural to assume that the probability that the first is a head and the second is also a head is the product $\frac{1}{2} \cdot \frac{1}{2}$. We will take this factorization condition as our definition.

Definition 1. Two events A and B are ***independent*** of each other if the probability that both occur factors into the product of the individual probabilities, that is:

$$P[A \cap B] = P[A] \cdot P[B]. \qquad (3.53)$$

More generally, several events $A_1, A_2, ..., A_n$, are ***mutually independent*** if the probability of the intersection of any subcollection of the events factors into the product of the probabilities of the events in that subcollection.

Notice that this gives us a new language to talk about the binomial branch process. In an n-step process, we are assuming that all events pertaining to transitions in different time steps are independent. For example,

$$P\left[U \text{ on } 1^{\text{st}} \cap U \text{ on } 2^{\text{nd}} \cap D \text{ on } 3^{\text{rd}} \cap D \text{ on } 4^{\text{th}}\right]$$

$$
\begin{aligned}
&= \ P\left[U \text{ on } 1^{\text{st}}\right] P\left[U \text{ on } 2^{\text{nd}}\right] P\left[D \text{ on } 3^{\text{rd}}\right] P\left[D \text{ on } 4^{\text{th}}\right] \\
&= \ p^2(1-p)^2.
\end{aligned}
$$

The fact that we have been doing this for awhile should give you a good comfort level with the idea of independent events.

But what about the intuition of independence? If A and B are independent events of probability greater than zero, then

$$P[B|A] = \frac{P[A \cap B]}{P[A]} = \frac{P[A]P[B]}{P[A]} = P[B], \qquad (3.54)$$

which shows that the knowledge that A occurs does not change the probability of B. Reversing the roles of A and B, the same derivation shows that, under independence, $P[A|B] = P[A]$. So the definition that we have taken gives us the very computationally useful factorization formula (3.53), and also implies the intuitive condition (3.54) that helps us better understand what independence really means.

It is time for a few examples.

Example 1. Four cards are selected in succession from an ordinary well-shuffled deck. Each card is replaced and the deck is shuffled again before the next card is drawn. A mentalist claims that he can use his extrasensory ability to correctly guess the cards. What do you think of his claim if he is able to get at least three of them right?

Solution. To get at the question, we can compute how likely it is to perform so well by simply guessing. If the mentalist is guessing randomly, he has a probability of $\frac{1}{52}$ of getting each draw right, and since cards are replaced and the deck is shuffled, it makes sense to assume that the events of guessing correctly on different draws are mutually independent. There are five disjoint cases that would result in at least three correct cards, which we could describe by

(R, R, R, W), (R, R, W, R), (R, W, R, R), (W, R, R, R), and (R, R, R, R) according to which card in the sequence, if any, was wrongly guessed. Each of these cases constitutes a sequence of independent events. By factorization,

$$
\begin{aligned}
P[(R, R, R, W)] &= \left(\frac{1}{52}\right)^3 \frac{51}{52} \\
&= P[(R, R, W, R)] \\
&= P[(R, W, R, R)] \\
&= P[(W, R, R, R)],
\end{aligned}
$$

and

$$
P[(R, R, R, R)] = \left(\frac{1}{52}\right)^4,
$$

hence,

$$
P[\text{at least 3 right}] = 4 \cdot \left(\frac{1}{52}\right)^3 \frac{51}{52} + \left(\frac{1}{52}\right)^4 = .000028.
$$

Thus, it is highly unlikely that the mentalist can get at least three correct answers by simply guessing. ∎

Example 2. In a group of four stocks whose prices change independently of each other, each goes up this week with probability .55. Find the probability that at least three of them go down this week.

Solution. The applied context is different, but the question turns out to be very similar to the one in Example 1. Since each stock goes up with probability .55, it goes down with probability .45. Let a list of four symbols such as (U, D, D, U) be used to indicate which of the stocks went up and which went down. The stocks behave independently, and so:

$$
\begin{aligned}
P[\text{at least 3 go down}] &= P[\{(D, D, D, U), (D, D, U, D), (D, U, D, D), \\
&\qquad (U, D, D, D), (D, D, D, D)\}] \\
&= P[(D, D, D, U)] + P[(D, D, U, D)] + P[(D, U, D, D)] \\
&\qquad + P[(U, D, D, D)\} + P[(D, D, D, D)] \\
&= 4 \cdot (.45)^3 + (.45)^4 \\
&= .241481. \ ∎
\end{aligned}
$$

Example 3. If events A, B, and C are mutually independent, show that A is also independent of the event $B \cup C$.

Solution. It suffices to verify the factorization condition (3.53) for the events A and $B \cup C$. By distributivity and the general sum rule, we have:

$$
\begin{aligned}
P[A \cap (B \cup C)] &= P[(A \cap B) \cup (A \cap C)] \\
&= P[A \cap B] + P[A \cap C] - P[(A \cap B) \cap (A \cap C)] \\
&= P[A]P[B] + P[A]P[C] - P[A \cap B \cap C] \\
&= P[A]P[B] + P[A]P[C] - P[A]P[B]P[C] \\
&= P[A](P[B] + P[C] - P[B]P[C]) \\
&= P[A](P[B] + P[C] - P[B \cap C]) \\
&= P[A] \cdot P[B \cup C].
\end{aligned}
$$

In line 3, factorization is used on each of the two-way intersections. In the fourth line, the mutual independence of the three events is used to factor the three-way intersection probability. In the sixth line, factorization is used in reverse to reintroduce $P[B \cap C]$ into the expression, which sets up the last line in which the general sum rule is applied to get $P[B \cup C]$. ∎

3.5.2   Independent Random Variables

Independence of random variables is defined along the lines of independence of events. We should have enough intuition built up already to proceed directly to the definition.

Definition 2. Two random variables X and Y are **independent** of each other if, for each subset A of the state space of X and each subset B of the state space of Y, the probability that both variables simultaneously take values in these subsets factors into the product of the individual probabilities, that is:

$$
P[(X \in A) \cap (Y \in B)] = P[X \in A] \cdot P[Y \in B]. \tag{3.55}
$$

Several random variables $X_1, X_2, ..., X_n$, are **mutually independent** if the probability of the intersection of any subcollection of the events $\{X_i \in B_i\}$ is the product of the probabilities of the events in that subcollection, where B_i is an arbitrary subset of the state space of X_i.

As with sets, it is also true that, if X and Y are independent random variables, then

$$
P[Y \in B | X \in A] = P[Y \in B], \text{ and } P[X \in A | Y \in B] = P[X \in A]. \tag{3.56}
$$

Thus, knowing the value taken on by one of the variables does not change the probability law of the other.

We know that the probability distribution of a random variable is characterized either by its probability mass function or its cumulative distribution function. In the case of two jointly distributed random variables, the joint c.d.f. is defined by:

$$F(x, y) = P[(X \leq x) \cap (Y \leq y)]. \qquad (3.57)$$

Of course, the joint probability mass function is:

$$f(x, y) = P[(X = x) \cap (Y = y)]. \qquad (3.58)$$

In view of how independence is defined, it should not be too surprising that the following are equivalent:

(a) X and Y are independent;

(b) The joint p.m.f. of X and Y factors into the product
of the individual p.m.f.'s: $f(x, y) = f(x) \cdot g(y)$; $\qquad (3.59)$

(c) The joint c.d.f. of X and Y factors into the product
of the individual c.d.f.'s: $F(x, y) = F(x) \cdot G(y)$.

We won't show this equivalence in general, but Exercise 10 asks you to prove that (a) implies both (b) and (c).

Example 4. We can reconsider the binomial branch process in the context of independence. Let $R_1, R_2, R_3, \ldots$ be a sequence of mutually independent random variables with common p.m.f.:

$$f(r) = \begin{cases} p & \text{if } r = b; \\ 1 - p & \text{if } r = a. \end{cases} \qquad (3.60)$$

In other words, each R_i takes value b with probability p, and a otherwise. These R_i are the random one-step rates of return in the price process if we construct the process by:

$$X_n = (1 + R_n) X_{n-1}, n = 1, 2, 3 \ldots, \qquad (3.61)$$

since, solving for R_n, we get:

$$X_n = X_{n-1} + R_n X_{n-1} \Longrightarrow R_n = \frac{X_n - X_{n-1}}{X_{n-1}}. \qquad (3.62)$$

The recursive equation (3.61) can be iterated down to the initial time 0 to get:

$$\begin{aligned} X_n &= (1 + R_n) X_{n-1} \\ &= (1 + R_n) (1 + R_{n-1}) X_{n-2} \\ &= (1 + R_n) (1 + R_{n-1}) (1 + R_{n-2}) X_{n-3} \\ &\vdots \qquad \vdots \\ &= (1 + R_n) (1 + R_{n-1}) (1 + R_{n-2}) \cdots (1 + R_1) X_0 \end{aligned} \qquad (3.63)$$

So either formula (3.61) or (3.63) can be viewed as *defining* the binomial branch process in terms of independent random variables R_i, each of which has the two-state distribution given by formula (3.60). ∎

Example 5. A lender has eight loans outstanding of value \$20,000 each. Suppose that defaults on these loans occur independently, each with probability .1. Find the joint distribution of the indicator random variables I_j, where I_j takes the value 1 if the loan defaults and 0 otherwise. Also find the probability distribution of the total number of defaults and the probability that the bank loses at least \$60,000 from defaults.

Solution. We are to assume that the indicator variables I_j are independent, so their joint distribution is the product of their individual distributions. We can write the p.m.f. of a particular indicator, say I_1, as follows:

$$f_1(i_1) = \begin{cases} .1 & \text{if } i_1 = 1 \\ .9 & \text{if } i_1 = 0 \end{cases} = (.1)^{i_1}(.9)^{1-i_1}, i_1 = 0, 1.$$

Defining each f_j analogously, we see that the joint distribution of the indicators has mass function given by:

$$\begin{aligned} f(i_1, i_2, ..., i_8) &= f_1(i_1) \cdot f_2(i_2) \cdots f_8(i_8) \\ &= (.1)^{i_1}(.9)^{1-i_1}(.1)^{i_2}(.9)^{1-i_2} \cdots (.1)^{i_8}(.9)^{1-i_8} \\ &= (.1)^{\sum_{j=1}^{8} i_j}(.9)^{8-\sum_{j=1}^{8} i_j}. \end{aligned}$$

The form of the joint mass function is interesting in the sense that it relates directly to the problem of finding the mass function of the total number of defaults. Any state $(i_1, i_2, ...i_8)$ that has a fixed number k of defaults gives a sum $\sum_{j=1}^{8} i_j = k$ and therefore contributes $(.1)^k(.9)^{8-k}$ to the total probability of k defaults. The number of such states is $\binom{8}{k}$, which is the number of ways of selecting k positions among eight to be default codes (i.e., 1s). So the total probability of exactly k defaults is:

$$P[\# \text{ defaults} = k] = \binom{8}{k}(.1)^k(.9)^{8-k}, k = 0, 1, 2, ..., 8. \qquad (3.64)$$

This is another case of the binomial probability distribution, with $n = 8$ and $p = .1$. Using it, we note that the bank will lose at least \$60,000 if and only if there are at least three defaults. Thus,

$$P[\text{bank loses at least } \$60,000] = P[\text{at least 3 defaults}]$$

$$= 1 - P[\text{2 or fewer defaults}]$$

$$= 1 - \left(\binom{8}{0} (.1)^0 (.9)^8 + \binom{8}{1} (.1)^1 (.9)^7 \right.$$

$$\left. + \binom{8}{2} (.1)^2 (.9)^6 \right)$$

$$= .0381. \blacksquare$$

Example 6. Suppose that an investment of $1000 earns random, independent rates of return R_1, R_2, and R_3 over three time periods, each with the probability mass function:

$$f(-.05) = \frac{1}{4}; f(0) = \frac{1}{2}; f(.05) = \frac{1}{4}.$$

Find the probability distribution and the mean of the final value of the investment.

Solution. There are $3 \cdot 3 \cdot 3 = 27$ possible cases for the final investment value depending on which of the three possible rates of return occurred in each period. We could list out all of the cases (see Exercise 11), but perhaps we can be wiser and combine cases in advance using a bit of combinatorial reasoning. The final investment value will be:

$$V = 1000 \left(1 + R_1 \right) \left(1 + R_2 \right) \left(1 + R_3 \right).$$

The R_is are independent. But it may be more useful to consider the dependent random variables:

$$N_1 = \#R's \text{ taking the value } -.05;$$
$$N_2 = \#R's \text{ taking the value } 0;$$
$$N_3 = \#R's \text{ taking the value } .05.$$

Then N_2 must equal $3 - N_1 - N_3$, and we can rewrite:

$$V = 1000(.95)^{N_1}(1.05)^{N_3}, 0 \le N_1 + N_3 \le 3. \tag{3.65}$$

There are fewer cases, namely, 10, of N_1 and N_3 jointly, summarized by the following, which tables the value of N_2 for each combination of possible values of the other two variables:

		n_3				
		0	**1**	**2**	**3**	
	0	3	2	1	0	
n_1	**1**	2	1	0	–	⟵ value of N_2
	2	1	0	–	–	
	3	0	–	–	–	

This table is an aid to computing the final investment values and their probabilities, which we do below. For example, in line 1 we have the event that $N_2 = 3$, that is, all three rates of return are 0, which by independence occurs with probability $\left(\frac{1}{2}\right)^3$. If this happens, then both N_1 and N_3 are 0, and by formula (3.65), the final value is 1000. In the second line is the case where two of the rates are 0 and one is .05, putting the final value at $1000(1.05)$. This occurs in three different orderings: rates of return $(0, 0, .05)$, $(0, .05, 0)$, and $(.05, 0, 0)$. Each has probability $\left(\frac{1}{2}\right)^2 \left(\frac{1}{4}\right)$, giving a total probability of $3 \cdot \left(\frac{1}{2}\right)^2 \left(\frac{1}{4}\right)$.

(n_1, n_2, n_3)	probability	final investment value
$(0,3,0)$	$P[R_i = 0, \text{all } i]$	1000
	$= \left(\frac{1}{2}\right)^3 = \frac{1}{8}$	
$(0,2,1)$	$P[2R_i = 0, 1R_i = .05]$	$1000(1.05) = 1050$
	$= 3\left(\frac{1}{2}\right)^2 \left(\frac{1}{4}\right) = \frac{3}{16}$	
$(0,1,2)$	$P[1R_i = 0, 2R_i = .05]$	$1000(1.05)^2 = 1102.50$
	$= 3\left(\frac{1}{4}\right)^2 \left(\frac{1}{2}\right) = \frac{3}{32}$	
$(0,0,3)$	$P[R_i = .05, \text{all } i]$	$1000(1.05)^3 = 1157.63$
	$= \left(\frac{1}{4}\right)^3 = \frac{1}{64}$	
$(1,2,0)$	$P[1R_i = -.05, 2R_i = 0]$	$1000(.95) = 950$
	$= 3\left(\frac{1}{2}\right)^2 \left(\frac{1}{4}\right) = \frac{3}{16}$	
$(1,1,1)$	$P[1R_i = -.05, 1R_i = 0, 1R_i = .05]$	$1000(1.05)(.95) = 997.50$
	$= 6\left(\frac{1}{4}\right)^2 \left(\frac{1}{2}\right) = \frac{3}{16}$	
$(1,0,2)$	$P[1R_i = -.05, 2R_i = .05]$	$1000(.95)(1.05)^2 = 1047.38$
	$= 3\left(\frac{1}{4}\right)^3 = \frac{3}{64}$	
$(2,1,0)$	$P[2R_i = -.05, 1R_i = 0]$	$1000(.95)^2 = 902.50$
	$= 3\left(\frac{1}{4}\right)^2 \left(\frac{1}{2}\right) = \frac{3}{32}$	
$(2,0,1)$	$P[2R_i = -.05, 1R_i = .05]$	$1000(.95)^2(1.05) = 947.625$
	$= 3\left(\frac{1}{4}\right)^3 = \frac{3}{64}$	
$(3,0,0)$	$P[R_i = -.05, \text{all } i]$	$1000(.95)^3 = 857.375$
	$= \left(\frac{1}{4}\right)^3 = \frac{1}{64}$	

You should check the rest of the entries, which use similar reasoning. Columns 2 and 3 describe the p.m.f. of the final value as demanded, and we can compute the expectation of this value routinely as:

$$
\begin{aligned}
E[V] &= \frac{1}{8}(1000) + \frac{3}{16}(1050) + \frac{3}{32}(1102.50) + \frac{1}{64}(1157.63) + \frac{3}{16}(950) \\
&\quad + \frac{3}{16}(997.50) + \frac{3}{64}(1047.38) + \frac{3}{32}(902.50) + \frac{3}{64}(947.625) \\
&\quad + \frac{1}{64}(857.375) \\
&= 1000. \ \blacksquare
\end{aligned}
$$

Perhaps you expected the answer of 1000 in the last example, because of the symmetry about 0 of the rates of return. To probe a little deeper, look again at the expression

$$V = 1000 \left(1 + R_1\right)\left(1 + R_2\right)\left(1 + R_3\right)$$

for the final value. If only one period was under consideration, then the expected final value would have been

$$
\begin{aligned}
E[V] &= E\left[1000\left(1 + R_1\right)\right] \\
&= 1000E\left[1 + R_1\right] \\
&= 1000\left(1 + E\left[R_1\right]\right) \\
&= 1000(1 + 0) = 1000.
\end{aligned}
$$

It appears as if this result carries over to two and three periods as well. Suppose that we had a theorem that expected values of products of functions of independent random variables factor into the product of the expectations. Then in the three-period case we would have:

$$
\begin{aligned}
E[V] &= E\left[1000\left(1 + R_1\right)\left(1 + R_2\right)\left(1 + R_3\right)\right] \\
&= 1000E\left[1 + R_1\right]\cdot E\left[1 + R_2\right]\cdot E\left[1 + R_3\right] \\
&= 1000\cdot 1\cdot 1\cdot 1 = 1000,
\end{aligned}
$$

which would explain the result we got.

Fortunately, there is such a theorem, and it will prove useful in a couple of other important ways too.

Theorem 1. Let X and Y be independent random variables, and let g and h be real-valued functions. Then:

$$E[g(X)h(Y)] = E[g(X)]E[h(Y)]. \tag{3.66}$$

This factorization also holds for expectations of products of several functions of mutually independent random variables.

Proof. (case of two random variables only) Let f_x and f_y be the probability mass functions of X and Y, respectively, and let $f(x, y)$ be the joint probability mass function of X and Y. By independence, $f(x, y) = f_x(x)f_y(y)$, as noted above. Then,

$$
\begin{aligned}
E[g(X)h(Y)] &= \sum_x \sum_y g(x)h(y)f(x, y) \\
&= \sum_x \sum_y g(x)h(y)f_x(x)f_y(y) \\
&= \sum_x g(x)f_x(x)\sum_y h(y)f_y(y) \\
&= E[g(X)]E[h(Y)].
\end{aligned}
$$

One of the most important consequences of Theorem 1 is the following, which in the independent case allows us to evaluate a total risk or variability in terms of the components from which it is built.

Theorem 2. Let X and Y be independent random variables, and let a and b be constants. Then:

$$\text{Var}(aX + bY) = a^2\text{Var}(X) + b^2\text{Var}(Y). \qquad (3.67)$$

This decomposition also holds for variances of linear combinations of several mutually independent random variables:

$$\text{Var}\left(\sum_{i=1}^{n} a_i X_i\right) = \sum_{i=1}^{n} a_i^2 \text{Var}(X_i). \qquad (3.68)$$

Proof. (two variable case only) Let X, Y, a, b be as in the hypothesis. By linearity,

$$E[aX + bY] = a\mu_x + b\mu_y,$$

where μ_x and μ_y are the means of X and Y, respectively. Then, by the definition of variance,

$$
\begin{aligned}
\text{Var}(aX + bY) &= E\left[((aX + bY) - (a\mu_x + b\mu_y))^2\right] \\
&= E\left[(a(X - \mu_x) + b(Y - \mu_y))^2\right] \\
&= E[a^2(X - \mu_x)^2 + 2ab(X - \mu_x)(Y - \mu_y) \\
&\quad + b^2(Y - \mu_y)^2] \\
&= a^2 E\left[(X - \mu_x)^2\right] + 2abE\left[(X - \mu_x)(Y - \mu_y)\right] \\
&\quad + b^2 E\left[(Y - \mu_y)^2\right].
\end{aligned}
\qquad (3.69)
$$

The first term on the right side of the last line is $a^2\text{Var}(X)$ and the third term is $b^2\text{Var}(Y)$, so the proof will be complete if we can show that the middle term vanishes. This is where the independence assumption comes into play. By Theorem 1,

$$
\begin{aligned}
E\left[(X - \mu_x)(Y - \mu_y)\right] &= E\left[X - \mu_x\right] E\left[Y - \mu_y\right] \\
&= (E[X] - \mu_x)(E[Y] - \mu_y) \\
&= 0 \cdot 0 = 0,
\end{aligned}
\qquad (3.70)
$$

which finishes the proof.

Example 7. Recall Example 6 from Section 3.3, in which a hotel room renting for \$100 per night is either occupied, with probability 2/3, or unoccupied, with probability 1/3. With a per night discount factor of .0005, we computed the expected present value of the total proceeds during the next year as \$22,236. Let us now assume that the event that a room is rented one night is independent of the event that it is rented another night. Compute the variance and standard deviation of the present value of the total proceeds over a year.

Solution. As we did in the previous example, define X_i to be 1 or 0 according to whether the room is occupied or not on day i. The mean of each X_i is clearly 2/3, and the variance of each X_i is:

$$\text{Var}\left(X_i\right) = E\left[X_i{}^2\right] - E\left[X_i\right]^2 = \frac{2}{3}\cdot 1^2 + \frac{1}{3}\cdot 0^2 - \left(\frac{2}{3}\right)^2 = \frac{2}{9}.$$

The present value of the revenue on day i is $.9995^i\,(\$100X_i)$. By Theorem 2, the variance of total present value over the 365 days is:

$$
\begin{aligned}
\text{Var}\left(\sum_{i=1}^{365}.9995^i\,(100X_i)\right) &= (100)^2\sum_{i=1}^{365}\left(.9995^2\right)^i\text{Var}\left(X_i\right)\\
&= (100)^2\cdot\left(\frac{2}{9}\right)\sum_{i=1}^{365}\left(.9995^2\right)^i\\
&= (100)^2\cdot\left(\frac{2}{9}\right)\left(\frac{1-\left(.9995^2\right)^{366}}{1-.9995^2}-1\right)\\
&= 679,194.
\end{aligned}
$$

The standard deviation is $\sigma = \sqrt{679,194}\approx\$824.13$. If you think back to Chebyshev's inequality, the impact of this computation is that the hotel can expect a large portion of the time that the room will produce a beginning-of-year present value in the interval $\mu\pm 2\sigma = \$22,236\pm 2\cdot 824.13$, or between about \$20,588 and \$23,884. ∎

3.5.3 Dependence: Covariance and Correlation

As you can well imagine, though independence is a convenient simplifying assumption to make, there are many situations in which random variables are not independent of one another. Although more complication ensues, this might actually be a good thing for investment situations. This is because, if one asset tends to go down as another goes up, a combination of the two might be found to reduce overall variability in value, which is the whole idea behind the concept of diversification. Our goal in this last subsection is to quantify the dependence between two random variables in hopes that we can measure risk in a wider variety of situations than Theorem 2 permits.

To begin, consider the two diagrams in Figure 3.21, which are state spaces for two discrete random vectors (X, Y). For simplicity, assume that equal probability mass of $1/5$ is placed on each of the five points (x, y) in each state space. In part (a), the points are so arranged that there is an increasing, roughly linear relationship between the two variables, and in part (b) there is no particular trend that relates X and Y to one another.

In the distribution of part (a) of Figure 3.21, the states are $(-1, -1)$, $(-.5, -.25)$, $(0, 0)$, $(.5, .25)$, and $(1, 1)$. The means of X and Y are:

$$\mu_x = \frac{1}{5}\cdot(-1) + \frac{1}{5}\cdot(-.5) + \frac{1}{5}\cdot(0) + \frac{1}{5}\cdot(.5) + \frac{1}{5}\cdot(1) = 0;$$

$$\mu_y = \frac{1}{5}\cdot(-1) + \frac{1}{5}\cdot(-.25) + \frac{1}{5}\cdot(0) + \frac{1}{5}\cdot(.25) + \frac{1}{5}\cdot(1) = 0.$$

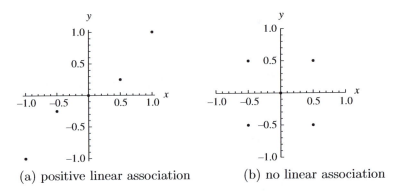

(a) positive linear association   (b) no linear association

Figure 3.21: State spaces of two random variables.

It is easy to check that the means are both 0 for the distribution of part (b) as well. So the two do not differ in their average values. Due to its narrower scale, the distribution in (b) might give smaller variances for both X and Y than the distribution in (a). But that is not the quality that we are really trying to measure. We want to quantify the strength and direction of the relationship between the two variables. This measure should tell us that there is a strong, increasing linear relationship for the distribution in (a), and no relationship for the distribution in (b). The following quantities do this job.

Definition 3. Let X and Y be two random variables with joint p.m.f. $f(x, y)$, means μ_x and μ_y, and standard deviations σ_x and σ_y. The **covariance** between X and Y is:

$$\begin{aligned} \sigma_{xy} = \text{Cov}(X, Y) &= E\left[(X - \mu_x)(Y - \mu_y)\right] \\ &= \sum_x \sum_y (x - \mu_x)(y - \mu_y) f(x, y), \end{aligned} \tag{3.71}$$

where the sum is taken over all joint states (x, y). The **correlation** between X and Y is:

$$\rho = \text{Corr}(X, Y) = \frac{\text{Cov}(X, Y)}{\sigma_x \cdot \sigma_y}. \tag{3.72}$$

To help us understand the meaning of covariance and correlation, let us compute them for the two example distributions in Figure 3.21. Starting with (a), since both means are zero, the expectation that defines the covariance is:

$$\sigma_{xy} = E\left[(X - \mu_x)(Y - \mu_y)\right]$$

$$= E[XY]$$

$$= \frac{1}{5} \cdot (-1)(-1) + \frac{1}{5} \cdot (-.5)(-.25) + \frac{1}{5} \cdot (0)(0)$$

$$+ \frac{1}{5} \cdot (.5)(.25) + \frac{1}{5} \cdot (1)(1)$$

$$= \frac{1}{5}(1 + (-.5)(-.25) + 0 + (.5)(.25) + (1)(1))$$

$$= 9/20.$$

Notice that all terms contributed positively to the covariance, because all products were either both negatives, zero, or both positives. That is, both states were below their means, at their means, or above their means for both random variables. Formula (3.71) suggests that the covariance can therefore indicate the direction of the relationship, if any. If Y tends to be higher (resp. lower) than its mean μ_y at the same time that X is higher (resp. lower) than its mean, then the terms of (3.71) will come out positive, at least if large mass $f(x, y)$ is given to such states. A decreasing relationship will be shown if the products $(x - \mu_x)(y - \mu_y)$ tend to be negative with high probability, meaning that large x corresponds to small y. So the sign of the covariance, positive or negative, seems to indicate the direction of the relationship, increasing or decreasing, respectively.

What about the distribution in (b)? Again the means are both zero, so we can compute:

$$E\left[(X - \mu_x)(Y - \mu_y)\right] = E[XY]$$

$$= \frac{1}{5} \cdot (-.5)(-.5) + \frac{1}{5} \cdot (-.5)(.5) + \frac{1}{5} \cdot (0)(0)$$

$$+ \frac{1}{5} \cdot (.5)(-.5) + \frac{1}{5} \cdot (.5)(.5)$$

$$= \frac{1}{5}((-.5)(-.5) + (-.5)(.5) + (0)(0) + (.5)(-.5)$$

$$+ (.5)(.5))$$

$$= 0.$$

All products of positives times negatives were offset exactly by products of negatives times positives. The random variable Y neither has a tendency to increase with X nor to decrease with it. A covariance of 0 indicates no (linear) relationship. (However, see Exercise 23.)

What does the correlation add that the covariance does not already tell us? By formula (3.72), the correlation between X and Y in distribution (b) must also be zero, indicating neither an increasing nor decreasing relationship. In distribution (a), the variances and standard deviations can be calculated as:

$$\sigma_x{}^2 = E\left[(X-0)^2\right]$$

$$= \frac{1}{5}\cdot(-1)^2 + \frac{1}{5}\cdot(-.5)^2 + \frac{1}{5}\cdot(0)^2 + \frac{1}{5}\cdot(.5)^2 + \frac{1}{5}\cdot(1)^2$$

$$= 1/2;$$

$$\sigma_x = \sqrt{\frac{1}{2}};$$

$$\sigma_y{}^2 = E\left[(Y-0)^2\right]$$

$$= \frac{1}{5}\cdot(-1)^2 + \frac{1}{5}\cdot(-.25)^2 + \frac{1}{5}\cdot(0)^2 + \frac{1}{5}\cdot(.25)^2 + \frac{1}{5}\cdot(1)^2$$

$$= 17/40;$$

$$\sigma_y = \sqrt{\frac{17}{40}}.$$

Therefore the correlation in part (a) is:

$$\rho = \mathrm{Corr}(X,Y) = \frac{9/20}{\sqrt{\frac{1}{2}}\cdot\sqrt{\frac{17}{40}}} \approx .976.$$

Which quantity, the covariance of $9/20 = .45$ or the correlation of about .976, has more meaning as a measure of the strength of the relationship between X and Y? The answer lies in the next theorem; a good measure of association should be invariant to changes in the unit of measurement. The covariance is not invariant, but the correlation is.

Theorem 3. If X and Y are random variables, a and c are positive constants, and b and d are constants, then:

$$\begin{aligned}\mathrm{Cov}(aX+b,cY+d) &= ac\,\mathrm{Cov}(X,Y) \text{ and}\\ \mathrm{Corr}(aX+b,cY+d) &= \mathrm{Corr}(X,Y).\end{aligned} \tag{3.73}$$

You are asked for a proof of this theorem, which is a straightforward application of linearity and algebra, in Exercise 20. Note that taking linear transformations $aX+b$ and $cY+d$ can be looked at as changing the units on X and Y. Displacements b and d do not matter to either measure, which is a good thing, but the covariance is affected by changes of scale a and c. Correlation is not, which makes it a better standard yardstick to measure strength of association.

The following important facts are also true.

(shortcut computational formula)

$$\mathrm{Cov}(X,Y) = E[XY] - \mu_x\mu_y. \tag{3.74}$$

(independence) If X and Y are independent, then

$$\mathrm{Cov}(X, Y) = 0 \text{ and } \mathrm{Corr}(X, Y) = 0. \tag{3.75}$$

(perfect correlation)

$$-1 \leq \rho \leq 1. \tag{3.76}$$

The extreme values of -1 and 1 occur if and only if all of the probability mass of the joint distribution lies on a single line, decreasing or increasing, respectively.

Formula (3.74) is easy to derive by expansion in the definition of covariance:

$$
\begin{aligned}
\mathrm{Cov}(X, Y) &= E\left[(X - \mu_x)(Y - \mu_y)\right] \\
&= E\left[XY - \mu_x Y - \mu_y X + \mu_x \mu_y\right] \\
&= E[XY] - \mu_x E[Y] - \mu_y E[X] + \mu_x \mu_y \\
&= E[XY] - \mu_x \mu_y - \mu_x \mu_y + \mu_x \mu_y \\
&= E[XY] - \mu_x \mu_y.
\end{aligned}
$$

Formula (3.75) was actually already established at the end of the proof of Theorem 2 (see formula (3.70)). We will not prove (3.76) here, but the proof can be found in most probability texts. Since, in our example distribution (a), the correlation came out to .976, which is near 1, fact (3.76) shows that there is a very strong linear relationship between the two random variables, as is apparent from Figure 3.21.

Incidentally, the converse of fact (3.75) is not true. Examples can be constructed of random variables that are not independent, but have correlation equal to 0.

Example 8. Suppose that two risky assets have rates of return as in the table below. Find $\mathrm{Cov}(R_1, R_2)$. If \$500 is invested in each of the two assets to make up a portfolio, find the variance of the profit on that portfolio.

value of (R_1, R_2)	$(-.04, -.04)$	$(0, -.01)$	$(0, .01)$	$(.04, .04)$
probability	$1/8$	$3/8$	$3/8$	$1/8$

Solution. To find the covariance, we will first need to find the individual means of R_1 and R_2. They are:

$$\mu_1 = \frac{1}{8}(-.04) + \frac{3}{8}(0) + \frac{3}{8}(0) + \frac{1}{8}(.04) = 0;$$

$$\mu_2 = \frac{1}{8}(-.04) + \frac{3}{8}(-.01) + \frac{3}{8}(.01) + \frac{1}{8}(.04) = 0.$$

Therefore, by the computational formula (3.74) for covariance,

$$\begin{aligned}
\text{Cov}(R_1, R_2) &= E[R_1 R_2] - \mu_1 \mu_2 \\
&= E[R_1 R_2] \\
&= \frac{1}{8}(-.04)(-.04) + \frac{3}{8}(0)(-.01) + \frac{3}{8}(0)(.01) + \frac{1}{8}(.04)(.04) \\
&= .0004.
\end{aligned}$$

The profit on the portfolio that is described in the problem is the total of the invested amounts times the rates of return, or $X = 500R_1 + 500R_2$. We can calculate the four possible values for each of the possible pairs (r_1, r_2) and use the table to get the probability mass function of X, from which the variance can be calculated. The values of X are:

$$\begin{aligned}
500(-.04) + 500(-.04) &= -40 \\
500(0) + 500(-.01) &= -5 \\
500(0) + 500(.01) &= 5 \\
500(.04) + 500(.04) &= 40.
\end{aligned}$$

By the symmetry of these values and their respective probabilities $\frac{1}{8}, \frac{3}{8}, \frac{3}{8}$, and $\frac{1}{8}$, we have that $E[X] = 0$. The variance of X is then

$$\sigma_x^2 = \frac{1}{8}(-40)^2 + \frac{3}{8}(-5)^2 + \frac{3}{8}(5)^2 + \frac{1}{8}(40)^2 = 418.75. \blacksquare$$

The example above suggests to us the need for a generalized formula to that of Theorem 2 for the variance of a linear combination of random variables, applicable to the general case of correlated random variables. This would allow us to bypass the calculation of the probability mass function of the linear combination. Fortunately, the extension is straightforward.

Theorem 4. Let X and Y be arbitrary random variables, and let a and b be constants. Then:

$$\text{Var}(aX + bY) = a^2 \text{Var}(X) + b^2 \text{Var}(Y) + 2ab\text{Cov}(X, Y). \qquad (3.77)$$

This decomposition also holds for variances of linear combinations of several mutually independent random variables:.

$$\text{Var}\left(\sum_{i=1}^{n} a_i X_i\right) = \sum_{i=1}^{n} a_i^2 \text{Var}(X_i) + \sum_{i=1}^{n} \sum_{j=1, j \neq i}^{n} a_i a_j \text{Cov}(X_i, X_j). \qquad (3.78)$$

Proof. (two variable case only) As derived in the proof of Theorem 2,

$$\text{Var}(aX + bY) = a^2 E\left[(X - \mu_x)^2\right] + 2ab E\left[(X - \mu_x)(Y - \mu_y)\right] + b^2 E\left[(Y - \mu_y)^2\right].$$

We can identify the first and third terms on the right side as $a^2\text{Var}(X)$ and $b^2\text{Var}(Y)$. By definition of covariance, the middle term is $2ab\text{Cov}(X, Y)$, which proves the theorem.

Theorem 4 will be of great use to us in the next chapter when we study the optimal selection of portfolios of risky assets. Already it gives us an alternative way to solve the problem posed in Example 8.

Example 8 (cont.) For the two rates of return R_1 and R_2 above we already know that the means are zero and the covariance is $\sigma_{12} = .0004$. To use Theorem 4 to find the variance of the profit variable $X = 500R_1 + 500R_2$, we will need the variances of R_1 and R_2. They are:

$$\sigma_1{}^2 = \frac{1}{8}(-.04)^2 + \frac{3}{8}(0)^2 + \frac{3}{8}(0)^2 + \frac{1}{8}(.04)^2 = .0004;$$

$$\sigma_2{}^2 = \frac{1}{8}(-.04)^2 + \frac{3}{8}(-.01)^2 + \frac{3}{8}(.01)^2 + \frac{1}{8}(.04)^2 = .000475.$$

Therefore, by formula (3.77),

$$
\begin{aligned}
\text{Var}(X) &= \text{Var}\,(500R_1 + 500R_2) \\
&= (500)^2\text{Var}\,(R_1) + (500)^2\text{Var}\,(R_2) + 2(500)^2\text{Cov}\,(R_1, R_2) \\
&= (500)^2(.0004 + .000475 + 2(.0004)) \\
&= 418.75. \ \blacksquare
\end{aligned}
$$

We close this section with a review of a topic from the previous section: Markov processes.

Example 9. Find the covariance between interest rates R_1 and R_2 on a variable-rate loan in two successive periods if the interest rate process is Markov with the transition matrix below. Assume that the initial rate R_0 is known to be .05.

$$
\begin{array}{cc}
 & \begin{array}{cc} .05 & .08 \end{array} \\
\begin{array}{c} .05 \\ .08 \end{array} & \left(\begin{array}{cc} 1/2 & 1/2 \\ 1/4 & 3/4 \end{array} \right)
\end{array}
$$

Solution. The calculation should be routine, if we can just find the joint distribution of R_1 and R_2. The possible states for each of these random variables are .05 and .08. By the multiplication rule, we have:

$$P\,[R_1 = r_1, R_2 = r_2] = P\,[R_1 = r_1]\,P\,[R_2 = r_2|R_1 = r_1].$$

(The probability measure P here is actually a conditional probability measure given the event $\{R_0 = .05\}$.) Since R_0 is .05, the mass function of R_1 is given by the first line of the transition matrix. The conditional probabilities in the second factor can be read off the transition matrix as well. Doing this for all four joint states (r_1, r_2) gives us:

$$P\left[R_1 = .05, R_2 = .05\right] = P\left[R_1 = .05\right] P\left[R_2 = .05 \,|\, R_1 = .05\right] = \frac{1}{2} \cdot \frac{1}{2} = \frac{1}{4};$$

$$P\left[R_1 = .05, R_2 = .08\right] = P\left[R_1 = .05\right] P\left[R_2 = .08 \,|\, R_1 = .05\right] = \frac{1}{2} \cdot \frac{1}{2} = \frac{1}{4};$$

$$P\left[R_1 = .08, R_2 = .05\right] = P\left[R_1 = .08\right] P\left[R_2 = .05 \,|\, R_1 = .08\right] = \frac{1}{2} \cdot \frac{1}{4} = \frac{1}{8};$$

$$P\left[R_1 = .08, R_2 = .08\right] = P\left[R_1 = .08\right] P\left[R_2 = .08 \,|\, R_1 = .08\right] = \frac{1}{2} \cdot \frac{3}{4} = \frac{3}{8}.$$

The individual means of R_1 and R_2 are:

$$\mu_1 = \frac{1}{4} \cdot (.05) + \frac{1}{4} \cdot (.05) + \frac{1}{8} \cdot (.08) + \frac{3}{8} \cdot (.08) = \frac{1}{2} \cdot (.05) + \frac{1}{2} \cdot (.08) = .065;$$

$$\mu_2 = \frac{1}{4} \cdot (.05) + \frac{1}{4} \cdot (.08) + \frac{1}{8} \cdot (.05) + \frac{3}{8} \cdot (.08) = \frac{3}{8} \cdot (.05) + \frac{5}{8} \cdot (.08) = .06875.$$

Therefore, by the computational formula, the covariance between R_1 and R_2 is:

$$
\begin{aligned}
\operatorname{Cov}\left(R_1, R_2\right) &= E\left[R_1 R_2\right] - \mu_1 \mu_2 \\
&= \frac{1}{4} \cdot (.05)(.05) + \frac{1}{4} \cdot (.05)(.08) + \frac{1}{8} \cdot (.08)(.05) \\
&\quad + \frac{3}{8} \cdot (.08)(.08) - (.065)(.06875) \\
&= .00005625. \ \blacksquare
\end{aligned}
$$

Important Terms and Concepts

Independent events - If events are independent of each other, then the probability of their intersection factors into the product of their individual probabilities.

Independent random variables - If random variables are independent of each other, then the joint probability that they fall into arbitrary subsets of their state spaces factors into the product of the individual probabilities.

Independence and the joint p.m.f. - If random variables are independent of each other, then the joint mass function factors into the product of the individual mass functions. The same holds for the cumulative distribution functions.

Covariance of two random variables - The quantity $E\left[(X - \mu_x)(Y - \mu_y)\right]$ measuring the degree of linear association between X and Y. But it is sensitive to the units of measurement. A negative value indicates negative association between X and Y, and a positive value indicates positive association.

Correlation between two random variables - The ratio $\frac{\text{Cov}(X,Y)}{\sigma_x \sigma_y}$, which also measures linear association, but does not depend on units of measurement. Its value must lie between -1 and 1, and it is 0 when the random variables are independent.

Expectation, independence, and dependence - When X and Y are independent, $E[g(X)h(Y)] = E[g(X)]E[h(Y)]$, from which it can be shown that $\text{Var}(aX + bY) = a^2\text{Var}(X) + b^2\text{Var}(Y)$. If X and Y are dependent, these formulas are not true, although $\text{Var}(aX + bY) = a^2\text{Var}(X) + b^2\text{Var}(Y) - 2a\,b\text{Cov}(X, Y)$.

Exercises 3.5

1. Two fair six-sided dice are rolled in such a way that each possible pair of faces (x, y) has likelihood $1/36$. Show that: (a) the event that 3 is rolled on the first die is independent of the event that 4 is rolled on the second; (b) the event that 2 is rolled on the first is not independent of the event that the sum of the two dice is 9.

2. Two stocks, A and B, are monitored over a period of 108 trading days. The number of times that they went up together, down together, or one went up and the other went down is counted. The table below displays these frequencies. If one trading day is selected at random from the 108 recorded here, is the event that stock A went up independent of the event that stock B went up? Is the event that A went up independent of the event that B went down?

		stock B	
		up	down
stock A	up	20	40
	down	16	32

3. Show that the empty set $\emptyset$ and the whole sample space Ω are independent of any other event.

4. Let events A, B, and C be mutually independent. Suppose that $P[A] = \frac{1}{4}$, $P[B] = \frac{1}{3}$, and $P[C] = \frac{1}{2}$. Find: (a) $P[A \cap B | C]$; (b) $P[A | B \cap C]$; and (c) $P[B \cup C | A]$.

5. Prove that if A and B are independent, then so are A^c and B^c.

6. Grandma Phyllis has a list of eight companies from which she will choose four to give 25% each of the fabulous wealth that she has stashed in the coffee can in her refrigerator. If she selects four companies at random and allows herself to reselect companies she has already chosen (such companies would get extra shares), is the event that company 1 is selected first independent of the event that company 5 is selected last? What if she excludes companies she has already picked; are these two events independent?

7. Two stocks act independently of each other, and transitions in separate time periods are also independent. Stock A goes up in value in a day by 1 with probability .6, otherwise it goes down by 1. Similarly, Stock B goes up in value by 1 with probability .5, otherwise it goes down by 1. If the pair starts at (10,10), what possible combined values are there two days from now, and with what probabilities do they occur?

8. A bank has five loans to families in a particular demographic category, in amounts $1000, $3000, $5000, $7000, and $9000. The probabilities of default for the loans are, respectively, .001, .01, .02, .04, and .05.   If default events are mutually independent, find the probability that the total loss for the bank is at least $9000.

9. A 3-year municipal bond of $10,000 is designed to pay semi-annual coupons at 4% per year and have a yield rate of 5% nominal per year. But there is a probability of .1 for each coupon that the coupon will not be paid. If non-payment events are independent, how should the bond be priced, and how different is that price from the case where the coupons are paid with certainty?

10. In the several random variables case, show that mutual independence implies factorization of both the joint p.m.f. and the joint c.d.f. into the products of the individual p.m.f's and c.d.f.'s of the random variables.

11. List all outcomes in Example 6 with their probabilities of occurrence.

12. Following Example 6, we commented on independence and the expected product $E\left[1000\left(1 + R_1\right)\left(1 + R_2\right)\left(1 + R_3\right)\right]$, assuming that the R_is were mutually independent. Show in general that if random variables X and Y are independent of each other, then $1 + X$ and $1 + Y$ are also independent of each other.

13. Find the distribution of the price X_n after n transitions in a general binomial branch process with initial value X_0, up probability p, and up and down rates b and a. For the particular case where the initial price is $X_0 = \$50$, $p = .4$, $b = .01$, and $a = -.005$, compute the probability that the price exceeds $51.50 at time 5.

14. Ralph sets up an annuity in which he is to contribute $200 at the end of each month for a total of 20 years. He receives an effective monthly rate of .8%. But

sometimes he runs out of money and does not contribute. This happens about $1/20^{\text{th}}$ of the time, and successive occurrences of this event can be assumed to be mutually independent. Find the mean, variance, and standard deviation of the future value of his annuity.

15. Three risky assets are such that the variances of their rates of return R_1, R_2, and R_3 are .1, .3, and .5, respectively. If the rates of return are independent random variables, find the variance of each of: (a) $\frac{1}{2}R_1 + \frac{1}{2}R_2$; (b) $\frac{1}{8}R_1 + \frac{3}{8}R_2 + \frac{1}{2}R_3$; (c) $R_1 - R_2 + R_3$.

16. The rates of return R_1 and R_2 in two successive time periods for an investment of \$2000 are independent random variables with means .05 each, and variances .2 each. Find the mean and variance of the value of this investment after these two time periods.

17. A two-variable probability distribution puts equal mass on the four points in the figure: $(-2, 1)$, $(0, 0)$, $(1, -1)$, and $(2, -1.5)$. Compute the covariance and correlation of the two random variables X and Y with this distribution.

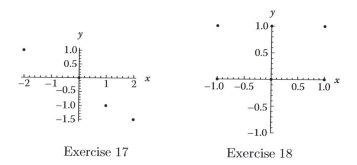

Exercise 17   Exercise 18

18. A two-variable probability distribution puts equal mass on the six points in the figure: $(-1, 0)$, $(0, 0)$, $(1, 0)$, $(-1, 1)$, $(0, 1)$, and $(1, 1)$. Compute the covariance and correlation of the two random variables X and Y with this distribution.

19. For the state space shown in Figure 3.21(b), suppose that the probability masses on the points $(-.5, .5)$ and $(.5, -.5)$ are just .05 each, and the other three points each has probability .3. Recompute the covariance and correlation. Explain what has happened.

20. Prove the change of units formulas (3.73).

21. In Exercise 15 above, instead of independence of the rates of return R_1, R_2,

and R_3, assume that R_1 is independent of R_3, R_2 is also independent of R_3, but the correlation between R_1 and R_2 is .8. Redo the calculations of variances in each of parts (a), (b), and (c).

22. Assume that indicator random variables I_1, I_2, and I_3 for default on three loans have the joint p.m.f. tabulated below. Find all of Cov (I_1, I_2), Cov (I_1, I_3), and Cov (I_2, I_3).

(i_1, i_2, i_3)	$(0,0,0)$	$(0,0,1)$	$(0,1,0)$	$(1,0,0)$
probability	.6	.1	.15	.05

(i_1, i_2, i_3)	$(1,1,0)$	$(1,0,1)$	$(0,1,1)$	$(1,1,1)$
probability	.04	.03	.02	.01

23. Suppose that two random variables X and Y have possible joint values $(-1, 1)$, $(0, 0)$, and $(1, 1)$, occurring with equal probability. Are X and Y independent? Find the covariance between X and Y. What does this exercise illustrate?

24. Find the correlation between the interest rate variables R_1 and R_2 in Example 9.

25. Prove Theorem 4 in the case of three random variables X_1, X_2, and X_3. Use the result to compute Var $(X_1 - 2X_2 + 3X_3)$ in the case where the individual variances are 4, 1, and 9, respectively, and the covariances are Cov $(X_1, X_2) = 1$, Cov $(X_1, X_3) = -1.8$, and Cov $(X_2, X_3) = -1.5$.

26. In Example 8, does the strategy of diversifying the $1000 investment by splitting it evenly between the two assets seem to help the investor or not? Explain.

27. In Example 9, suppose that R_0, R_1, and R_2 are the interest rates on a loan at simple interest of $1000 over the first three periods. Find the mean and variance of the total interest that is assessed.

28. What is the covariance of a random variable with itself? The correlation of a random variable with itself? How does this simplify formula (3.78)?

3.6   Estimation and Simulation

In all of the previous sections we have illustrated financial situations into which randomness entered, and we were frequently concerned with means, variances, and covariances of groups of random variables such as interest rates on loans or accounts, rates of return on risky assets, and indicators of default on loans or bonds. But we have had to assume that we know the parameters of these random phenomena. Therefore, an important question is: how do we use historical data to estimate parameters? Moreover, in some financial situations, such as the valuation of certain kinds of options to be studied in Chapter 5, the quantity that we must model is some function $h(X_0, X_1, ..., X_n)$ of the prices through time n of an underlying asset. Historical data may not be available, but we may be able to **simulate** the price path a large number of times in order to get an estimate of the important features of this function (such as its expected value). In this spirit, we will close the chapter with a brief treatment of estimation of the parameters of a probability distribution using random samples of real-world data, and an illustration of algorithms for simulating random systems and thereby estimating their key characteristics.

3.6.1   The Sample Mean

To begin, we call a sequence of random variables $X_1, X_2, ..., X_n$ a **random sample of size n from a probability distribution** if the Xs are mutually independent and are identically distributed with the given distribution. The abbreviation "i.i.d." for "independent and identically distributed" is in common use. For example, if we repeatedly choose a number randomly from the universe $\{1, 2, 3, 4\}$ and allow reuse of numbers, then, under the assumption that all lists of n numbers have the same likelihood, the random variables X_i, $i = 1, 2, 3, 4$ that record the numbers that were chosen are independent and identically distributed, and hence they form a random sample. (See Exercise 1.)

A caution is in order right away. In the small example of the last paragraph, we were literally sampling at random, basically picking numbers out of a hat, replacing and mixing before sampling again. But in financial situations that commonly arise, you are only able to obtain a list of data values after the fact; no randomization has been prearranged. It can be a severe assumption that such a list actually constitutes the observed values of an i.i.d. sequence of random variables. One price or rate of return may well depend upon its predecessors, and the probability distribution may not be stable as time passes. There are more advanced statistical techniques to deal with some of these difficulties, but in this book we will start with the basics.

If we are interested in estimating the mean μ of a probability distribution, and a random sample $X_1, X_2, ..., X_n$ from that distribution is available, then a reasonable estimator is the most obvious one.

Definition 1. The **sample mean** $\bar{X}$ of a random sample of size n is the arithmetical average of the sample values:

$$\bar{X} = \frac{1}{n} \sum_{i=1}^{n} X_i. \tag{3.79}$$

Since the distributional mean μ is a kind of weighted average of the possible values of a generic random variable X with the given distribution, it certainly makes intuitive sense that the average of randomly sampled values from the distribution should estimate μ well. But we can actually be more precise about the good qualities of $\bar{X}$ as an estimator. By linearity of expectation,

$$E\left[\bar{X}\right] = E\left[\frac{1}{n}\sum_{i=1}^{n}X_i\right] = \frac{1}{n}\sum_{i=1}^{n}E\left[X_i\right] = \frac{1}{n}\sum_{i=1}^{n}\mu = \frac{1}{n}\cdot n\cdot\mu = \mu. \qquad (3.80)$$

The fact that $E\left[\bar{X}\right] = \mu$ is a profound property called **unbiasedness**. We say that the sample mean is an **unbiased estimator** of the distributional mean. This requires some reflection to truly understand. The sample values X_i are random variables; they cannot be predicted in advance and can only be known after the data is observed. Thus, the sample mean $\bar{X}$, being a function of them, is also a random variable. It has its own probability distribution, and unbiasedness says that the mean of the distribution of $\bar{X}$ is the same as μ, the mean of the distribution from which the sample was taken. We might say that $\bar{X}$ is "correctly aimed"; its average value is the quantity that we are trying to estimate. The sample mean neither consistently underestimates nor overestimates μ. You should reread this paragraph a few times to be sure you understand this concept.

Another valuable property of the sample mean is true: it has increasing **precision** as the sample size increases. To see this, let us compute $\text{Var}\left(\bar{X}\right)$ using the properties of variance and the assumption of independence of the sample values X_i. Assume that the variance of the distribution being sampled from is σ^2. Then:

$$\begin{aligned}
\text{Var}\left(\bar{X}\right) &= \text{Var}\left(\frac{1}{n}\sum_{i=1}^{n}X_i\right) \\
&= \frac{1}{n^2}\sum_{i=1}^{n}\text{Var}\left(X_i\right) \\
&= \frac{1}{n^2}\sum_{i=1}^{n}\sigma^2 \\
&= \frac{1}{n^2}\cdot n\cdot\sigma^2 = \frac{\sigma^2}{n}.
\end{aligned} \qquad (3.81)$$

The presence of n in the denominator is crucial. This says that, as n grows, the variance of the estimator $\bar{X}$ approaches 0, so that most of the probability mass of the distribution of $\bar{X}$ is concentrated tightly around μ. So, for large sample sizes, the sample mean is very likely to take a value close to the true μ.

Example 1. Below are data on 26 consecutive weekly price values for Kellogg Company stock on the New York Stock Exchange (ticker symbol "K") beginning in January 2014. You may find such data in print newspapers, although it is far easier now to download it from financial sites such as Yahoo Finance.

58.82, 59.01, 58.71, 58.09, 56.3, 56.66, 58.44, 58.21, 58.93, 60.27, 59.58, 60.25, 60.79, 62.39, 64.26, 64.9, 65.27, 64.18, 65.76, 66.08, 66.1, 67.95, 67.88, 66.2, 66.23, 65.44

The associated 25 weekly rates of return can be calculated as:

$$\frac{\text{next week price} - \text{current week price}}{\text{current week price}}.$$

These are displayed below. For example, for the first week, the rate of return on Kellogg stock would be $(59.01 - 58.82)/58.82 \approx .00323$.

0.00323, −0.005084, −0.010560, −0.030814, 0.006394, 0.031416, −0.003936, 0.012369, 0.022739, −0.011449, 0.011245, 0.008963, 0.026320, 0.029973, 0.009960, 0.005701, −0.016700, 0.024618, 0.004866, 0.000303, 0.027988, −0.001030, −0.024750, 0.000453, −0.011928

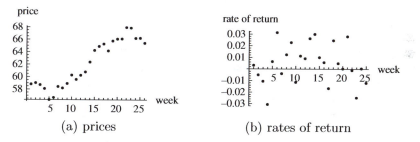

(a) prices   (b) rates of return

Figure 3.22: Time series plots of weekly Kellogg data, January-June 2014.

We can graph both the price data and the rate of return data as a function of week, as in Figure 3.22. In part (a) of the figure, it is apparent that there is a trend in price, so that the price list $X_1, X_2, ..., X_{26}$ cannot be considered to be a random sample. But in part (b), the rate of return sequence $R_1, R_2, ..., R_{25}$ is relatively featureless, suggesting no time trends nor changes in location or spread as the weeks go by. Roughly, we do not see obvious red flags resulting from treating the Rs as a random sample. There are of course statistical techniques to check more carefully for lack of independence or shifts in the probability distribution, but we do not discuss those here.

Figure 3.23 shows a data histogram of the rate of return list. The bar heights are the proportions of the 25 observations falling into each of the intervals of length .01. This gives us a visual estimate of the distribution of a generic rate of return random variable R for Kellogg during this time period. Its center seems to be slightly above 0. We can compute that center by calculating the sample mean $\bar{R}$ of the rate of return sample:

$$\bar{R} = \frac{0.00323 + (-0.005084) + \cdots + (-0.011928)}{25} \approx .00441.$$

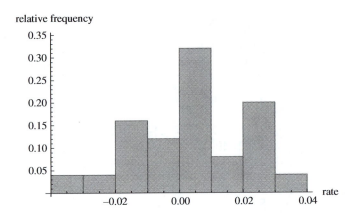

Figure 3.23: Data histogram of weekly Kellogg rates of return, January–June 2014.

   This means that we have an unbiased estimate, $\mu \approx \bar{R} \approx .00441$, of the mean of the distribution of weekly rates of return on Kellogg during this period. How good is that estimate? Remember that the random variable $\bar{R}$ has mean μ and standard deviation $\sigma / \sqrt{n} = \sigma / \sqrt{25} = \sigma / 5$, where σ is the standard deviation of the underlying distribution of rates of return. We would like to be able to cite Chebyshev's inequality, which says that most of the probability mass of the distribution of $\bar{R}$ lies within 2 standard deviation units of its mean, μ. But we don't know σ yet. We will have a better way of estimating σ shortly, but for now note that Figure 3.23 indicates that the highest and lowest values of the distribution of the rate of return are roughly .04 units from the mean. We could estimate that 2 standard deviations of R are about .04, so that:

$$2\sigma \approx .04 \Longrightarrow \sigma \approx .02.$$

In terms of the sample mean $\bar{R}$, its standard deviation would be about $\sigma / 5 = .004$, so two standard deviations of $\bar{R}$ would be about .008. We are highly confident that the true mean rate of return μ is in the range $.00441 \pm .008$, which is still not a very precise estimate. A larger sample size could help, although then we incur the risk that the distribution of the rate of return could have shifted over a longer time frame. In general, estimation of the parameters of a rate of return distribution is fraught with peril. ■

Example 2. In situations where the state space of the probability distribution of interest consists only of the values 0 and 1, as in the case of an indicator random variable of default on a loan, the distributional mean and sample mean take on an interesting form. Suppose that a bank is interested in the chance of default on a certain class of loan, and it consults its historical records and finds the following background data:

$$0, 0, 0, 1, 0, 0, 0, 0, 0, 0, 0, 1, 0, 0, 0, 0, 0, 0, 0, 0, 0, 1, 0,$$
$$0, 0, 0, 0, 0, 0, 0, 0, 1$$

Here the 0s indicate a loan that did not default, and the 1s stand for a default. There are 30 observations in total, which we boldly assume constitute a random sample from a distribution with p.m.f.:

$$f(x) = \begin{cases} p & \text{if } x = 1; \\ 1 - p & \text{if } x = 0. \end{cases} \qquad (3.82)$$

So the parameter p is the probability of default. We have seen such two-point distributions before, and it is easy to check that:

$$\mu = E[X] = p; \;\; \sigma^2 = \text{Var}(X) = p(1 - p). \qquad (3.83)$$

In this case the mean μ of the distribution is the probability parameter p. The sample mean $\bar{X} = \sum_{i=1}^{n} X_i / n$ consists of the total of a list of 1s and 0s, divided by n, which is nothing more than the number of 1s divided by the size of the list. We call this the **sample proportion** of 1s, and denote it by:

$$\hat{p} = \frac{\text{total } \#1\text{s}}{\text{sample size}} = \bar{X}. \qquad (3.84)$$

The estimation of the default probability p by the sample proportion of defaults $\hat{p}$ (which is $4/30 = 2/15$ for this data) is just an instance of the estimation of μ by $\bar{X}$. Thus, $\hat{p}$ is an unbiased estimator of p and its variance is:

$$\text{Var}(\hat{p}) = \frac{\sigma^2}{n} = \frac{p(1 - p)}{n}. \qquad (3.85)$$

The variance of the sample proportion approaches 0 as the sample size becomes large, which means once again that $\hat{p}$ becomes more and more likely to be close to p as the sample size increases. ∎

The convergence of the sample mean to the distributional mean as the sample size goes to infinity is a major theorem in probability called the **Law of Large Numbers**. More is true than we have implied so far: for an open-ended sample $X_1, X_2, X_3, ...$, not only is the sample mean of the first n values likely to be close to μ, but in fact with probability 1 it will converge to μ as more and more sample values are included in the calculation. To illustrate this, we can simulate values from a given distribution whose mean we know, and plot the sequence of updated sample means as n grows. Specifically, let us apply the idea to a random sample from the distribution that places equal mass of $\frac{1}{5}$ on all points in the set $\{1, 2, 3, 4, 5\}$. The mean μ is clearly 3. The algorithm that we use is:

1. Simulate the first value from the distribution;

2. Begin a running sum with this value;

3. Initialize the first sample mean $\bar{X}_1$ as the first sample value and place it into the front of a list of means;

4. While more sample values are desired,

    (a) Simulate a new sample value;

    (b) Add the new sample value to the running sum;

    (c) Increment the sample size n;

    (d) Recalculate $\bar{X}$, including the new value, and adjoin it to the list of sample means;

5. Return the list of sample means.

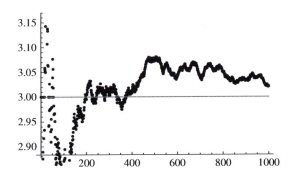

Figure 3.24: Plot of updated sample means; sample of 1000 from uniform distribution on $\{1, 2, 3, 4, 5\}$.

In Figure 3.24 we see the result of such a simulation of 1000 data values using a program in *Mathematica*. The Law of Large Numbers says that any time you run such an experiment, the sample means will approach the distributional mean of 3, although it makes no guarantees about how long that might take.

3.6.2   Sample Variance, Covariance, and Correlation

The mean is certainly not the only parameter of a probability distribution in need of estimation. Among all applications, finance is particularly concerned with modeling and estimating risk, and variance is one measure of that risk.

Again let $X_1, X_2, ... X_n$ be a random sample from a distribution with mean μ and variance σ^2. The distributional variance is defined by $\sigma^2 = E\left[(X - \mu)^2\right]$ in order to summarize an average or expected square distance of an observation from the mean of its distribution. To estimate σ^2 from a random sample, we imitate this idea.

Definition 2. The **sample variance** S^2 of a random sample of size n is the quantity:

$$S^2 = \frac{1}{n-1} \sum_{i=1}^{n} (X_i - \bar{X})^2. \tag{3.86}$$

The **sample standard deviation** is the square root of the sample variance: $S = \sqrt{S^2}$.

This is a kind of average square difference between the sample values and the sample mean, and so it has the same nature as the distributional variance. But why divide by $n-1$ rather than by n, as we would do in a true average? The answer is that, by defining the sample variance as in formula (3.86), we allow S^2 to be an unbiased estimate of σ^2, that is:

$$E\left[S^2\right] = \sigma^2. \tag{3.87}$$

If we had used n in the denominator, the average would have been slightly smaller, so that the sample estimate would have systematically underestimated the parameter σ^2. You are led through a proof of the unbiasedness of S^2 in Exercise 15.

Example 3. Referring back to Example 2, we had a random sample of size 30 from the two-point distribution in Formula (3.82). Since four of the observations were equal to 1 and the rest were 0, the unbiased estimator $\hat{p}$ of the probability p was $4/30 = .1333$. Now the distributional variance for this probability distribution is $\sigma^2 = p(1-p)$. For the random sample that was given, since there were 26 0s and 4 1s in the sample, we can calculate the sample variance by:

$$
\begin{aligned}
S^2 &= \frac{1}{29}\left(\left(0 - \frac{4}{30}\right)^2 + \left(0 - \frac{4}{30}\right)^2 + \cdots + \left(1 - \frac{4}{30}\right)^2 + \right) \\
&= \frac{1}{29}\left(26\left(0 - \frac{4}{30}\right)^2 + 4\left(1 - \frac{4}{30}\right)^2\right) = .11954.
\end{aligned}
$$

This is one way of estimating σ^2; another is to estimate $p(1-p)$ by $\hat{p}\,(1-\hat{p}) = \frac{4}{30} \cdot \frac{26}{30}$. This product turns out to be about .11556, so the two estimates are similar. We had also determined that $\text{Var}\,(\hat{p}) = \frac{\sigma^2}{n} = \frac{p(1-p)}{n}$, so we can estimate the variance and standard deviation of the estimator $\hat{p}$ by:

$$\text{Var}\,(\hat{p}) \approx \frac{\hat{p}\,(1-\hat{p})}{n} = \frac{.1333(1 - .1333)}{30} = .00385;$$

$$\sigma_{\hat{p}} = \sqrt{.00385} = .06206.$$

You can check that, if we had used $S^2 = .11954$ to estimate σ^2 in the formula $\frac{\sigma^2}{n}$, we would have obtained a similar estimate of $\text{Var}\,(\hat{p})$ of .00398. ∎

As with the distributional variance, the sample variance has an equivalent form that is sometimes easier to compute. It is:

$$S^2 = \frac{1}{n-1}\left(\sum_{i=1}^{n}X_i^2 - n\cdot\bar{X}^2\right),\tag{3.88}$$

which Exercise 14 asks you to prove.

The standard deviation of an estimator like $\bar{X}$ or $\hat{p}$ is an important quantity to measure the variability of that estimator and its precision as an estimate of the corresponding distributional parameter. (Remember that, by Chebyshev's inequality, a random variable like these is quite likely to be within 2 of its standard deviations from its mean.) This standard deviation goes by the name of the **standard error** of the estimate. For $\bar{X}$, the standard error is $\sqrt{\sigma^2/n} = \sigma/\sqrt{n}$, and for $\hat{p}$ the standard error is $\sqrt{\frac{p(1-p)}{n}}$. These standard errors involve distributional parameters that are usually unknown, however. But we can estimate them. Doing so gives:

$$\text{approximate standard error of } \bar{X} = \frac{S}{\sqrt{n}};\tag{3.89}$$

$$\text{approximate standard error of } \hat{p} = \sqrt{\frac{\hat{p}(1-\hat{p})}{n}}.\tag{3.90}$$

Example 4. In Example 1, we looked at 25 weekly rates of return on Kellogg stock, reproduced below.

0.00323, −0.005084, −0.010560, −0.030814, 0.006394, 0.031416,
−0.003936, 0.012369, 0.022739, −0.011449, 0.011245, 0.008963,
0.026320, 0.029973, 0.009960, 0.005701, −0.016700, 0.024618,
0.004866, 0.000303, 0.027988, −0.001030, −0.024750, 0.000453,
−0.011928

The sample mean of these was $\bar{R} \approx .00441$, and we were concerned about the precision of this estimate of the distributional mean. We can compute the sample variance using this data:

$$
\begin{aligned}
S^2 &= \frac{1}{24}\sum_{i=1}^{n}\left(X_i - \bar{X}\right)^2 \\
&= \frac{1}{24}((.00335 - .00441)^2 + (-.00501 - .00441)^2 + \cdots \\
&\quad + (-.01205 - .00441)^2) \\
&= .000285.
\end{aligned}
$$

Then the approximate standard error of $\bar{R}$ is $\frac{S}{\sqrt{n}} = \frac{\sqrt{.000285}}{\sqrt{25}} = .00337$. Two standard errors of .0067 is an improvement upon the very rough estimate of .008 obtained earlier. ∎

Besides estimating the variability of individual random variables, it is important to be able to estimate the degree of association between two random variables. In Section 3.5 we defined the distributional covariance and correlation between random variables X and Y as:

$$\sigma_{xy} = \text{Cov}(X,Y) = E\left[(X - \mu_x)(Y - \mu_y)\right] = \sum_x \sum_y (x - \mu_x)(y - \mu_y) f(x,y),$$

$$\rho = \text{Corr}(X,Y) = \frac{\text{Cov}(X,Y)}{\sigma_x \cdot \sigma_y}.$$

It is intuitively reasonable to define a **sample covariance** to be an average of products $(X_i - \bar{X})(Y_i - \bar{Y})$ using a sample of pairs $(X_1, Y_1), ..., (X_n, Y_n)$ from a joint distribution with mass function $f(x,y)$. The **sample correlation** would then be the sample covariance divided by the product of the sample standard deviations of the individual variables. We define these formally now, and we show a slightly simplified form of the sample correlation.

Definition 3. Let $(X_1, Y_1), (X_2, Y_2), ... (X_n, Y_n)$ be a random sample from a joint distribution. The **sample covariance** between X and Y is:

$$S_{xy} = \frac{1}{n-1} \sum_{i=1}^{n} (X_i - \bar{X})(Y_i - \bar{Y}), \qquad (3.91)$$

where $\bar{X}$ and $\bar{Y}$ are the means of the individual samples $X_1, X_2, ..., X_n$ and $Y_1, Y_2, ..., Y_n$, respectively. The **sample correlation** between X and Y is:

$$R = \frac{S_{xy}}{S_x \cdot S_y} = \frac{\sum_{i=1}^{n} (X_i - \bar{X})(Y_i - \bar{Y})}{\sqrt{\sum_{i=1}^{n} (X_i - \bar{X})^2} \cdot \sqrt{\sum_{i=1}^{n} (Y_i - \bar{Y})^2}}, \qquad (3.92)$$

where S_x and S_y are the individual sample standard deviations.

In Formula (3.92), both S_{xy} and $S_x \cdot S_y$ have denominators of $n - 1$, which divide away to yield the rightmost expression. As with the distributional covariance and correlation, these quantities will tend to be large and positive if the Y sample values exceed their sample mean when the X values exceed theirs, and negative if there is a decreasing relationship between the Ys and their corresponding Xs. It is also true that

$$|R| \leq 1, \text{ and } |R| = 1 \text{ only if all sample points lie perfectly on a line.} \qquad (3.93)$$

Hence R is a normalized measure of the strength of the linear relationship between sample X and Y observations, and its sign indicates the direction of that relationship. The sample covariance also has a convenient simplified formula that you are asked to prove in Exercise 20:

$$S_{xy} = \frac{1}{n-1} \left(\sum_{i=1}^{n} X_i Y_i - n \cdot \bar{X} \cdot \bar{Y} \right). \tag{3.94}$$

Example 5. To illustrate the definitions, let us make a couple of quick calculations for small data sets. We will compute the sample covariance and correlation if the data is: (a) $(0, 2), (1, .75), (2, .5)$; (b) $(-1, 0), (0, 1), (1, 2), (2, 3)$.

Solution. (a) We will need the sample means, variances, and standard deviations first. This time we will use the computational formula (3.88) for the sample variance:

$$\bar{X} = \frac{0+1+2}{3} = 1; \bar{Y} = \frac{2+.75+.5}{3} = 1.08333;$$

$$S_x^2 = \frac{1}{2} \left(0^2 + 1^2 + 2^2 - 3 \cdot 1^2 \right) = 1;$$

$$S_y^2 = \frac{1}{2} \left(2^2 + .75^2 + .5^2 - 3 \cdot 1.08333^2 \right) = .645844;$$

$$S_x = \sqrt{1} = 1; S_y = \sqrt{.645844} = .803644.$$

Now we compute the average product, using the original definition of the covariance.

$$
\begin{aligned}
S_{xy} &= \frac{1}{n-1} \sum_{i=1}^{n} \left(X_i - \bar{X} \right) \left(Y_i - \bar{Y} \right) \\
&= \frac{1}{2} ((0-1)(2 - 1.08333) + (1-1)(.75 - 1.08333) \\
&\quad + (2-1)(.5 - 1.08333)) \\
&= -.75.
\end{aligned}
$$

This implies that the sample correlation is:

$$R = \frac{S_{xy}}{S_x \cdot S_y} = \frac{-.75}{1 \cdot .803644} = -.93325.$$

We see that these two samples are highly negatively correlated.

(b) Doing similar computations for the data set of size 4 gives:

$$\bar{X} = \frac{-1+0+1+2}{4} = \frac{1}{2}; \bar{Y} = \frac{0+1+2+3}{4} = \frac{3}{2};$$

$$S_x^2 = \frac{1}{3} \left((-1)^2 + 0^2 + 1^2 + 2^2 - 4 \cdot \left(\frac{1}{2} \right)^2 \right) = \frac{5}{3};$$

$$S_y^2 = \frac{1}{3} \left(0^2 + 1^2 + 2^2 + 3^2 - 4 \cdot \left(\frac{3}{2} \right)^2 \right) = \frac{5}{3};$$

$$S_x = \sqrt{5/3} = S_y.$$

The sample covariance, by formula (3.94), is

$$
\begin{aligned}
S_{xy} &= \frac{1}{n-1}\left(\sum_{i=1}^{n} X_i Y_i - n \cdot \bar{X} \cdot \bar{Y}\right) \\
&= \frac{1}{3}\left((-1)(0) + (0)(1) + (1)(2) + (2)(3) - 4 \cdot \frac{1}{2} \cdot \frac{3}{2}\right) \\
&= 5/3.
\end{aligned}
$$

Hence the sample correlation turns out to be:

$$R = \frac{S_{xy}}{S_x \cdot S_y} = \frac{5/3}{\sqrt{5/3} \cdot \sqrt{5/3}} = 1.$$

We should have expected this perfect positive correlation, because, looking back at the data set $(-1,0), (0,1), (1,2), (2,3)$, we see that all points lie directly on the positively sloped line $y = x + 1$. ∎

In general, for any but very small data sets it is preferable to use technology to calculate the sample covariance and correlation. Scientific calculators with statistical functions will be able to do this, and of course dedicated statistical programs such as SAS, SPSS, R, and Minitab can make these computations. Excel has functions COVAR(xrange, yrange) and CORREL(xrange, yrange) that also can compute these quantities, if the X data has been placed in a range of cells (such as A1:AN), and the corresponding Y data is in an associated range (such as B1:BN). However, in Excel the COVAR function uses n rather than $n-1$ in the denominator of (3.91), so, if the more traditional definition of sample covariance that we have used is desired, then the Excel result would have to be multiplied by n and divided by $n - 1$ to correct.

It is far more interesting to go beyond small, concocted examples to study larger, real financial data sets. The next example does this.

Example 6. Here we present results on rates of return for two real stocks, Ford (ticker symbol "F") and General Motors (ticker symbol "GM"), which we would suspect would be correlated due to the fact that they are competitors in the same market. It will be interesting to see if they are positively correlated because of global market trends for the auto industry, or negatively correlated because the success of one tends to be associated with the failure of the other. We will show the nature of the computations, but we will have to rely on a form of technology to do the tedious arithmetic for us. Below are 52 weekly closing prices for 2013 (a period of economic recovery, though certainly not a boom period).

Ford prices: $13.08, 13.18, 12.78, 12.25, 12.33, 12.25, 11.74, 11.87,$
$12.21, 12.66, 12.48, 12.37, 11.71, 12.73, 12.17, 12.86, 13.11,$
$13.38, 14.29, 14.02, 14.86, 14.91, 14.57, 14.22, 14.66, 15.83, 16.22,$
$15.89, 16.13, 16.69, 16.23, 15.54, 15.68, 15.44, 16.21, 16.54, 16.58,$
$16.26, 16.3, 16.31, 16.71, 16.78, 16.2, 16.16, 16.37, 16.31, 16.38,$
$16.01, 15.91, 14.79, 14.67, 14.8;$

GM prices: $29.32, 28.28, 28.08, 27.21, 27.59, 26.81, 26.18, 26.28, 27.04,$
$27.25, 27.26, 26.87, 26.58, 28.61, 28.15, 29.46, 31., 30.35, 32.28,$
$31.75, 32.73, 33.83, 33.12, 31.11, 32.17, 33.49, 35.16, 35.36, 35.42, 35.69,$
$34.79, 33.21, 33.86, 32.92, 34.92, 34.83, 35.57, 35.13, 34.48, 34.14, 34.66,$
$34.37, 36.11, 35.41, 37.45, 36.35, 37.41, 38.8, 38.67, 39.59, 39.54, 39.47.$

From these prices, we compute weekly rates of return. For example, starting with the first week for Ford,

$$R_1 = \frac{13.18 - 13.08}{13.08} = .00765; R_2 = \frac{12.78 - 13.18}{13.18} = -.03035;$$

etc., and for GM,

$$R_1 = \frac{28.28 - 29.32}{29.32} = -.03547; R_2 = \frac{28.08 - 28.28}{28.28} = -.00707;$$

etc. The full lists of 51 rates of return are:

Ford rates: $0.00765, -0.03035, -0.04147, 0.00653, -0.00649, -0.04163,$
$0.01107, 0.02864, 0.03686, -0.01422, -0.00881, -0.05335, 0.08711, -0.04399,$
$0.05670, 0.01944, 0.02060, 0.06801, -0.01889, 0.05991, 0.00336, -0.02280,$
$-0.02402, 0.03094, 0.07981, 0.02464, -0.02035, 0.015103, .03472, -0.02756,$
$-0.04251, 0.00901, -0.01531, 0.04987, 0.02036, 0.00242, -0.01930, 0.00246,$
$0.00061, 0.02452, 0.00419, -0.03457, -0.00247, 0.01300, -0.00367, 0.00429,$
$-0.02259, -0.00625, -0.07040, -0.00811, 0.00886;$

GM rates: $-0.03547, -0.00707, -0.03098, 0.01397, -0.02827, -0.02350,$
$0.00382, 0.02892, 0.00777, 0.00037, -0.01431, -0.01079, 0.07637, -0.01608,$
$0.04654, 0.05227, -0.02097, 0.06359, -0.01642, 0.03087, 0.03361, -0.02099,$
$-0.06069, 0.03407, 0.04103, 0.04987, 0.00569, 0.00170, 0.00762, -0.02521,$
$-0.04542, 0.01957, -0.02776, 0.06075, -0.00258, 0.02125, -0.01237, -0.01850,$
$-0.00986, 0.01523, -0.00837, 0.05063, -0.01939, 0.05761, -0.02937, 0.02916,$
$0.03716, -0.00335, 0.02379, -0.00126, -0.00177.$

Identify the Ford rates of return as the X variable and GM as the Y variable. The individual summary statistics are:

$$\bar{X} = \frac{.00765 + \cdots + .00886}{51} \approx .00297;$$

$$\bar{Y} = \frac{-.03547 + \cdots + -.00177}{51} \approx .00633;$$

$$S_x{}^2 = \frac{(.00765 - .00297)^2 + \cdots + (.00886 - .00297)^2}{50} \approx .00113;$$

$$S_y{}^2 = \frac{(-.03547 - .00633)^2 + \cdots + (-.00177 - .00633)^2}{50} \approx .00098;$$

$$S_x = \sqrt{.00113} \approx .03357; \ S_y = \sqrt{.00098} \approx .03137.$$

For this particular year's worth of data, we see that GM is a strictly better individual investment than Ford, because it has a higher mean return and a slightly lower standard deviation. It tends to be the exception rather than the rule for one stock to dominate another in this way. The sample covariance between the two rates of return is:

$$\begin{aligned} S_{xy} &= \frac{1}{50}((.00765 - .00297)(-.03547 - .00633) + \cdots \\ &\quad + (.00886 - .00297)(-.00177 - .00633)) \\ &\approx .00064. \end{aligned}$$

This makes the correlation come out to be:

$$R = \frac{S_{xy}}{S_x S_y} \approx \frac{.00064}{.03357 \cdot .03137} = .60622.$$

We have found a moderately positive correlation of about .6 between weekly rates of return. In Figure 3.25 we show a scatter plot of the (Ford, GM) rate of return pairs in the plane, and, although there is variability, you can make out an increasing and roughly linear dependence of the GM rates on the Ford rates, in line with our correlation computation. Perhaps the rising economic tide of 2013 floated both of these boats, overcoming any competitive tendency for one to move in the opposite direction to the other. ∎

3.6.3 Simulation for Estimation

Some of the most important applications of probability, particularly in the area of finance, involve complex systems that do not easily allow for closed form answers to significant questions. For example, some financial contracts have a value that depends upon the full path of a process of prices of an underlying asset. Specifically, such contracts may earn a profit that is a function $Y = h(X_0, X_1, ..., X_k)$ of the (random) values of this underlying asset between the initial time 0 and the final step k. Expectations of functions like this may be

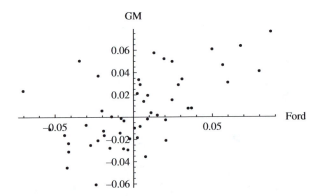

Figure 3.25: Plot of GM weekly rates of return vs. Ford weekly rates, 2013.

difficult or impossible to compute. Can anything be done to estimate them, or in general to gain insight into the workings of the complex system?

One way of obtaining useful approximate information is through computer simulation of the system. We are interested here in probabilistic systems, so this will involve designing and implementing an algorithm to simulate appropriate random variables that model what the system does. Then, we run the algorithm a large number of times to replicate values of whatever function Y, as described in the last paragraph, that we are interested in. If the computer system on which we run the algorithm has good random number generation methods, then we essentially produce a random sample of observations $Y_1, Y_2, ..., Y_n$ from the distribution of Y. Notice that Y is itself a random variable, and its mean $\mu_y = E[h(X_0, X_1, ..., X_n)]$ is the quantity that we are looking to estimate. So the Law of Large Numbers applies, which tells us that the sample mean $\bar{Y}$ of our simulated observations converges to μ_y. It does not, however, tell us how quickly that happens, so in simulation estimation we need to pay attention to the variability of the estimate.

On the issue of variability, at this point we have two facts available to us. First, $\text{Var}(\bar{Y}) = \frac{\sigma_y^2}{n}$, where σ_y^2 is the distributional variance of Y. The latter is normally unknown, but can be estimated by the sample variance S_y^2 of the simulated sample. Second, Chebyshev's inequality tells us that most of the probability mass of the distribution of Y lies within 2 standard deviations, that is, about $2 \cdot \frac{S_y}{\sqrt{n}}$ of $\mu_y = E[\bar{Y}]$, which gives us a reasonable range

$$\bar{Y} \pm 2 \cdot \frac{S_y}{\sqrt{n}} \tag{3.95}$$

of possible values of $\mu_y = E[h(X_0, X_1, ..., X_n)]$.

The preceding paragraphs lay out a general plan for simulation estimation of quantities in probabilistic financial problems. There is much more to the subject than can be done in this brief introduction. You can find further information in Ross [12] and Wang [15]. But we will be able to show how to bring that plan to fruition in a particular interesting and reasonably complex case study.

Our approach in the printed text will be to explain a little background about how computers simulate random values, and to describe the solution of the case study in some detail but to stop at the algorithm level for problem solving. In the electronic version of this text, the details of implementing the algorithm in *Mathematica* will be shown. The algorithm can easily be translated to other platforms, such as Java, C, and popular computer algebra systems.

First, how do we go about simulating a value X from a given discrete distribution with p.m.f. $f(x) = P[X = x]$ on a finite space of states? For specificity, suppose that there are j states $x_1, x_2, ..., x_j$, which have probabilities $p_1, p_2, ..., p_j$, respectively. Most computer systems have a way of simulating streams of **uniformly distributed real numbers** on $[0, 1]$, that is, sequences of numbers $U_1, U_2, U_3, ...$ with values in $[0, 1]$ such that the probability that each U takes a value in a subinterval $[a, b]$ of $[0, 1]$ is just the length $b - a$ of the subinterval. No particular portion of $[0, 1]$ is favored over any other portion, except by how much length that portion has. For example, $P\left[U \in \left(\frac{1}{2}, \frac{3}{4}\right]\right] = \frac{3}{4} - \frac{1}{2} = \frac{1}{4}$, and $P[U \in [0, .3]] = .3 - 0 = .3$.

Now refer to Figure 3.26. Suppose that we simulate a uniformly distributed U on $[0, 1]$. If $U \leq p_1$, then return $X = x_1$ as the simulated value of X. Otherwise, if $p_1 < U \leq p_1 + p_2$, then return $X = x_2$. Otherwise, if $p_1 + p_2 < U \leq p_1 + p_2 + p_3$, then return $X = x_3$, etc. As the figure shows, the probability that $X = x_i$ is the length of the interval $(p_1 + p_2 + \cdots + p_{i-1}, p_1 + p_2 + \cdots + p_i]$, which is p_i. Thus, the simulated X has the desired discrete distribution.

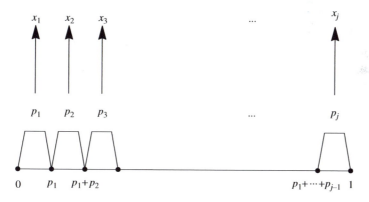

Figure 3.26: Simulating from a discrete distribution.

Now to formulate the example that we will consider, suppose that a financial asset follows a **random walk** on a set of prices $x_1, x_2, ..., x_k$ arranged in ascending order, which means that, independently of previous moves, if the current price is x_i, then at the next time instant the price can only move to one of the adjacent states. The next price will be x_{i+1} with probability p, or x_{i-1} with probability $1 - p$. The model is displayed in Figure 3.27. We must decide how to handle the left and right boundary states x_1 and x_k, however. We will suppose for this example that the price bounces off x_1 and goes with certainty back to x_2, and we will suppose that, once the price hits the right-hand boundary x_k, it stays there forever. Our question: what is the expected value of the time that it takes, starting from some particular state x_i, to reach the right-hand boundary x_k?

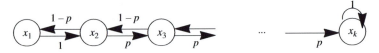

Figure 3.27: Random walk with left-hand reflecting boundary and right-hand absorbing boundary.

It is not immediately obvious how to answer the question, although Exercise 22 suggests a way of setting up a system of linear equations that can be solved. Let us use simulation to approximate the mean time to absorption at price x_k.

Following the general strategy outlined at the beginning of the subsection, we should repeatedly simulate a path of the random walk starting at a given initial state, in each replication noting the time, that is, the number of steps that it took the process to reach the boundary state x_k. The estimate of the mean time to absorption would be the sample mean of the times taken among all replications. As post hoc analysis, we should compute the sample variance, sample standard deviation, and the 2-standard error interval in formula (3.95) in order to understand the precision of the estimate.

The random aspects of the model are simple. Wherever the process is, we need to make it move to the adjacent state to the right with probability p and to the left with probability $1 - p$. This is not hard to do using the observations above: generate a uniform random number on $[0, 1]$ and if it is less than p then return a right move, otherwise return a left move. But the boundary states x_1 and x_k are special cases. If the process is at x_1, it must reflect back to x_2. If the process is at x_k, then it must stay there. This results in an easy function whose algorithm is:

OneMove[input parameters: p, the current state x_i, and the ordered list of all states $x_1, ..., x_k$]

{local variable: U}

If $x_i = x_1$ then return x_2,

Else if $x_i = x_k$ then return x_k,

Else
{Simulate U, a uniform$[0,1]$ random variable;
If $U \leq p$ return x_{i+1}, otherwise return x_{i-1}}

We have used a relatively formal notation to describe the algorithm, one that lends itself easily to being translated to a computer program, but which could benefit from some explanation. The algorithm has a name, OneMove, and input variables as listed. A variable U is used locally inside the algorithm to hold the results of a call for a uniformly distributed number. When two or more things are to be done inside a clause such as If or Else, then we group them with braces to indicate that the whole group of statements is to be done in that case. Statements that are to be executed in succession are separated by semicolons.

Next we would like to simulate one full path and determine how many steps the process took to get to x_k. One way would be to simulate a complete path, say $\{y_1 = \text{initial } x, y_2, y_3, ..., y_l\}$, and then return $l - 1$, that is, the length of the list minus 1, because the initial element of the list does not count as a step. Another way would be to only keep track of the current state and how many steps have been taken so far, and update those as the random walk moves. We will use the second approach. Here is an algorithm for a function called SimulateOneTime.

SimulateOneTime[input parameters: p, the initial state x_i, and the ordered list of all states $x_1, ..., x_k$]

{local variables: currentstate, currenttime}

Initialize currentstate = x_i;

Initialize currenttime = 0;

While currentstate is not x_k, do the following:

{Update currenttime = currenttime + 1;

Update currentstate = OneMove[p, currentstate, list of all states]};

Return currenttime

In this algorithm, we do initializations of the state and time first, in light of the fact that the random walk has not moved yet. Then we continue to get a new state using the OneMove function, and increment the time by 1. This is done until the local variable *currentstate* is the right-hand boundary x_k. After the While loop is done, we return the variable *currenttime*, which equals the number of steps taken. If the walk begins at x_k initially, the While loop will never be entered at all, and a value of 0 will be returned for *currenttime*, which is appropriate in that case.

All that remains is to build a main program to wrap around SimulateOne-Time which calls it a desired number of times, tabulating a list of its return values, and reports the important statistics, namely, the mean, variance, and standard deviation of the list of simulated times, and the 2-standard deviation interval. We call this algorithm SimulateAllTimes.

SimulateAllTimes[input parameters: p, the initial state x_i, the ordered list of all states $x_1, ..., x_k$ and the number n of replications desired]

{local variables: timelist, time}

Initialize timelist to an empty list;

Do the following n times:

    {Compute the next time = SimulateOneTime[p, x_i, list of all states];

    Append time to the timelist};

Return the mean of timelist;

Return the variance of timelist;

Return the standard deviation of timelist;

Return the interval $\left(\text{mean} \pm 2 \cdot \dfrac{\text{standard deviation of timelist}}{\sqrt{n}}\right)$

The result of executing the simulation of 1000 replications of the random walk using our algorithm and *Mathematica* program is displayed below. We get an estimate of a little under 12 for the mean time, and the width of the 2-standard error interval is around 1.5. The simulation has succeeded well in estimating the correct answer, which turns out to be exactly 12.

Mean time to absorption: 11.684

Variance: 145.069

Standard deviation: 12.0445

Two standard error interval: {10.9222,12.4458}

The exercise set contains a few problems for you to practice your modeling and algorithm development skills. As time goes by and financial situations become more and more intricate, simulation estimation may be the best hope of coping with these complications.

Important Terms and Concepts

Sample mean - The arithmetical average of sample values $X_1, X_2, ..., X_n$: $\bar{X} = \frac{1}{n}\sum_{i=1}^{n} X_i$.

Unbiasedness of sample mean - The sample mean satisfies $E[\bar{X}] = \mu$, where μ is the mean of the underlying distribution being sampled from.

Variance of sample mean - $\mathrm{Var}\left(\bar{X}\right) = \frac{\sigma^2}{n}$, where n is the sample size.

Strong Law of Large Numbers - If sample values X_i have common distribution with mean μ, then as more and more sample values are observed, the sample mean $\bar{X}$ will converge to μ.

Sample proportion - In a random sample of dichotomous observations encoded as 0 and 1, the proportion $\hat{p}$ of 1s in the sample, that is, the number of 1s divided by the size of the sample. The sample proportion is just a special case of a sample mean.

Unbiasedness of sample proportion - The sample proportion satisfies $E\left[\hat{p}\right] = p$, where p is the probability that 1 occurs.

Variance of sample proportion - $\mathrm{Var}\left(\hat{p}\right) = \frac{p(1-p)}{n}$, where n is the sample size.

Sample variance - The quantity $S^2 = \frac{1}{n-1}\sum_{i=1}^{n}\left(X_i - \bar{X}\right)^2$, which estimates the distributional variance σ^2 of the distribution from which the X_is were sampled. S^2 is an unbiased estimate of σ^2.

Sample covariance - The quantity $S_{xy} = \frac{1}{n-1}\cdot\sum_{i=1}^{n}\left(X_i - \bar{X}\right)\left(Y_i - \bar{Y}\right)$, which estimates the distributional covariance $\sigma_{xy} = \mathrm{Cov}(X,Y)$ between two random variables X and Y based on a random sample of pairs (X_i, Y_i) of size n.

Sample correlation - The quantity $R = \frac{S_{xy}}{S_x \cdot S_y}$, which estimates the distributional correlation $\rho_{xy} = \mathrm{Corr}(X,Y)$ between two random variables X and Y based on a random sample of pairs (X_i, Y_i) of size n.

Simulation estimation of a mean - If Y is a quantitative characteristic of a random phenomenon depending on one or more random variables, that is, $Y = h\left(X_1, X_2, ..., X_k\right)$, replicate the phenomenon a large number n of times by simulating the input variables X_i for each replication and computing the appropriate Y from them. Then $\bar{Y}$ estimates the expectation $E[Y] = E\left[h\left(X_1, X_2, ..., X_k\right)\right]$.

Exercises 3.6

1. Show, as in the example at the start of the section, that if we choose a number randomly and with replacement from $\{1, 2, 3, 4\}$ three successive times, denoting our choices by X_1, X_2, and X_3, that these Xs form a random sample of size 3 from a distribution. What distribution? (You may assume that all outcomes are equally likely.)

2. Below is a list of 20 consecutive monthly prices of a risky asset. Find the associated monthly rates of return; does it seem as if the rates of return are

independent? Estimate the mean and standard deviation of the monthly rate of return.

$$40, 40.69, 43.44, 43.19, 46.32, 50.22, 49.58, 53.11, 52.03, 52.04, 56.36,$$
$$59.05, 61.33, 61.81, 61.01, 62.83, 64.88, 62.75, 67.86, 67.69$$

3. The list below follows the monthly variable interest rate on a credit card over a 2-year span. Does it seem as if the interest rates are a random sample? If not, where does a change seem to have taken place? Estimate the amount by which the mean interest rate changed.

$$0.0325, 0.0293, 0.0494, 0.0340, 0.0448, 0.0492, 0.0223, 0.0435, 0.0357, 0.0385,$$
$$0.0514, 0.0387, 0.0467, 0.0831, 0.0725, 0.0607, 0.0857, 0.1074, 0.0995, 0.0597,$$
$$0.0781, 0.0713, 0.0825, 0.0660.$$

4. For a random sample of size 2 from a distribution with probability mass function below, find the joint probability $P[X_1 > 1, X_2 < 2]$.

$$f(x) = \begin{cases} 1/8 & \text{if } x = 0; \\ 3/8 & \text{if } x = 1; \\ 3/8 & \text{if } x = 2; \\ 1/8 & \text{if } x = 3. \end{cases}$$

5. For a random sample of size 3 from a distribution with probability mass function below, find the joint probability $P[X_1 = 1, X_2 < 4, X_3 > 2]$.

$$f(x) = \begin{cases} 1/3 & \text{if } x = 1; \\ 1/3 & \text{if } x = 2; \\ 1/6 & \text{if } x = 3; \\ 1/6 & \text{if } x = 4. \end{cases}$$

6. Let X_1, X_2, X_3 be a random sample of size 3 from a distribution with mean μ and variance σ^2. Show that $Y = \frac{1}{2}X_1 + \frac{1}{4}X_2 + \frac{1}{4}X_3$ is an unbiased estimator of μ. Which has the smaller variance, Y or $\bar{X}$?

7. Below is actual rate of return data for IBM stock on the New York Stock Exchange for June 2012. Assume that it does constitute a random sample from the distribution of the random variable R, the daily rate of return on IBM. Estimate the mean and variance of the distribution of rates of return, and the mean and variance of the sample mean rate of return.

$$-0.00288, 0.00350, 0.02532, 0.00232, 0.00360, -0.01345, 0.01060,$$
$$-0.00747, 0.01035, 0.02053, -0.00410, 0.00322, -0.00074, -0.02710,$$
$$0.00157, -0.00432, -0.00472, 0.00545, -0.00824, 0.02181$$

8. Below is a data set of indicator variables for loan defaults in a certain class. Estimate the probability of default, and the mean and variance of the sample proportion.

$0,0,1,0,1,0,0,0,0,0,0,0,0,0,0,0,1,0,0,0,1,0,0,0,0,0,1,0,0,0,0,0,0$

9. For a random sample of size 16 from the distribution of Exercise 1, find the mean and variance of the sample mean $\bar{X}$.

10. For a random sample of size 25 from the distribution of Exercise 5, find the mean and variance of the sample mean $\bar{X}$.

11. In Example 1, how would you reestimate σ if you assumed that the histogram in Figure 3.24 was actually showing a 3-standard deviation interval? What would be the corresponding 2-standard deviation of $\bar{R}$ range?

12. Suppose that you are trying to estimate the average 30-year fixed-rate mortgage interest rate throughout the country, and a small preliminary sample resulted in an average of 4.8% with a sample standard deviation of 1.5%. How large a random sample should you take so that it is highly likely (at least 75%) that your estimate of the average interest rate differs by less than .1% from the true average?

13. A bank wishes to estimate the default probability of a certain category of loans. A small pilot sample showed that 2.3% of such loans defaulted. How large a random sample of loans should be taken so that it is highly likely (at least 75%) that the bank estimate of the default probability is within .2% of the true probability?

14. Verify the computational formula for the sample variance:

$$S^2 = \frac{1}{n-1}\left(\sum_{i=1}^{n} X_i{}^2 - n \cdot \bar{X}^2\right).$$

15. Show that the sample variance S^2 is an unbiased estimator of the distributional variance σ^2. (Hint: use the result of the previous exercise and the computational formula $\mathrm{Var}(Y) = E\left[Y^2\right] - (E[Y])^2$ to compute the expected squares.)

16. Compute the sample covariance and correlation for the following data sets:

(a) $(2,1),(3,0),(5,-1)$;

(b) $(-1,-1),(-1,0),(0,0),(0,-1)$.

17. Use any form of technology to compute the sample covariance and correlation for the list of pairs of prices of two stocks below.

$$(19.72, 29.10), (20.06, 31.46), (22.19, 30.45), (21.11, 29.88), (20.26, 31.21),$$
$$(21.19, 31.07), (20.82, 33.00), (21.80, 34.43), (20.30, 27.23), (19.08, 30.54)$$

18. Referring to Exercise 17, estimate the mean and variance of $.3P_1 + .7P_2$, where P_1 and P_2 are the two prices.

19. In the case of a random sample of n pairs of indicator random variables $(I_1, J_1), \ldots, (I_n, J_n)$, suppose that the number of times that $I = 1$ is n_1, the number of times that $J = 1$ is n_2, and the number of times that both I and $J = 1$ is n_{12}. Find expressions for the sample covariance and correlation.

20. Derive formula (3.94), $S_{xy} = \frac{1}{n-1} \left(\sum_{i=1}^{n} X_i Y_i - n \cdot \bar{X} \cdot \bar{Y} \right)$, for the sample covariance.

21. For the Ford and GM data in Example 6, create a histogram and a time series plot of the rates of return for each company. Make note of the salient features of the plots.

22. For the random walk in Figure 3.27, if $g(x)$ is defined to be the expected time to reach x_k starting from state x, argue that:

$$g(x_k) = 0;$$

$$g(x_1) = 1 + g(x_2);$$

$$g(x_i) = 1 + p \cdot g(x_{i+1}) + (1 - p) \cdot g(x_{i-1}) \text{ for } 1 < i < k.$$

Solve the system and find the value of each $g(x_i)$ in the case where the state space is $\{1, 2, 3, 4\}$ and the right-hand probability is $p = .5$.

23. Write an algorithm for simulating a path of a binomial branch process with initial price x and parameters p, a, and b with the usual meanings.

24. Use the algorithm in Exercise 23 to write another algorithm that estimates the proportion of times that the price process ever drops below a boundary B among a desired number of replications of the path.

25. How could you use a computer's ability to simulate uniformly distributed random real observations on the interval $[0, 1]$ to simulate random and uniform observations on an arbitrary interval $[a, b]$? How about equally likely (i.e., discrete uniform) values on a discrete set of integers $\{a, a + 1, ..., b\}$?

26. In the SimulateAllTimes algorithm, can you think of an alternative approach so that the entire list of simulated times does not have to be kept? (Hint: you can define a variable that keeps track of the running sum of times.)

27. Suppose that the distribution of Exercise 5 gives the dollar value per $100 owned of a quarterly dividend issued by a company to a stockholder. Write a simulation algorithm to estimate the expected total dividend per $100 over 4 years.

Chapter 4

Portfolio Theory

One of the main thrusts in the subject of financial mathematics is the study of the optimal distribution of wealth among collections of risky assets called *portfolios*. We will consider mostly portfolios of stocks in this chapter, and also limit our attention to single-time step problems. In such problems, there is a current time "now" and a time "later" at which we observe the performance of our investment. We have a finite collection of stocks in which we are interested in investing. Our task is to decide how to allocate money among the stocks, in view of the tradeoffs between return and riskiness that are natural in a financial market.

We will keep our model very simple for the sake of this text: besides the simplification of a single time step, we will pay no fees for transacting, no dividends are issued on stocks, our purchase prices now will be known and fixed rather than bid upon, we pay cash in full for our purchases rather than borrowing, and the parameters that we need to know in order to make our decisions are constant and estimable. These are fairly severe simplifying assumptions, but we need them to make a first pass at solving the problem.

To partially counterbalance all the simplifications, in Section 4.1 we give a short introduction to what actually goes on when investors buy stocks in the real world. A particular focus there is the opportunity to *buy on margin*, which means that the investor only pays a part of the full transaction, borrowing the rest from the broker. Section 4.2 quickly reviews what was done in Chapter 3 with regard to modeling and estimating key parameters of rates of return on risky assets, and then goes on to explore the *rate of return on a portfolio* of risky assets. It closes by introducing *risk aversion*, which is a means of quantifying an investor's attitude toward risk of a single asset or a portfolio of assets. Finally, Section 4.3 formulates and solves the *mean-variance portfolio optimization problem* using standard calculus techniques. We will keep the number of assets small in our examples in order to not require too much mathematical prerequisite material, but in slightly more advanced work, techniques and ideas from linear algebra can be brought to bear to simplify both presentation and computation.

By no means does this chapter tell the whole story about portfolios. One of the most important generalizations is to allow the investment process to take place over multiple time periods. This raises the possibility that investors may choose to withdraw (or even invest more) money into the system at each time. The more general **portfolio investment-consumption problem** in discrete time emerges, which requires background in a mathematical subject called **dynamic programming**. At a much more advanced level, time can be continuous, and partial differential equations can be necessary.

Also, the problem in this chapter will depend entirely on the means, variances, and covariances of the risky assets. The exact probability distributions of the assets do not figure in. In **utility theory**, one tries to construct portfolios that optimize a **utility function** of final wealth, whose expected value may well depend on the probability distributions of the rates of return.

As another direction of study, the results of mean-variance optimization of risky assets in this chapter, in the presence of a risk-free asset like a government bond, imply something very deep about the full market of assets. The **capital market theory** in economics says roughly that there is a **market portfolio** of risky assets so that all investors keep those in the same proportions to each other, differing only in what proportion of wealth to reserve for the risk-free asset. There are interesting consequences of capital market theory having to do with enforced relationships between asset means and variances and those of the market.

All of these references to extensions are meant to whet your appetite for advanced material for which we set the stage with the basic background in this chapter.

4.1   Stocks

4.1.1   Common Stocks, Preferred Stocks, and Stock Indices

There are two main ways in which a corporation raises capital for its regular operation and special projects. The first way is to incur debt by issuing bonds. This book dealt extensively earlier with the valuation of such financial instruments. The other way to raise money is to sell shares of ownership of the company to investors, who then have **equity** in the company. We refer to these shares as **stock**. Whereas a bondholder has first rights to repayment, the bondholder is not a partial owner, does not benefit from increased company value, and also does not get to participate in any way in the decision making of the corporation. A stockholder does have the right to elect board members and participate in other referenda involving the company. In addition, the investor in stock hopes that the company will pay **dividends**, that is, small percentages of the company's profits, from time to time, as well as profiting from the growth in value of the company stock.

If, for example, a stockholder owns 100 shares of stock in company A, purchased at $30 per share, and the stock market bids up the price to $35 per share, then there is a net gain of $100(\$35 - \$30) = \$500$. If at this time the company decided to pay stockholders a dividend of 2% per share, then the stockholder receives a payment of:

$$100(.02)(\$35) = \$70.$$

Stocks are risky, however. The price per share could also have gone down to $25 since it was purchased, in which case there would be a net loss in value of $100(\$30 - \$25) = \$500$. This chapter is all about achieving an appropriate balance between non-risky (or, we should probably say "less risky," given the small possibility of defaulted coupon payments or redemption value) investment instruments such as bonds and risky assets such as stocks.

Example 1. Find the internal rate of return expressed monthly for Elizabeth, whose financial situation is as follows. At the beginning of the year, she buys 10 shares of the stock of Company X, at $60 per share, and 20 shares of Company Y at $30 per share. At the end of the first quarter, she examines her holdings and finds that the value of her Company X shares has increased to $65, but the Company Y shares have decreased to $28. So she sells half of her Company Y shares and invests in two more shares of Company X, consuming the remainder. Meanwhile, Company X has issued a 2% per share dividend (on her original 10 shares), which she also consumes. At the end of the third quarter, Company X shares are still selling for $65 per share, but Company Y has rallied to a price of $32 per share, and has issued a 4% per share dividend, which she consumes. She does not transact. At the end of the year, Company X has increased again to $70 per share, and Company Y is at $35 per share. Assume that, since Elizabeth's brother Dudley is a stockbroker, she pays no fees for transacting.

Solution. Recall that the present value equation for the internal rate of return i is:

$$P_0 + P_1(1 + i)^{-1} + \cdots + P_n(1 + i)^{-n} = B(1 + i)^{-n},$$

where the P_js are payments into the system (with negative values for consumptions from the system), and B is the ending balance. We take months as our time unit for the calculation, so that the i we will solve for will be the monthly internal rate of return. Multiplication by 12 will then give the nominal yearly rate.

Then $n = 12$, transactions take place at times 0, 3, and 9, and we observe the final balance at time 12. At time 0, Elizabeth pays $P_0 = \$60(10) + \$30(20) = \$1200$ for her shares. At time 3 she rearranges her holdings, selling 10 shares of Company Y stock for a total of $\$28(10) = \$280$, purchasing 2 shares of Company X at $65 each for a total of $130, and consuming the residual $\$280 - \$130 = \$150$. She also consumes

$$(10 \text{ shares }) \times (2\% \text{ per dollar invested }) \times (\$65 \text{ per share }) = \$13$$

as a result of the dividend. So $P_3 = -\$163$ and she goes into the next period owning 12 shares of Company X and 10 of Company Y. At time 9 months, she consumes the Company Y dividend, which means that:

$$P_9 = -(10 \text{ shares}) \times (4\% \text{ per dollar invested}) \times (\$32 \text{ per share}) = -\$12.80.$$

Elizabeth did not buy or sell any shares, so she ends the year still owning 12 and 10 shares, respectively, of Company X and Y, whose value is:

$$
\begin{aligned}
B &= (12 \text{ shares }) \times (\$70 \text{ per share }) + (10 \text{ shares}) \times (\$35 \text{ per share}) \\
&= \$840 + \$350 \\
&= \$1190.
\end{aligned}
$$

Therefore, her monthly internal rate of return satisfies the equation:

$$1200 - 163(1+i)^{-3} - 12.80(1+i)^{-9} = 1190(1+i)^{-12}.$$

Numerical solution yields the value $i = .012$ per month, which scales out to a successful $12(.012) = .144$ nominal per year. ■

There are many physical markets, called **stock exchanges**, across the world where stocks are traded, some of the most important being London, Tokyo, Hong Kong, Shanghai, and Toronto. By far the largest exchange in terms of the monetary value of transactions is the **New York Stock Exchange** (NYSE). At the time of this writing there were roughly 3500 companies listed in the NYSE. There is also a computer network called **NASDAQ** (National Association of Security Dealers Automated Quotation System) operated out of New York, where trades are processed electronically (referred to as **over-the-counter trading**). On-line trading has grown enormously in recent years with the emergence of web-based services such as *etrade.com* and *scottrade.com*.

To estimate the overall trend of the value of stocks in a market, various indices have been created that are essentially weighted averages of the values of the included stocks. The **Dow Jones Industrial Average** (DJIA) is an old but still much monitored index that is proportional to the sum of the prices of 30 specific NYSE stocks in very large publicly held companies, including Boeing, McDonald's, AT&T, and General Electric. The **NYSE Composite Index** is a fairly recent creation, which consists of many more companies (more than 2000 at this writing) in various segments of the economy, such as Alcoa, Bank of America, IBM, and J.C. Penney. Another widely reported stock market index is the **Standard & Poor's 500** (S&P 500), made up of a selection of about 500 large companies comprising about 75% of the total value of all shares in the market. Examples of companies in this index include Adobe, Costco, and Delta Air Lines. Financial institutions also offer **index funds**, which are essentially portfolios of the assets included in these indices, whose values mirror the changes in the index values. Investing in an index fund gives investors the capability of creating a very diversified portfolio that moves in lockstep with the whole market.

Example 2. The S&P 500 index is an example of a ***market-weighted index***, where the assets' prices are weighted according to their total value in the market. To illustrate how this might work, consider the Frostbite Falls Exchange, a very small market that consists of five companies. Company 1 has 50 total shares outstanding, priced at $40 each; Company 2 has 100 shares, priced at $20 each; Company 3 has 100 shares, priced at $60 each; Company 4 has 50 shares, priced at $30 each; and Company 5 has 40 shares, priced at $15 each. Suppose that the FF index is defined as the weighted average per share price, where the weight given to a company is the proportion of total market value that the company's shares make up. The dollar values of shares are as follows:

Company 1: $2000; Company 2: $2000; Company 3: $6000;
Company 4: $1500; Company 5: $600;

Total value : $2000 + $2000 + $6000 + $1500 + $600 = $12,100.

Then the value of the FF index would be:

$$\frac{\$2000}{\$12100}(\$40) + \frac{\$2000}{\$12100}(\$20) + \frac{\$6000}{\$12100}(\$60) + \frac{\$1500}{\$12100}(\$30)$$

$$+\frac{\$600}{\$12100}(\$15) = \$44.13.$$

Notice that Company 3 contributes most to the index with a weight of around 1/2, and Company 5 contributes the least with a weight of around 1/20. ∎

A phenomenon that can complicate the computation of an index occurs when a stock ***splits***. A 2-for-1 split, for example, means that the value of each share suddenly is cut in half, but the shareholders receive twice as many shares so that the total value of their investment does not change. The total outstanding value of all shares, called the ***market capitalization*** of the company, therefore also does not change. One reason why a company may decide to split stock is to increase access to ownership of the company to a wider audience. Some experts believe that a stock split tends to lead to increased trading volume and higher prices for the stock later, although this is not certain. Note that, if a Dow Jones company splits, there is a radical change to the price and the numerator of the DJIA index value, so that the denominator in the average that defines the index must somehow compensate. We will not attempt to clarify the arcane details of the Dow Jones index computation here.

We have been speaking as if there is only one kind of stock issued by corporations; the conditions described so far apply to so-called ***common stock***. There is a second main category called ***preferred stock***. This is also an equity rather than debt issue, meaning that it implies partial ownership of the company, but it has some of the features of a bond. In particular, preferred stock has an agreed upon liquidation value, like the face value of a bond, and dividends are fixed in advance as a percentage of this value. It is possible that the company may not be able to pay a particular dividend, but owners of preferred stock receive their dividends before owners of common stock, and some preferred stock contracts allow for a backlog of unpaid dividends to accumulate and be paid later. Voting rights may be limited to certain situations; in fact, companies can try to guard against hostile takeovers by issuing large amounts of preferred stock which have heavy voting rights only in cases where management is about to change. In this book we will confine ourselves to common stock, whose value changes frequently in the market.

4.1.2   Stock Transactions

There are two traditional ways to acquire stock, or to sell it: first, a ***dealer*** maintains a large inventory of assets, and will sell to a purchaser for a slight premium above the market price, or buy from a seller for a slight discount below the market price. These dealer ***spreads*** allow this kind of agent to stay in business. We won't talk much about stock dealers in this book. The other way to transact in stocks is to work through a ***broker***, who is an intermediary with privileges to trade in a market. The broker may charge a fixed fee and also an additional fee proportional to the value of the transaction. Stockbrokers are typically not individuals working on their own, but rather employees of large financial institutions called ***brokerages***.

Stock transactions may be made partially on credit with a brokerage, but there are restrictions imposed by federal agencies, stock exchanges, and the brokerage itself. Buying stocks using credit is called buying ***on margin***. Interest is also generally paid by the stockholder on the borrowed value. The ***equity*** that a stockholder owns is the current market value of the assets owned minus the amount that is owed to the brokerage. The ***margin rate*** m for the stockholder is the proportion of the stock's current value that is equity. Currently the Federal Reserve Board requires m to be initially 50% or more. For example, if you open up an account with a brokerage in which you purchase 500 shares of a stock at $200 per share, paying $60,000 of the total value of $100,000, then your total equity stands at $60,000, and the margin rate is $m = \$60,000/\$100,000 = .6$, which would more than satisfy the initial margin requirement of 50%.

In addition to initial margin requirements, there may be other maintenance requirements on your margined amount as time passes and the values of assets change. Suppose, in the example of the last paragraph, there is a continuous 40% maintenance margin requirement, and the stock's price has plunged to $100 per share. Ignoring interest on the margin account for a moment for simplicity, the investor has borrowed $40,000, and the total value of the asset is (500 shares)($100 per share) $= \$50,000$, hence the investor only has $\$50,000 - \$40,000 = \$10,000$ of equity. The margin rate has sunk to $\$10,000/\$50,000 = 20\%$, which does not satisfy the maintenance margin requirement. The brokerage then issues a ***margin call***, which is notification to pay cash to decrease the debt sufficiently, or to sell off assets to do so.

Example 3. Consider the case of an investor Si, who has purchased 25 shares of stock through a broker. The stock begins at a price of $40 per share. Si takes maximum advantage of margining the purchase at an initial margin rate of 50%. He also pays the broker's commission of 4% of the total transaction value by cash. His brokerage has a 40% maintenance margin requirement, and he is called upon to either pay cash or sell sufficiently many shares of stock if his equity should fall below that level. Interest of 2% per period is charged on margined funds. Suppose that one time period goes by, and the price of the stock falls to $30 per share. Is there a margin call? How much cash will he have to pay to bring his margin rate up to 40%? If Si chooses to sell shares to restore his margin rate to the acceptable level, how many should he sell?

Solution. Initially, Si pays half of the total value of ($40)(25) = $1000 for the shares. So he has $500 of equity and $500 of debt to the brokerage. He will also pay 4% of the value of the purchase transaction, which is $40. One period goes by, and, because of interest on the margined amount, he owes the brokerage $500(1 + .02) = $510. The total value of his shares is now ($30)(25) = $750, hence his margin rate is

$$\frac{\$750 - \$510}{\$750} = .32.$$

A margin call will be issued by the brokerage, since the margin rate has fallen below the required value of .40. In the scenario that he chooses to pay an amount of cash c to the brokerage, his debt will be reduced to $510 − c$ while the total value of the shares is still $750. The value c must therefore satisfy:

$$m \geq .4 \Longrightarrow \frac{\$750 - (\$510 - c)}{\$750} \geq .4 \Longrightarrow \$240 + c \geq \$750(.4) \Longrightarrow c \geq \$60.$$

So Si can choose to pay $60 cash to resolve the problem. The other way to do it is to sell off equity, to apply to the debt. If s shares are sold, then ($30)$s$ comes off both the debt and the total value of shares. Then s satisfies:

$$m \geq .4 \implies \frac{(\$750 - \$30s) - (\$510 - \$30s)}{\$750 - \$30s} \geq .4$$
$$\implies \frac{\$750 - \$510}{\$750 - \$30s} \geq .4$$
$$\implies \$240 \geq (\$750 - \$30s)(.4).$$

Solution of the linear inequality gives $s \geq 5$. Thus, Si can sell off 5 shares for a total of $150, and then be back in compliance with the minimum maintenance margin. He would own 20 shares for a total value of $600, but he would owe the brokerage $510 − $150 = $360. Notice that this has not been a particularly good deal for Si, since he paid $500 initially, and another $40 for the transaction fee, and his net account value with the brokerage is just $600 - $360 = $240. The 25% drop in share value has cost him $300. ■

We now consider a more complicated example that evolves over several time periods.

Example 4. Let us trace the financial position of an investor, Bill, over 3 months. Bill is interested in stocks A and B, and will initially buy 100 shares of A and 200 shares of B at their current prices of $20/share and $10/share. Bill utilizes the 50% margin requirement to buy the stocks. He pays interest to the brokerage on his margin account at the rate of .5% per month. Each of A and B individually has a maintenance margin: 30% for A and 40% for B. He instructs his broker that, if either stock violates its margin condition and a margin call is made, he will sell off enough of stock B to put him back in compliance. He makes no transactions other than the ones he is required to make during the 3 months. Also, the brokerage generously charges no transaction fee for a reaction to a margin call. If the share prices for A at the end of months 1, 2, and 3 are, respectively, $18, $19, and $22, and the share prices for B are $7, $9, and $10, find the value of his equity at the end of the third month.

Solution. Initially, Bill buys $20(100) = $2000 worth of stock A, paying half himself and putting a (borrowed) balance of $1000 into a margin account. He buys $10(200) = $2000 of stock B, again paying $1000 himself and leaving an additional $1000 that he owes in his margin account. Here is his initial situation:

month 0	market value	margin	equity	margin rate
stock A				
100 shares	$2000	$1000	$1000	50%
$20/share				
stock B				
200 shares	$2000	$1000	$1000	50%
$10/share				

At month 1, A has gone down to $18/share and B has also gone down to $7/share. We must check to see if Bill must sell some of stock B to meet margin requirements. The price declines put him in the following situation:

month 1 pre-call	market value	margin	equity	margin rate
stock A				
100 shares	$1800	$1000(1.005)	$1800 - $1005	$\frac{\$795}{\$1800}$
$18/share		= $1005	= $795	= 44.2%
stock B				
200 shares	$1400	$1000(1.005)	$1400 - $1005	$\frac{\$395}{\$1400}$
$7/share		= $1005	= $395	= 28%

A margin call goes out because of the decline in the equity percentage for stock B. Bill must sell off enough stock B value, say x, to reduce his margin account so that the equity percentage is at least 40%. The new market value will be $1400 - x$, and the new margin will be $1005 - x$, so the equity remains at $395. Thus, we require:

$$\frac{395}{1400 - x} \geq .40 \implies 395 \geq .40(1400 - x).$$

Solving the inequality yields that at least $x = \$412.50$ worth of stock B must be sold. Since the price per share is $7, and $412.50/$7 = 58.93$ approximately, he will have to sell 59 shares, reducing both his market value and his margin account by $59 \cdot (\$7) = \$413$.

The new situation is:

month 1 post-call stock A	market value	margin	equity	margin rate
100 shares $18/share	$1800	$1000(1.005) = $1005	$1800 - $1005 = $795	$\frac{\$795}{\$1800}$ = 44.2%
stock B				
141 shares $7/share	$987	$1005 - $413 = $592	$987 - $592 = $395	$\frac{\$395}{\$987}$ = 40%

In month 2, stocks A and B rebound to, respectively, $19/share and $9/share. Bill's financial position is now:

month 2 stock A	market value	margin	equity	margin rate
100 shares $19/share	$1900	$1005(1.005) = $1010.20	$1900 - $1010.20 = $889.80	$\frac{\$889.80}{\$1900}$ = 46.8%
stock B				
141 shares $9/share	$1269	$592(1.005) = $594.60	$1269 - $594.60 = $674.40	$\frac{\$674.40}{\$1269}$ = 53.1%

Bill is not required to sell off stock B, and does not do so. Therefore, since the prices per share of the two stocks move to $22 and $10, at the end of month 3 this is his situation:

month 3 stock A	market value	margin	equity	margin rate
100 shares $22/share	$2200	$1010.20(1.005) = $1015.25	$2200 - $1015.25 = $1184.75	$\frac{\$1184.75}{\$2200}$ = 53.9%
stock B 141 shares $10/share	$1410	$594.60(1.005) = $597.57	$1410 - $597.57 = $812.43	$\frac{\$812.43}{\$1410}$ = 57.6%

Now Bill originally paid $2000 out of pocket. If he closes his account at the end of the third month, his equity is $1184.75 + $812.43 = $1997.18, despite the fact that ultimately the stock price for A went up from $20 to $22, and the price for B remained at $10. What happened? Some money went into interest on the borrowed amount in the margin account, but the bulk of the loss happened because he was forced to sell 59 shares of stock B at only $7. ■

4.1.3   Long and Short Positions

To have a **long position** in a stock means that you own shares, hoping that they will go up in value. To have a **short position** in a stock means that you owe shares, i.e., you have borrowed them from a broker and sold them. You can collect the proceeds (or part of them, because the broker holds onto part for collateral); however, you are obliged to rebuy the shares later to restore them to their original owner. You are anticipating that the share price will decrease, so that you will have to pay less than you received in order to reacquire the shares and give them back to the owner. Think of a short position as owning negative numbers of shares (i.e., owing them instead of owning them).

Because, in a long position, shares can be bought on margin, an investor can **leverage** his investment, potentially earning higher rates of return than someone who pays cash (but risking higher loss rates as well). The next example illustrates how this can occur.

Example 5. Derrick and his sister Rose are both interested in buying a particular stock, now priced at $40 per share. They each have $2000 to invest, and will pay the initial brokerage fees for puchase (3% of the purchase value) in cash. Rose pays in full for 50 shares, and so must also pay $2000(.03) = $60 in fees. Derrick, who believes it to be very likely for the stock to go up substantially and who is something of a daredevil, uses his $2000 to buy 100 shares with 50% on margin. The total value of the transaction upon which the brokerage fee is based is then $4000; therefore he pays $120 in fees. The brokerage charges 1% per period on borrowed money. Find the value of the investments and rates of return for these two people at the end of one time period under two scenarios: (a) the price of the stock rises to $45/share; (b) the price of the stock falls to $35/share.

Solution. (a) First, Rose spent $2060 initially for her 50 shares, which are worth $45(50) = \$2250$ at the end. The rate of return for Rose is:

$$\frac{\$2250 - \$2060}{\$2060} = .092.$$

For Derrick, the initial investment cost $2120 including the fees. We must compute his final position carefully, taking account of what he owes. His brokerage account shows 100 shares, which at the end are worth $4500. The $2000 that was borrowed lands on the debit side of the equation, and with interest he now owes $2000(1 + .01) = \$2020$. Thus, Derrick's equity at the end of this period is $4500 - \$2020 = \$2480$. His rate of return is:

$$\frac{\$2480 - \$2120}{\$2120} = .170.$$

Using the leverage effect created by margin buying, Derrick has enabled himself to earn almost twice the rate of return of Rose.

(b) Lest that you think buying on margin is a foolproof technique for making money, suppose instead that the price per share falls to $35. The 50 shares that Rose purchased are now worth $35(50) = \$1750$, and so her rate of return is:

$$\frac{\$1750 - \$2060}{\$2060} = -.150.$$

Derrick's 100 shares are worth $3500, and again he owes the brokerage $2020, hence his equity is $3500 - \$2020 = \$1480$. This changes Derrick's rate of return to:

$$\frac{\$1480 - \$2120}{\$2120} = -.302.$$

This time, Derrick's gamble did not pay off, and his rate of return is more than twice as bad as that of his sister. We see that leverage is a two-edged sword. By using it, an investor can magnify both profits and losses. ∎

Both the benefits and the dangers of short-selling are illustrated in the next example.

Example 6. Suppose that Clara decides to short-sell 100 shares of a stock currently priced at $50, in hopes that the stock price will go down. Her broker requires her to put down a 50% cash margin deposit (per Federal Reserve Board regulation), but gives a modest .2% interest rate per time period on that deposited money. Find Clara's possible profits and rates of return per period, if after one time period the share price of the stock is each of: (a) $45; (b) $48; (c) $50; (d) $52; (e) $55.

Solution. Consider the original cash deposit of $.5(100)(\$50) = \$2500$ as her initial investment for the purpose of calculating a rate of return. At the end she will have the proceeds of the short sale of the stock, plus interest earned on the margin deposit, minus the amount that she has to repay to rebuy the stock after one period. In all cases, she realizes $5000 from the short sale, and she earns interest in the amount $2500(.02) on her cash margin deposit.

(a) If the stock price falls to $45, then Clara's profit and rate of return are:

$$\text{profit} = \$5000 + .002(2500) - 100(\$45) = \$505;$$

$$\text{rate of return} = \frac{\text{profit}}{\$2500} = .202.$$

Even though the stock price itself went down by 10%, she achieved a rate of return of just over 20%.

(b) In this case, where the stock decreases by $\frac{2}{50} = 4\%$, her rate of return is over 8%:

$$\text{profit} = \$5000 + .002(2500) - 100(\$48) = \$205;$$

$$\text{rate of return} = \frac{\text{profit}}{\$2500} = .082.$$

(c) If the stock price stays the same, her rate of return is just what she got from the margin deposit account:

$$\text{profit} = \$5000 + .002(2500) - 100(\$50) = \$5;$$

$$\text{rate of return} = \frac{\text{profit}}{\$2500} = .002.$$

(d) An increase in the value of the stock is not what Clara was hoping for. In this scenario the stock price rises by $\frac{2}{50} = 4\%$, and her rate of return is now negative, at -7.8%:

$$\text{profit} = \$5000 + .002(2500) - 100(\$52) = -\$195;$$

$$\text{rate of return} = \frac{\text{profit}}{\$2500} = -.078.$$

(e) Finally, her situation is even worse if the stock price rises by $\frac{5}{50} = 10\%$; her rate of return in this case is -19.8%:

$$\text{profit} = \$5000 + .002(2500) - 100(\$55) = -\$495;$$

$$\text{rate of return} = \frac{\text{profit}}{\$2500} = -.198. \quad \blacksquare$$

There is a general formula that can be derived in the short-selling scenario of the last example. Let r be the proportion of the short sale proceeds that must be kept on margin ($r = .5$ in the example), let i be the brokerage interest rate per period on the margin account ($i = .002$ in the example), let V_0 be the total value of all shares that were short sold at the initial time, and let V_1 be the total value of the shares after one period. Do not consider any investment earnings that the short seller may have accumulated with the proceeds of the short sale. Then rV_0 is the initial investment, the difference $V_0 - V_1$ plus the interest earned $i \cdot r \cdot V_0$ is the profit earned by the short seller, and therefore the rate of return per period on the short sale is:

$$\frac{V_0 - V_1 + i\,(rV_0)}{rV_0} = \frac{1}{r}\frac{V_0 - V_1}{V_0} + i.$$

Exercise 16 considers the additional benefits of leveraging an investment in an asset that is expected to go up by short-selling a second asset expected to fall, and investing the proceeds in the first asset.

Important Terms and Concepts

Debt vs. equity - A company may raise money by issuing bonds (debt) or stock (equity).

Common stock - Shares of ownership of a company. Value is determined by market forces.

Dividends - Payments made to stockholders at the discretion of the company.

Preferred stock - Shares of ownership of a company with some characteristics of bonds: fixed claim value, priority in dividend awards, but sometimes limited privileges in company decision-making.

Stock indices - Weighted averages of a number of stocks in the particular market, meant to estimate the performance of the whole market.

Index funds - Assets that are actually portfolios of assets that make up the index.

Brokers and dealers - Buyers and sellers of stock: dealers have inventories of assets and charge spreads for the transaction, while brokers are agents performing transactions in the market and charging commissions.

Buying on margin - Purchasing stock by borrowing a portion of the transaction value from the broker.

Initial margin requirements - Proportion of the net initial value of a stock purchase that must be paid, rather than borrowed.

Equity (in a margin account) - The difference Equity $=$ Market value of shares $-$ Net amount owed.

Maintenance margin - Threshhold proportion of total value of shares that must be in the form of equity.

Margin call - Notification to stockholders to correct a violation of margin requirements.

Long position, short position - In a long position, the investor owns shares of stock, while in a short position the investor borrows stock to sell and is required to restore those shares at a later date. This is called ***short-selling***.

Exercises 4.1

1. Recalculate the internal rate of return in Example 1 if Elizabeth's brother charges a 3% transaction fee.

2. Jim buys 20 shares of a new stock issue, MegaMan energy drink, at $25 per share, and 20 shares of a more tested second stock, Boring Aircraft, at $30 per share. (For simplicity, we will ignore transaction costs in this problem.) At the end of a month, MegaMan has gone down to $20 per share, while Boring is still at $30. Fearing that he might have chosen badly, Jim sells half of his MegaMan, consumes $50, and invests the remainder of his proceeds in Boring. After another month, MegaMan increases to $30 per share, while Boring has increased to $32. Feeling more confident in that stock, Jim buys his 10 shares back, selling 5 shares of Boring Aircraft and using cash to make up the difference. Finally, after one more month, both MegaMan and Boring return to their original prices of $25 per share and $30 per share. How much did Jim gain or lose, and what is his internal rate of return (expressed on a monthly basis)?

3. Mario invests with a brokerage that charges a 2% transaction fee on all transactions. At the beginning of a year, he invests $10,000 in a stock whose price is $50 per share and another $5000 in a 1-year certificate-of-deposit earning 1.5% nominal interest per year, compounded semi-annually. At the end of 6 months the price per share of the stock is $52, he receives a dividend of 3% per share, and buys another 50 shares, leaving the CD alone as per bank regulations. Then at the end of the year, there is a second dividend of 2% per share, and the price of his stock is $55. Find the internal rate of return on Mario's investment, expressed as a semi-annual rate.

4. To have a barometer of its economic health, the council of the town of Galesville sets up a market-weighted index of its four publicly traded companies: A, B, C, and D. Suppose that, at the beginning of the year, A has 1000 shares outstanding selling at a price of $40 each, B has 500 shares selling at $30 per share, C has 1000 shares at $50 per share, and D has 800 shares at $25 each. Find the value of this index initially, and after each of the first three months, assuming that the numbers of shares in the market stay the same, and the prices per share behave as follows:

	A	B	C	D
month 1	$42	$25	$51	$27
month 2	$44	$30	$49	$29
month 3	$40	$32	$48	$30

5. Suppose, as in Example 2, that Company 3 undergoes a 2-for-1 split. What happens to the weight of Company 3 in the FF index? Devise a way to reweight the Company 3 price such that the split does not result in a change in the index.

6. In Exercise 4, suppose that a local bank offers a Galesville Index Fund asset. A local resident, Gale, purchases 10 shares of the fund. Find Gale's rate of return during each of the first three months, and find her overall rate of return.

7. In Example 3, what increase in share value for the next time period would be necessary in order for Si's net losses to be zero? Assume that he pays cash to resolve the margin call. (Do not forget that he pays interest again on what he owes to the brokerage.)

8. Monica deals with a brokerage that demands that its customers continually maintain the 50% margin rate that is required initially on margin buying. The brokerage also charges 1% interest per period on borrowed amounts. She invests $2000 in a stock priced at $50 per share, taking maximum advantage of the initial margin. (There is a broker's fee that she pays in cash, not relevant to this computation.) What is the lowest price to which the stock can fall without resulting in a margin call? If the stock falls to $45 per share, how much cash will she have to pay to resolve the margin call? In this case, if she chooses to sell shares instead of paying cash, how many will she have to sell and what will be her new margin rate?

9. Jeff buys 40 shares of an asset currently priced at $50 per share from a broker, using the highest possible intial margin of 50%. The broker charges interest on the margined amount each week at the nominal yearly rate of 4%. After the initial purchase, the margin requirement is 40%, that is, equity must remain at least 40% of the full value of the shares. If the price of the asset at the end of one week is $48, and the second is $47, is a margin call necessary in either week? (Assume that, if it is made after the first week, Jeff's strategy is to pay the broker cash in order to reduce his margined amount sufficiently to meet the

requirement.)

10. Phil buys $10,000 of stock, fully utilizing the 50% initial margin requirement provided by his broker. There is a 40% maintenance margin requirement, and there is also 1% per time period interest charged on the amount borrowed. At the end of one time period, the stock decreases in value by 20%. Is there a margin call? If so, how much value in new securities must be added in order to meet the maintenance margin? If he chooses to pay off some of his margined amount instead, how much would he have to pay?

11. Steve buys 400 shares of stock C at a price of $20/share from a broker who charges a commission fee of 2% of the transaction value. The stock has an initial margin requirement of 50%, but Steve actually pays just enough to be able to put the commission fee on margin. There is a subsequent margin requirement of 45%, which is checked every month. The broker charges interest on the margin account monthly, at a yearly nominal rate of 5%. Steve instructs the broker in the case of a margin call to sell off just enough of his shares to comply with the monthly margin requirement. In the first 4 months the stock price moves from its original value through the values $22/share, $21, $19, and $18. Trace Steve's account activity through these 4 months. What is the final value of his equity? Is a margin call ever necessary?

12. Laverne senses a great deal in stock D, currently priced at $5.20 per share. Having just received a tax refund, she wants to invest part of it in 100 shares of this stock, using a 50% margin which is charged interest of .4% each month. Laverne's broker charges 2% of the purchase price for the transaction for which Laverne pays cash. Her more cautious roommate Shirley has some money saved and will buy 100 shares using cash and no margin, and she pays a fixed dealer fee of $20. If the price of stock D at the end of the year is $6.50 per share, find the effective yearly rate of return on the whole transaction for both Laverne and Shirley.

13. Ben believes that the stock of company "TanksAMillion" is going to go down. The stock is currently priced at $36, and he decides to sell short 50 shares. But his broker charges a 4% commission for this action. The proceeds will be invested in an interest-bearing account earning nominal yearly interest at rate 3% compounded monthly. Find the future value of the profit on his investment if the price at the end of the year is $32, and he is required to conclude the short sale. Plot a graph of Ben's profit as a function of the final price of the stock, and find the final price at which he breaks even.

14. Beau and his brother Jeff are interested in investing in a stock whose current price is $40 per share. The conservative Beau, who has read in the financial section of his newspaper that the stock is expected to go up soon, buys 100 shares, paying the full amount. Jeff, who plays his instincts, has a feeling that the stock will go down. If Jeff sells 100 shares of the stock short with a 50% cash

margin requirement and 1% per period interest on this deposited money, find an expression for the profit that each brother earns as a function of the price x of the stock after one period. Evaluate the profits if the final price is each of $37, $40, $43. Graph these two profit functions on the same set of axes. At what price x will the brothers' profits be identical?

15. An investor short sells a stock priced at $25 and is required to put down 40% of the proceeds as a deposit with the broker, which does not earn interest.

(a) How much of the stock should be short sold in order that the rate of return will be at least 10%, in the case that the final price is $22? How much should be short sold to yield a net profit of at least $100?

(b) If $10,000 is sold short, at most how much can the final price be in order that the rate of return will be at least 15%?

16. Investor Robin hopes to gain leverage in Stock A that she suspects is going to go up by borrowing Stock B from her sister Isabella and selling it. Robin thinks that Stock B will go down, so that when she has to return the shares to Isabella, she will not have to pay as much as she collected from her short sale. Stock A is currently priced at $40 per share, and Stock B is now $20 per share. Robin buys 10 shares of A with cash (ignore brokerage costs). What is her rate of return if Stock A rises to $45 per share? What will Robin's rate of return be if she sells short 10 shares of B to buy 5 additional shares of A, and A finishes at $45 while the final price of B is $15?

17. Consider the two stocks described in Exercise 16. Write a general formula for Robin's rate of return as a function of n if she sells short n shares of B to buy additional shares of A, where n is chosen so that the proceeds from the short sale are exactly enough to purchase an integer number of shares of A.

4.2   Portfolios of Risky Assets

The investor who puts all of his money into a single asset is certainly not being very wise, given that the asset may come tumbling down. It has long been known that advantage can be taken of **diversification**, that is, spreading wealth among several different assets, in order to reduce the risk of investing all wealth in one asset. It is our goal in this section to quantify the ideas of risk and reward for a portfolio of assets. Then we define a quantity that an investor may use to make personalized decisions about the tradeoffs between risk and reward, called the **risk aversion measure**. This will set up the mathematical model for the problem of mean-variance portfolio optimization in the following section.

4.2.1 Asset Rates of Return: Modeling and Estimation

Certain elements of Chapter 3 are worth reviewing here. A **random variable** X is a function from the space of all possible outcomes of a random experiment to some subset of the real line. Its **probability distribution** is characterized by the **probability mass function** $f(x) = P[X = x]$. This function can be used as a system of weights in the quantity we call the **mean**, which measures the center of the probability distribution:

$$\mu = E[X] = \sum_x x \cdot f(x). \tag{4.1}$$

The spread of the probability distribution is measured by the **variance**:

$$\sigma^2 = E\left[(X - \mu)^2\right] = \sum_x (x - \mu)^2 \cdot f(x). \tag{4.2}$$

A computational formula for variance can simplify the arithmetic at times:

$$\sigma^2 = E\left[X^2\right] - \mu^2 = \sum_x x^2 \cdot f(x) - \mu^2. \tag{4.3}$$

If a random sample of size n is taken, that is, a sequence $X_1, X_2, ..., X_n$ of independent and identically distributed random variables from the probability distribution, then the distributional mean μ can be estimated by the **sample mean**:

$$\bar{X} = \frac{1}{n} \cdot \sum_{i=1}^{n} X_i, \tag{4.4}$$

and the distributional variance σ^2 can be estimated by the **sample variance**:

$$S^2 = \frac{1}{n-1} \sum_{i=1}^{n} \left(X_i - \bar{X}\right)^2. \tag{4.5}$$

These two estimators have the desirable property that they are **unbiased** for their respective parameters, that is, $E\left[\bar{X}\right] = \mu$ and $E\left[S^2\right] = \sigma^2$.

The random variable that we will mostly be interested in is the rate of return on a risky asset. Recall that, if an asset has value $A_0 = A(t_0)$ at time t_0, and its value at time t_1 is $A_1 = A(t_1)$, then the **rate of return** of the asset over the time interval $[t_0, t_1]$ is:

$$R = \frac{A_1 - A_0}{A_0} = \frac{A(t_1) - A(t_0)}{A(t_0)}. \tag{4.6}$$

From now on, most assets that we will consider will be random in the sense that the price $A(t_1)$ at the later time is not perfectly known at the earlier time t_0, but rather has a probability distribution with mean rate of return μ and variance σ^2.

Example 1. If a common stock has price \$20 now, and at the end of the week its price can be \$19, \$21, or \$22 with probabilities $1/8$, $1/2$, and $3/8$, respectively, then there are three possible values of the rate of return R, namely, the possible prices minus 20, divided by 20. The mean rate of return would be:

$$\mu = E[R] = \frac{1}{8} \cdot \frac{19 - 20}{20} + \frac{1}{2} \cdot \frac{21 - 20}{20} + \frac{3}{8} \cdot \frac{22 - 20}{20} = .05625.$$

The variance of the rate of return would be $\sigma^2 = E\left[R^2\right] - \mu^2$, which is calculated as:

$$
\begin{aligned}
\sigma^2 &= \frac{1}{8} \cdot \left(\frac{19 - 20}{20}\right)^2 + \frac{1}{2} \cdot \left(\frac{21 - 20}{20}\right)^2 + \frac{3}{8} \cdot \left(\frac{22 - 20}{20}\right)^2 - \mu^2 \\
&= .0053125 - (.05625)^2 \\
&= .00214844. \quad \blacksquare
\end{aligned}
$$

As we move toward formulating optimization problems in which we decide the proportion of wealth to devote to each of several assets, we must face the reality that the probability distributions of the rates of return on the assets will not be known. Fortunately, we will not need the full distributions, but we will need to estimate the means and the variances from available historical data. (*A cautionary note*: such parameters are subject to change, so one should not count on using rate of return data for long historical periods. This can be a problem for estimation, since a smaller sample size of rates of return over a shorter historical period results in less precise estimates of the mean and variance parameters.)

Our work so far suggests an approach to estimation of the mean and variance of the rate of return on an asset. Observe asset values $A_0 = A(t_0)$, $A_1 = A(t_1)$, $A_2 = A(t_2)$, ... $A_n = A(t_n)$ at evenly spaced times $t_0 < t_1 < t_2 < \cdots < t_n$. Form a sample of rates of return:

$$R_1 = \frac{A_1 - A_0}{A_0}, R_2 = \frac{A_2 - A_1}{A_1}, \ldots, R_n = \frac{A_n - A_{n-1}}{A_{n-1}}. \tag{4.7}$$

Estimate the mean of the distribution of rate of return by:

$$\mu = \bar{R} = \frac{1}{n} \sum_{i=1}^{n} R_i, \tag{4.8}$$

and the variance of the distribution of rate of return by:

$$S^2 = \frac{1}{n-1} \sum_{i=1}^{n} \left(R_i - \bar{R}\right)^2. \tag{4.9}$$

Example 2. Data on asset prices is widely available from many print and web sources, including Yahoo! Finance and Nasdaq.com. Below are daily closing prices of Yahoo! stock from 2/3/14 through 3/20/14 obtained from the Nasdaq web site.

$$34.90, 35.66, 35.49, 36.24, 37.23, 37.76, 38.50, 38.11, 38.52,$$
$$38.23, 38.31, 37.81, 37.79, 37.29, 37.42, 37.26, 37.62, 38.47,$$
$$38.67, 38.25, 39.63, 39.50, 39.66, 38.70, 38.05, 37.56, 37.50,$$
$$37.23, 37.60, 39.11, 39.45, 38.61, 37.77$$

The rates of return can be easily calculated as:

$$\frac{35.66 - 34.90}{34.90} = .0217765, \quad \frac{35.49 - 35.66}{35.66} = -.00476725, \ldots,$$

$$\frac{37.77 - 38.61}{38.61} = -.021756.$$

Such rates of return are easy to calculate in a spreadsheet. For instance, if the data were located in column C starting in row 1, the first rate of return could be found by setting the formula

$$= (C2 - C1)/C1$$

into cell D1. Then just copy the formula down to the adjacent cells in column D. The entire list of rates of return is below.

$$0.0217765, -0.00476725, 0.0211327, 0.0273179, 0.0142358, 0.0195975,$$
$$-0.0101299, 0.0107583, -0.00752856, 0.0020926, -0.0130514, -0.000528961,$$
$$-0.013231, 0.00348619, -0.00427579, 0.00966184, 0.0225944, 0.00519886,$$
$$-0.0108611, 0.0360784, -0.00328034, 0.00405063, -0.0242057, -0.0167959,$$
$$-0.0128778, -0.00159744, -0.0072, 0.00993822, 0.0401596, 0.00869343,$$
$$-0.0212928, -0.021756$$

The sample mean and variance of this set of daily rates turn out to be $\bar{X} = .00260603$ and $S^2 = .000277241$. Spreadsheets also have built-in functions to calculate these quantities. In Excel, you would use the =AVERAGE(datarange) and =STDEV(datarange) functions to compute the sample mean and sample standard deviation. ■

It will be important in the assessment of the risk of a portfolio of assets to know about the association between the rates of return, if any. This may be characterized by the covariance or correlation between the rates. Recall that this is the expected product of the standardized variables; if X and Y are the two rates of return, μ_x and μ_y are their means, and σ_x and σ_y are their standard deviations, the **correlation** between X and Y is:

$$\rho = \frac{\text{Cov}(X, Y)}{\sigma_x \sigma_y} = \frac{E\left[(X - \mu_x)(Y - \mu_y)\right]}{\sigma_x \sigma_y}. \tag{4.10}$$

The more linearly associated the two rates of return are, the closer ρ is to 1 in magnitude. Negative correlation shows a decreasing relationship between Y and X, while positive correlation indicates an increasing relationship.

Example 3. Suppose that two rates of return X and Y are random variables such that only four values of the pair are possible: $(-.01, -.015)$, $(0, -.002)$, $(0, .002)$, and $(.01, .02)$ with equal probabilities. Find the correlation between X and Y.

Solution. Several quantities must be calculated before calculating ρ: μ_x, μ_y, σ_x^2, σ_y^2, σ_x, σ_y, and $E\left[(X - \mu_x)(Y - \mu_y)\right]$.

$$\mu_x = \frac{1}{4}(-.01 + 0 + 0 + .01) = 0.$$

$$\mu_y = \frac{1}{4}(-.015 - .002 + .002 + .02) = .005/4 = .00125.$$

$$\sigma_x^2 = E\left[(X - 0)^2\right] = \frac{1}{4}\left((-.01)^2 + 0^2 + 0^2 + (.01)^2\right) = .00005.$$

$$
\begin{aligned}
\sigma_y^2 &= E\left[(Y - .00125)^2\right] \\
&= \frac{1}{4}((-.015 - .00125)^2 + (-.002 - .00125)^2 + (.002 - .00125)^2 \\
&\quad + (.02 - .00125)^2) \\
&= .000156658.
\end{aligned}
$$

$$\sigma_x = \sqrt{\sigma_x^2} = \sqrt{.00005} = .007071.$$

$$\sigma_y = \sqrt{\sigma_y^2} = \sqrt{.000156658} = .0125175.$$

$$
\begin{aligned}
E[(X - 0)(Y - .00125)] &= \frac{1}{4}((-.01)(-.015 - .00125) + 0(-.002 - .00125) \\
&\quad + 0(.002 - .00125) + (.01)(.02 - .00125)) \\
&= .0000875.
\end{aligned}
$$

Therefore, the correlation is:

$$\rho = \frac{E[(X - 0)(Y - .00125)]}{\sigma_x \sigma_y} = \frac{.0000875}{.007071 \cdot .0125175} = .988575. \ \blacksquare$$

The correlation ρ, which would typically be unknown, can be estimated by the **sample correlation** R between the sampled rates of return $X_1, X_2, ..., X_n$ and $Y_1, Y_2, ..., Y_n$. Recall from Section 3.6 that this is defined by the **sample covariance** S_{xy} divided by the product of the two sample standard deviations.

$$R = \frac{S_{xy}}{S_x S_y} = \frac{\frac{1}{n-1}\sum_{i=1}^{n}\left(X_i - \bar{X}\right)\left(Y_i - \bar{Y}\right)}{S_x S_y}. \tag{4.11}$$

For instance, if asset 1 has rates of return .05, .08, and .05, and for the same time intervals asset 2 has rates of return .03, .02, .04, then:

$$\bar{X} = \frac{.05 + .08 + .05}{3} = .06; \bar{Y} = \frac{.03 + .02 + .04}{3} = .03;$$

$$S_x^2 = \frac{1}{2}\left((.05 - .06)^2 + (.08 - .06)^2 + (.05 - .06)^2\right) = .0003;$$

$$S_y^2 = \frac{1}{2}\left((.03 - .03)^2 + (.02 - .03)^2 + (.04 - .03)^2\right) = .0001;$$

$$
\begin{aligned}
S_{xy} &= \frac{1}{n-1}\sum_{i=1}^{n}\left(X_i - \bar{X}\right)\left(Y_i - \bar{Y}\right) \\
&= \frac{1}{2}((.05 - .06)(.03 - .03) + (.08 - .06)(.02 - .03) \\
&\quad + (.05 - .06)(.04 - .03)) \\
&= -.00015;
\end{aligned}
$$

$$R = \frac{-.00015}{\sqrt{.0003}\sqrt{.0001}} = -.866025.$$

For larger data sets, technology would be necessary to calculate the sample correlation.

4.2.2   Portfolio Rate of Return

Now we want to characterize the rate of return on a portfolio of risky assets. To fix the ideas we consider again the case of a single time period and just two possible risky assets between which the investor will split his wealth. Let the initial wealth of the investor be W and let the rates of return on the two assets be random variables R_1, R_2. Their means and variances are denoted by $\mu_1, \mu_2, \sigma_1{}^2, \sigma_2{}^2$ and their correlation is ρ. Also, denote:

$$x_1 = \text{proportion of wealth in asset 1}.$$

Thus, $x_2 = 1 - x_1$ is the proportion of wealth invested in the second asset. Since the change in value of any investment is the rate of return times the amount invested, it is clear that the final total portfolio value as a function of x_1 is:

initial wealth + gain in value of asset 1 + gain in value of asset 2

$$= W + x_1 W R_1 + (1 - x_1) W R_2.$$

The rate of return on the whole portfolio investment is:

$$R_p = \frac{\text{final wealth} - \text{initial wealth}}{\text{initial wealth}}$$
$$= \frac{W + x_1 W R_1 + (1 - x_1) W R_2 - W}{W} \qquad (4.12)$$
$$= x_1 R_1 + (1 - x_1) R_2.$$

Notice that the portfolio rate of return R_p does not depend on the initial wealth invested W. If, for instance, asset 1 gave a rate of return of .025, asset 2 had a .01 rate of return, and the investor devoted 2/3 of wealth to asset 1 and 1/3 to asset 2, the rate of return on the portfolio would be the weighted average:

$$\frac{2}{3}(.025) + \frac{1}{3}(.01) = \frac{.05 + .01}{3} = .02.$$

The analysis above easily generalizes to the case of more than two assets, as in Theorem 1 below. (You should be able to prove this theorem easily for, say, three assets, and if you know mathematical induction you can construct a general proof.)

Theorem 1. Let there be n assets with rates of return $R_i, i = 1, 2, ..., n$. If proportions $x_i, i = 1, 2, ..., n$ of wealth are invested in the assets, respectively, where $\sum_{i=1}^{n} x_i = 1$, then the rate of return on the portfolio of assets is

$$R_p = \sum_{i=1}^{n} x_i R_i. \qquad (4.13)$$

Now if the rates of return R_i on the assets contained in a portfolio are random variables, then so will be the portfolio rate of return R_p. We can study the distribution of R_p if we know the joint distribution of the constituent rates R_i, as in the following example.

Example 4. Suppose that there are three assets that are being considered in a portfolio and four possible combinations of their rates of return, as listed in the following table.

$$(R_1, R_2, R_3): \quad (-.01, 0, .04) \quad (0, .02, .02) \quad (.01, .04, .01) \quad (.02, .05, 0)$$
$$\text{probability}: \qquad 1/8 \qquad\quad 1/4 \qquad\qquad 1/4 \qquad\qquad 3/8$$

For a portfolio that devotes equal weight to each of the three assets, find the possible values of the portfolio rate of return and their probabilities.

Solution. For each of the four possible scenarios, we find the portfolio rate of return:

$$(-.01, 0, .04): R_p = \frac{1}{3}(-.01) + \frac{1}{3}(0) + \frac{1}{3}(.04) = .01$$

$$(0, .02, .02): R_p = \frac{1}{3}(0) + \frac{1}{3}(.02) + \frac{1}{3}(.02) = .01\bar{3}$$

$$(.01, .04, .01): R_p = \frac{1}{3}(.01) + \frac{1}{3}(.04) + \frac{1}{3}(.01) = .02$$

$$(.02, .05, 0): R_p = \frac{1}{3}(.02) + \frac{1}{3}(.05) + \frac{1}{3}(0) = .02\bar{3}$$

So the portfolio rate of return has a discrete probability distribution summarized in the table below.

value of R_p:	.01	.01$\bar{3}$	.02	.02$\bar{3}$
probability:	1/8	1/4	1/4	3/8

Notice that, with the distribution of R_p in hand, we can compute its mean and variance as well.

$$\mu_p = E[R_p] = \frac{1}{8}(.01) + \frac{1}{4}(.01\bar{3}) + \frac{1}{4}(.02) + \frac{3}{8}(.02\bar{3}) = .018333;$$

$$\begin{aligned}
\sigma_p^2 &= \mathrm{Var}(R_p) \\
&= E[R_p^2] - \mu_p^2 \\
&= \frac{1}{8}(.01)^2 + \frac{1}{4}(.01\bar{3})^2 + \frac{1}{4}(.02)^2 + \frac{3}{8}(.02\bar{3})^2 - .018333^2 \\
&= .000025. \ \blacksquare
\end{aligned}$$

Although, if we are equipped with information about the joint distribution of the asset rates of return R_i in a portfolio, we could in theory find the probability distribution of the portfolio rate of return, as in the previous example, it is not very practical to go this route if there are many possible joint values of the R_is. Besides, we usually don't even know the individual distributions of asset rates of return, much less how they behave jointly. But it would still be highly desirable to have formulas for the mean and variance of the portfolio rate of return. With luck, such formulas will depend on parameters that are estimable from background data.

We are indeed in luck in this regard. For simplicity, consider first the two-asset case with rate of return variables R_1 and R_2, whose means are μ_1 and μ_2. The portfolio mean rate of return is easy; by linearity of expectation:

$$\mu_p = E[R_p] = E[x_1 R_1 + (1 - x_1) R_2] = x_1 \mu_1 + (1 - x_1) \mu_2. \qquad (4.14)$$

The variance of R_p can also be computed as follows:

$$\begin{aligned}
\sigma_p^2 &= \operatorname{Var}(R_p) \\
&= \operatorname{Var}(x_1 \, R_1 + (1 - x_1) \, R_2) \\
&= E\left[((x_1 R_1 + (1 - x_1) \, R_2) - (x_1 \mu_1 + (1 - x_1) \, \mu_2))^2\right] \\
&= E\left[(x_1 \, (R_1 - \mu_1) + (1 - x_1) \, (R_2 - \mu_2))^2\right] \\
&= x_1^2 E\left[(R_1 - \mu_1)^2\right] + (1 - x_1)^2 \, E\left[(R_2 - \mu_2)^2\right] \\
&\quad + 2x_1 \, (1 - x_1) \, E\left[(R_1 - \mu_1) \, (R_2 - \mu_2)\right] \\
&= x_1^2 \sigma_1^2 + (1 - x_1)^2 \, \sigma_2^2 + 2x_1 \, (1 - x_1) \operatorname{Cov}(R_1, R_2).
\end{aligned}$$
(4.15)

Since it follows from the definition of covariance that $\operatorname{Cov}(R_1, R_2) = \rho_{12} \sigma_1 \sigma_2$, where ρ_{12} is the correlation between R_1 and R_2, both the portfolio mean and variance can be estimated using the sample means, variances, and correlation.

The extension of formulas (4.14) and (4.15) to many assets is straightforward.

Theorem 2. Let there be n assets with rates of return $R_i, i = 1, 2, ..., n$. Let the ith asset mean be denoted by μ_i, and let the ith asset variance be $\sigma_i{}^2$. If proportions $x_i, i = 1, 2, ..., n$ of wealth are invested in the assets, respectively, where $\sum_{i=1}^n x_i = 1$, then the mean and variance of the rate of return on the portfolio of assets are:

$$\mu_p = \sum_{i=1}^n x_i \mu_i \quad , \text{ and}$$
(4.16)

$$\sigma_p^2 = \sum_{i=1}^n x_i^2 \sigma_i^2 + 2 \sum_{i=1}^n \sum_{j=1}^{i-1} x_i x_j \operatorname{Cov}(R_i, R_j).$$
(4.17)

(The double sum in formula (4.17) can also be written $\sum_{i=1}^n \sum_{j \neq i} x_i x_j \operatorname{Cov}(R_i, R_j)$, since $\operatorname{Cov}(R_i, R_j) = \operatorname{Cov}(R_j, R_i)$. Another form is $2 \sum_{i=1}^n \sum_{j<i} x_i x_j \rho_{ij} \sigma_i \sigma_j$, where ρ_{ij} is the correlation between R_i and R_j.)

Proof. For formula (4.16), linearity implies that:

$$\mu_p = E[R_p] = E\left[\sum_{i=1}^n x_i R_i\right] = \sum_{i=1}^n x_i E[R_i] = \sum_{i=1}^n x_i \mu_i.$$

You are asked for a general proof of formula (4.17) for the portfolio variance in Exercise 9; here is a derivation in the case of three assets:

$$\begin{aligned}
\sigma_p^2 &= \text{Var}\,(R_p) \\
&= \text{Var}\,(x_1 R_1 + x_2 R_2 + x_3 R_3) \\
&= E\left[\left((x_1 R_1 + x_2 R_2 + x_3 R_3) - (x_1\mu_1 + x_2\mu_2 + x_3\mu_3)\right)^2\right] \\
&= E\left[\left(x_1\,(R_1 - \mu_1) + x_2\,(R_2 - \mu_2) + x_3\,(R_3 - \mu_3)\right)^2\right] \\
&= x_1^2 E\left[(R_1 - \mu_1)^2\right] + x_2^2 E\left[(R_2 - \mu_2)^2\right] + x_3^2 E\left[(R_3 - \mu_3)^2\right] \\
&\quad + 2x_1 x_2 E\left[(R_1 - \mu_1)(R_2 - \mu_2)\right] + 2x_1 x_3 E\left[(R_1 - \mu_1)(R_3 - \mu_3)\right] \\
&\quad + 2x_2 x_3 E\left[(R_2 - \mu_2)(R_3 - \mu_3)\right] \\
&= x_1^2\sigma_1^2 + x_2^2\sigma_2^2 + x_3^2\sigma_3^2 + 2x_1 x_2 \text{Cov}\,(R_1, R_2) + 2x_1 x_3 \text{Cov}\,(R_1, R_3) \\
&\quad + 2x_2 x_3 \text{Cov}\,(R_2, R_3)\,. \quad\blacksquare
\end{aligned}$$

Example 5. Suppose that there are three assets under consideration, and we have estimated the means, variances, and correlations between the rates of return as follows:

$$\mu_1 = .025;\ \mu_2 = .03;\ \mu_3 = .04;\ \sigma_1^2 = .001;\ \sigma_2^2 = .0015;\ \sigma_3^2 = .002;$$

$$\rho_{12} = .35;\ \rho_{23} = -.28;\ \rho_{13} = -.10.$$

In choosing between a portfolio with equal weight on all assets and one with weights $x_1 = 1/2, x_2 = 1/4, x_3 = 1/4$, is the decision clear?

Solution. For the portfolio with equal weights, the mean rate of return is:

$$\mu_p = \frac{1}{3}(.025) + \frac{1}{3}(.03) + \frac{1}{3}(.04) = .03167.$$

For the other portfolio, the mean rate of return is:

$$\mu_p = \frac{1}{2}(.025) + \frac{1}{4}(.03) + \frac{1}{4}(.04) = .03.$$

So the first portfolio has the higher expected rate of return.

But let us check the portfolio variances. For the equally weighted portfolio, we have:

$$\begin{aligned}
\sigma_p^2 &= \sum_{i=1}^{n} x_i^2\sigma_i^2 + 2\sum_{i=1}^{n}\sum_{j<i} x_i x_j \text{Cov}\,(R_i, R_j) \\
&= \left(\frac{1}{3}\right)^2 (.001 + .0015 + .002) + 2\left(\frac{1}{3}\right)^2 (.35\sqrt{.001}\sqrt{.0015} \\
&\quad - .28\sqrt{.0015}\sqrt{.002} - .10\sqrt{.001}\sqrt{.002}) \\
&= .000456.
\end{aligned}$$

Notice that this is substantially smaller than any of the individual asset variances. For the second portfolio we have:

$$\sigma_p^2 = \sum_{i=1}^{n} x_i^2 \sigma_i^2 + 2 \sum_{i=1}^{n} \sum_{j<i} x_i x_j \text{Cov}(R_i, R_j)$$

$$= \left(\frac{1}{2}\right)^2 (.001) + \left(\frac{1}{4}\right)^2 (.0015) + \left(\frac{1}{4}\right)^2 (.002)$$

$$+ 2\left(\left(\frac{1}{2}\right)\left(\frac{1}{4}\right).35\sqrt{.001}\sqrt{.0015} - \left(\frac{1}{4}\right)\left(\frac{1}{4}\right).28\sqrt{.0015}\sqrt{.002}\right.$$

$$\left. - \left(\frac{1}{2}\right)\left(\frac{1}{4}\right).10\sqrt{.001}\sqrt{.002}\right)$$

$$= .000480.$$

Therefore the second portfolio has a slightly higher variance than the first, while also giving a lower mean rate of return. Because investors will value mean return and avoid risk, the first portfolio appears to be more attractive than the second regardless of the degree to which the investor may be averse to risk. ■

4.2.3   Risk Aversion

The analysis of Example 5 should begin to give you an idea about how to handle the portfolio optimization problem. Investors will value high mean return and dislike high variance, which is a measure of the risk of the investment. If one possible investment has a higher mean and lower variance than another, the choice is clear. But what if two investments were such that the first had a higher mean than the second, but also had higher variance? How could the investor make the choice? It would depend how a particular investor quantifies the tradeoff between increased variance and increased mean.

Suppose that we are able to question an investor about how many extra units of expected return she would expect to receive in exchange for an extra unit of variance. The answer, call it a, is called her (mean-variance) **risk aversion**.

$$\text{risk aversion } a = \frac{\text{change in mean return expected}}{\text{change in variance}}. \qquad (4.18)$$

If $a = 2$, for example, and one investment has mean return .03 and variance .01, then, for her to invest in another opportunity with variance .02, (.01 more), she would require .02 more units of mean return, i.e. a mean return of .05 (because $\frac{.05-.03}{.02-.01} = 2$). In general, for mean returns of μ_1 and μ_2, and variances σ_1^2 and σ_2^2, an investor with risk aversion a would then be indifferent between investment 1 and investment 2 as long as

$$a = \frac{\mu_2 - \mu_1}{\sigma_2^2 - \sigma_1^2} \iff a\left(\sigma_2^2 - \sigma_1^2\right) = \mu_2 - \mu_1 \iff \mu_2 - a\sigma_2^2 = \mu_1 - a\sigma_1^2.$$

It starts to make sense to model the investor's decision as picking an investment, or portfolio of several assets, to maximize $\mu - a\sigma^2$.

Example 6. For an investor with risk aversion 5, which investment would be preferred: one with mean .08 and variance .02 or one with mean .09 and variance .025? How about risk aversion 20? At what risk aversion would the investor be indifferent between the two?

Solution. An investor with risk aversion 5 would demand $5(.005) = .025$ additional units of mean return in exchange for tolerating the $.025 - .02 = .005$ units of additional variance in the second investment. Since the increase in mean is only $.09 - .08 = .01$, this investor's demands are not met, and so the first investment would be preferred. The same would hold for an investor with a higher risk aversion of 20, who would demand $20(.005) = .1$ extra units of mean return. The investor would consider these two the same if her risk aversion is:

$$a = \frac{.09 - .08}{.025 - .02} = \frac{.01}{.005} = 2. \ \blacksquare$$

We have argued that, if a is the risk aversion constant for an investor, then the investor is indifferent between one investment and another if:

$$\mu_2 - a\sigma_2^2 = \mu_1 - a{\sigma_1}^2 \tag{4.19}$$

and since high mean return is good and high variance is bad, a sensible (not necessarily the only) criterion for a best investment is to maximize $\mu - a\sigma^2$. But with respect to what? What are the choices? Our next example uses this idea of penalizing variance according to risk aversion to choose between two assets. But in the next section, we will formulate the portfolio optimization problem as one of maximizing

$$\mu_p - a\sigma_p^2 \tag{4.20}$$

with respect to the choice of weights x_i. We now have all the elements necessary to do this.

Example 7. Suppose an investor has risk aversion $a = 10$, and asset 1 has mean and variance of return $\mu_1 = .05, \sigma_1^2 = .04$, and asset 2 has mean and variance of return $\mu_2 = .07, {\sigma_2}^2 = .05$. If the investor must choose either one or the other, then we can compare as follows.

$$\mu_1 - a\sigma_1^2 = .05 - 10(.04) = -.35.$$

$$\mu_2 - a\sigma_2^2 = .07 - 10(.05) = -.43.$$

So investment 1 is better for this investor. $\blacksquare$

Finally, it is reasonable to ask how one might determine what a particular investor's risk aversion is. One simple method can be described as follows. Suppose that you present the investor with two investment scenarios. One asset gives a possible range of rates of return from $-.01$ through $.03$, with $.01$ being the likeliest. Another gives a possible range of $-.03$ through $.06$, with $.015$ being the likeliest. The investor prefers the second investment. What does that say about the investor's risk aversion a?

Using the data in the last paragraph and the fact that the risk aversion value determines the preference via computation of the quantity $\mu - a\sigma^2$, we can set up an inequality that a must satisfy:

$$\mu_2 - a\sigma_2^2 > \mu_1 - a\sigma_1^2.$$

The investor is likely to interpret the information about the middle value of the given range as the mean rate of return, so $\mu_1 = .01$ and $\mu_2 = .015$ are good estimates to use. Although the casual investor may not know much about the meaning of standard deviation and Chebyshev's inequality, with our background we might interpret the ranges as intervals of length 4σ about the mean. This would mean that:

$$4\sigma_1 \approx .03 - (-.01) = .04 \Longrightarrow \sigma = .01;$$

$$4\sigma_2 \approx .06 - (-.03) = .09 \Longrightarrow \sigma = .0225.$$

Therefore the inequality for a becomes:

$$.015 - a(.0225)^2 > .01 - a(.01)^2 \Longrightarrow .005 > a\left((.0225)^2 - (.01)^2\right)$$

$$\Longrightarrow 12.31 > a.$$

It should be apparent that a series of such questions can be asked of the investor in order to derive several other inequalities for a so as to zero in on an appropriate value. We follow up on this idea in Exercise 17.

Important Terms and Concepts

Portfolio rate of return - The rate of return $R_p = \sum_{i=1}^{n} x_i R_i$ for a portfolio of assets with rates of return R_i in which a proportion of total wealth x_i is invested in asset i, $i = 1, 2, ..., n$.

Portfolio mean rate of return - The expected value $\mu_p = \sum_{i=1}^{n} x_i \mu_i$ for a portfolio of assets with mean rates of return μ_i in which a proportion of total wealth x_i is invested in asset i, $i = 1, 2, ..., n$.

Portfolio variance of rate of return - The variance of the rate of return

$$\sigma_p^2 = \sum_{i=1}^{n} x_i^2 \sigma_i^2 + 2\sum_{i=1}^{n}\sum_{j=i}^{i} x_i x_j \text{Cov}\left(R_i, R_j\right)$$

for a portfolio of assets with rates of return R_i in which a proportion of total wealth x_i is invested in asset i, $i = 1, 2, ..., n$.

Risk aversion - The ratio between difference in mean return and difference in variance for two investments for which an investor is indifferent between the two. Also interpreted as the additional mean return required to bear an additional unit of variance. Formula: $a = \frac{\mu_2 - \mu_1}{\sigma_2^2 - \sigma_1^2}$.

Exercises 4.2

1. One asset begins at a value of \$100 and after one time unit can be worth either \$102 or \$99. A second asset begins at a value of \$50 and after one time unit follows the first asset in lock step, worth \$51 in the first scenario and \$49 in the second. Find the possible portfolio rate of return values if a portfolio has: (a) equal weights on the two assets; (b) weights of 3/4 on the first and 1/4 on the second.

2. Assume that the rate of return on an asset over a particular time period is a random variable R with four possible values: $-.01, 0, .005, .015$, occurring with probabilities $\frac{1}{6}, \frac{1}{6}, \frac{1}{3}$, and $\frac{1}{3}$, respectively. If the initial price of the asset is \$20, find the distribution, mean, and variance of the price A_1 at the end of the time period.

3. Let X and Y be two random rates of return, such that the possible values of the pair are $(0, .01), (.01, 0)$, and $(.02, -.01)$ with probabilities $\frac{1}{4}, \frac{1}{2}$, and $\frac{1}{4}$, respectively. Compute the individual means and variances of X and Y, and the correlation between them.

4. Below are sample data on weekly rates of return for two stocks. Use the data to estimate the mean and variance of the portfolio rate of return on a portfolio that combines the two stocks in equal proportions.

stock 1: $.001, -.003, .007, .008, -.002, -.01, .04, .005, .003, -.001, .002$

stock 2: $.004, .003, -.002, -.005, 0, .01, 0, .001, .03, -.007, -.001$

5. Suppose that an asset starts at a price A_0 at time 0, and moves at time 1 either to a price of $1.01 A_0$ or $.98 A_0$ with probability p and $1 - p$, respectively. Is there a probability p that forces $E[A_1] = A_0$? Is it unique?

6. Suppose that an asset starts at a price A_0 at time 0, and moves at time 1 either to a price of $u \cdot A_0$ or $d \cdot A_0$ with probabilities $1/2$ each, where u and d are positive constants with $u > d$. Are there values of u and d that imply $E[A_1] = 1.05 A_0$? Are these unique?

7. Suppose that there are two assets, the first with initial price $20 and the second with intial price $10. The two assets have just two possible values for their price at time 1, as indicated in the table below.

values of (A_1, A_2): ($22, $13.50) ($19, $4.50)
probability: 2/5   3/5

Assume that it is possible to sell short, that is, to hold a negative proportion of wealth on one asset (essentially selling it without owning it in order to invest extra money in the other asset). What portfolio weights would result in a rate of return of .05 on the portfolio regardless of which of the two outcomes happens?

8. Suppose that there are two assets that are being considered in a portfolio, and three possible combinations of their rates of return, as listed in the following table.

values of (R_1, R_2): $(0, .03)$ $(.02, .02)$ $(.03, .05)$
probability: 1/4   1/2   1/4

Find the distribution, mean, and variance of the portfolio rate of return for: (a) the portfolio with weights 2/3 and 1/3, respectively, on assets 1 and 2; (b) the portfolio with equal weights on assets 1 and 2.

9. Prove formula (4.17) in Theorem 2 for the variance of a portfolio for general n.

10. Consider three independent assets with mean rates of return .03, .02, and .04 and variances .002, .001, and .003. Find the mean and variance of the portfolio that has proportions of wealth: (a) 1/3 for each asset; (b) 1/2 for assets 2 and 3, and nothing invested in asset 1. Is one portfolio clearly better than the other?

11. Suppose that there are four possible outcomes w_1, w_2, w_3, w_4 for which the joint rates of return on two assets R_1 and R_2 are given in the table below. Find the mean and variance of the rate of return on the portfolio that combines assets 1 and 2 in equal proportions.

outcome	$P[w]$	value of R_1	value of R_2
w_1	1/3	−.02	−.01
w_2	1/6	0	.02
w_3	1/3	.03	0
w_4	1/6	.05	.04

12. Consider a two-asset problem in which $\mu_1 = .05$, $\mu_2 = .08$, $\sigma_1 = .4$, $\sigma_2 = .6$. In the case where $\rho = -1$, find the portfolio weights such that $\mu_p = .07$. What is the corresponding standard deviation σ_p in this case?

13. Consider two assets with mean rates of return .04 and .05, variances .0003 and .0004, and correlation .2. (a) Write expressions in terms of x for the mean and variance of the rate of return on a portfolio that devotes a proportion x of

wealth to the first asset. (b) Suppose the investor decides to choose a portfolio that maximizes the mean return minus a penalty coefficient of five times the portfolio variance. Find the optimal x for this investor's criterion.

14. Derive a general expression for mean and variance of a two-asset portfolio if asset 1 is non-risky with sure rate of return .045. With this asset, the second asset of the previous problem, and the investor's strategy described in part (b) of that problem, find the best proportion of wealth x to invest in the non-risky asset.

15. What risk aversion is necessary for the investor in Example 7 to be indifferent between the two investments?

16. Suppose that, over a given time interval, we estimate the rate of return parameters for Walmart and Commonwealth Edison stock as:

Walmart: mean = .00225, variance = .0002258

ComEd: mean = .001125, variance = .000148517

Does an investor with risk aversion 2 prefer Walmart or ComEd? How about an investor with risk aversion 8? Find a value for the risk aversion such that investors above that value prefer ComEd and investors below that value prefer Walmart.

17. In the example of the investor's preferences at the end of the section, suppose that the investor also prefers asset 2 to another asset whose range of values is $-.05$ through .09, with .07 being likeliest. What more can be said about the investor's risk aversion?

4.3 Optimal Portfolio Selection

At the end of the last section we suggested that a good criterion for an investor to use when selecting a portfolio is to maximize:

$$\mu_p - a\sigma_p^2, \qquad (4.21)$$

where μ_p is the mean portfolio rate of return, a is the investor's risk aversion constant, and σ_p^2 is the variance of the portfolio rate of return. In the case where there are n assets, the expression in (4.21) is a function of all of the proportions x_i of wealth to be devoted to all of the assets. In light of Theorem 2 of Section 4.2, the objective that we must maximize becomes:

$$f(x_1, x_2, ..., x_n) = \begin{array}{l} \sum_{i=1}^{n} x_i\mu_i - a \cdot (\sum_{i=1}^{n} x_i^2\sigma_i^2 \\ + 2\sum_{i=1}^{n} \sum_{j<i} x_i x_j \text{Cov}(R_i, R_j)) \end{array} \qquad (4.22)$$

where the μ_is and σ_is are the means and standard deviations of the individual asset rates of return. We have also seen how to estimate all of the problem parameters from historical data. Thus, we are now ready to solve portfolio problems. In all of the cases that we will consider, the objective function to be maximized in formula (4.22) is quadratic as a function of each of the x_is with a negative leading coefficient, so that there is a critical point that is a global maximum.

4.3.1 Two-Asset Problems

Let us first do a few simple examples in the case of two assets.

Example 1. Consider an investor with risk aversion 15 who will invest in a risk-free asset with rate of return .03 and a risky asset with mean rate of return .06 and standard deviation .04. What portfolio should the investor choose?

Solution. Since the first asset is risk free, it has variance 0, and its covariance with the risky asset is also 0. This simplifies the objective function in (4.22) quite a bit. The goal is to maximize $f(x_1, x_2) = x_1 \cdot .03 + x_2 \cdot .06 - 15 \cdot x_2{}^2 \cdot .04^2$. Also, $x_1 = 1 - x_2$ can be eliminated, yielding the following function of x_2 to maximize:

$$\begin{aligned} g(x_2) &= (1 - x_2) \cdot .03 + x_2 \cdot .06 - 15 \cdot x_2^2 \cdot .04^2 \\ &= .03x_2 + .03 - 15(.04)^2 x_2^2. \end{aligned}$$

The negative coefficient on the square term assures us that the critical point will be a maximum. Differentiation gives $.03 - 30(.04)^2 x_2 = 0$, and therefore the optimal proportion in asset 2 is:

$$x_2 = \frac{.03}{30(.04)^2} = .625.$$

This means that a proportion $1 - .625 = .375$ of the wealth is held in the risk-free asset. ∎

Observe that the ability to eliminate one portfolio variable in these two-asset problems gives us a simple differential calculus optimization problem, as we see again in the next example.

Example 2. Assume a market with two independent risky assets whose rates of return have means μ_1 and μ_2 and variances σ_1^2 and σ_2^2. Derive general expressions for the optimal portfolio weights.

Solution. Since the assets are independent, the covariance between them is 0. Denote the proportion in the first asset by x, so that the proportion in the second is $(1 - x)$. Then we can write the expression for the portfolio objective function as:

$$\text{maximize}: E\left[x\ R_1 + (1-x)R_2\right] - a \cdot \text{Var}\left(x\ R_1 + (1-x)R_2\right)$$
$$= x\mu_1 + (1-x)\mu_2 \ - a\left(x^2\sigma_1^2 + (1-x)^2\sigma_2^2\right).$$

This is a quadratic function of the variable x, with negative leading coefficient $-a\left(\sigma_1^2 + \sigma_2^2\right)$, so there will be a unique maximum. The derivative of the objective function is:

$$\mu_1 - \mu_2 - a\left(2x\sigma_1^2 - 2(1-x)\sigma_2^2\right) = \mu_1 - \mu_2 - a\left(2x\left(\sigma_1^2 + \sigma_2^2\right) - 2\sigma_2^2\right).$$

Setting this to zero yields:

$$\mu_1 - \mu_2 + 2a\sigma_2^2 = 2a\left(\sigma_1^2 + \sigma_2^2\right)x$$

$$\implies x = \frac{\mu_1 - \mu_2 + 2a\sigma_2^2}{2a\left(\sigma_1^2 + \sigma_2^2\right)}. \tag{4.23}$$

It follows that the optimal proportion in the second asset is:

$$\begin{aligned}
1 - x &= 1 - \frac{\mu_1 - \mu_2 + 2a\sigma_2^2}{2a\left(\sigma_1^2 + \sigma_2^2\right)} \\
&= \frac{2a\left(\sigma_1^2 + \sigma_2^2\right) - \left(\mu_1 - \mu_2 + 2a\sigma_2^2\right)}{2a\left(\sigma_1^2 + \sigma_2^2\right)} \\
&= \frac{\mu_2 - \mu_1 + 2a\sigma_1^2}{2a\left(\sigma_1^2 + \sigma_2^2\right)}. \ \blacksquare
\end{aligned} \tag{4.24}$$

Formula (4.24) in the last example allows us to look at the dependence of the solution on parameters. Suppose for concreteness that asset 2 has larger mean and also larger variance than asset 1. Certainly the form of the numerator in the formula implies that, as the mean return μ_2 increases, all other things remaining the same, the proportion invested in asset 2 increases. Similarly, as μ_1 decreases, this proportion also increases. Now look at the dependence of $1 - x$ on σ_2^2. Since σ_2^2 appears only in the denominator, $1 - x$ decreases as σ_2^2 increases. Similarly, formula (4.23) implies that, as σ_1^2 increases, the proportion x in asset 1 decreases, so the proportion in asset 2 increases. All of these observations are consistent with the principle that, as asset 2 becomes more desirable compared to asset 1 from either the improved mean or variance perspective, a greater proportion of wealth will be invested in it. The dependence of $1 - x$ on a is not so apparent from formula (4.24). But we can look at a graph of the proportion of wealth $1 - x$ in asset 2 as a function of the risk aversion. If, for instance, $\mu_1 = .04$, $\mu_2 = .06$, $\sigma_1 = .02$, and $\sigma_2 = .05$, then we have that:

$$1 - x = \text{prop of wealth in asset 2} = \frac{.06 - .04 + 2a(.02)^2}{2a\left(.02^2 + .05^2\right)}.$$

Figure 4.1 shows the relationship. As we might think, the larger is the investor's aversion to risk, the smaller is the proportion of wealth that the investor gives to asset 2, whose variance is higher than that of asset 1. Notice that, for very low risk aversions, it is optimal for the investor to put more than 100% of her wealth in asset 2, short-selling asset 1 in order to do so.

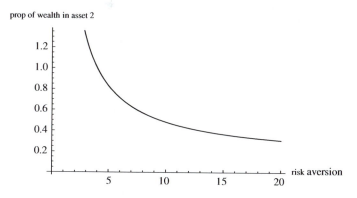

prop of wealth in asset 2

Figure 4.1: As risk aversion grows, the proportion of wealth in the riskier asset decreases.

Example 3. Suppose that two stocks, Walmart and Commonwealth Edison, have estimated weekly means and variances over a certain period of time as follows:

Walmart: mean = .0022, variance = .00023

ComEd: mean = .0011, variance = .00015

If the correlation between the two rates is estimated to be .34, find the optimal portfolio of the two stocks for an investor with risk aversion $a = 6$. What are the portfolio mean and variance for this optimal portfolio?

Solution. Recall the identity that the covariance between two random variables is the correlation times the product of the standard deviations: $\sigma_{xy} = \rho \sigma_x \sigma_y$. Therefore, if x_1 is the proportion of wealth in Walmart and $x_2 = 1 - x_1$ is the proportion in ComEd, then the objective function becomes:

$$x_1 \mu_1 + (1 - x_1) \mu_2 - a \left(x_1^2 \sigma_1^2 + (1 - x_1)^2 \sigma_2^2 + 2x_1 (1 - x_1) \rho \sigma_1 \sigma_2 \right).$$

Estimating the parameters with the sample information in the problem statement, we get:

$$.0022x_1 + .0011 (1 - x_1) - 6(.00023x_1^2 + .00015 (1 - x_1)^2$$
$$+ 2(.34) \left(\sqrt{.00023} \right) \left(\sqrt{.00015} \right) x_1 (1 - x_1))$$
$$= .0002 + .00214x_1 - .00152x_1^2.$$

Setting the derivative equal to zero to find the critical point, we obtain:

$$.00214 - .00304x_1 = 0 \implies x_1 = \frac{.00214}{.00304} = .704.$$

Again, the fact that the objective function is quadratic with leading coefficient less than zero implies that the critical point is a maximum. So about 70% of the wealth is dedicated to Walmart and 30% to Commonwealth Edison. Therefore, the mean return on the optimal portfolio is:

$$.0022(.704) + .0011(.296) = .00187,$$

and the variance of the rate of return on the optimal portfolio is:

$$.00023\left(.704^2\right) + .00015(.296)^2 + 2(.34)(\sqrt{.00023})(\sqrt{.00015})(.704)(.296)$$
$$= .000153.$$

It is interesting that the diversified portfolio is able to have a mean return almost as high as Walmart and a variance almost as small as ComEd. ∎

4.3.2   Multiple-Asset Problems

It is fairly easy to solve for the optimal portfolio in the case of n independent assets using elementary tools. The objective function in this case is:

$$f(x_1, x_2, ..., x_n) = \sum_{i=1}^{n} x_i \mu_i - a \cdot \left(\sum_{i=1}^{n} x_i^2 \sigma_i^2\right). \qquad (4.25)$$

However, remember that the portfolio weights must add to 1. So one of the variables, say x_n, may be expressed in terms of the others as:

$$x_n = 1 - \sum_{i=2}^{n-1} x_i. \qquad (4.26)$$

Substitution of (4.26) into (4.25) produces an unconstrained optimization problem in the variables $x_1, x_2, ..., x_{n-1}$. One just computes partial derivatives with respect to all variables and sets these partials equal to zero, solving the resulting system of linear equations. Recall that a partial derivative of a function with respect to a variable x_i is computed by differentiating the expression defining the function with respect to that variable and treating all other variables as constants in the process.

(To solve the problem properly and in generality really needs the method of Lagrange multipliers for constrained optimization problems, and it also benefits from some techniques from linear algebra. We will not attempt to be this ambitious in this book, however, in order to keep the level more accessible.)

Example 4. Suppose that an investor of risk aversion 8 is interested in three independent risky assets whose mean rates of return are .03, .05, and .07, respectively, and whose variances are .001, .002, and .003. Let us find the optimal portfolio.

Solution. As a function of all three portfolio weight variables, our objective is:

$$f(x_1, x_2, x_3) = .03x_1 + .05x_2 + .07x_3 - 8\left(.001x_1^2 + .002x_2^2 + .003x_3^2\right).$$

Substitute $x_3 = 1 - x_1 - x_2$, and write the objective as:

$$g(x_1, x_2) = \begin{aligned}[t] &.03x_1 + .05x_2 + .07(1 - x_1 - x_2)\\ &- 8\left(.001x_1^2 + .002x_2^2 + .003(1 - x_1 - x_2)^2\right).\end{aligned}$$

Routine, but tedious, expansion and combination of like terms results in the simpler expression:

$$g(x_1, x_2) = .046 + .008x_1 + .028x_2 - .032x_1^2 - .048x_1x_2 - .04x_2^2.$$

The partial derivative of g with respect to x_1 is:

$$\frac{\partial g}{\partial x_1} = .008 - .064x_1 - .048x_2,$$

and the partial with respect to x_2 is:

$$\frac{\partial g}{\partial x_2} = .028 - .048x_1 - .08x_2.$$

Setting both partial derivatives to zero gives us the system of equations:

$$\begin{cases} .064x_1 + .048x_2 = .008 \\ .048x_1 + .08x_2 = .028. \end{cases}$$

The solution is $x_1 = -.25$, $x_2 = .5$, and therefore $x_3 = 1 - x_1 - x_2 = .75$. The interpretation of the results is that an investor with $100 to spend would sell short the first asset to raise an additional $25, then invest $50 in the second asset and $75 in the third asset. ∎

To optimize the complete portfolio objective function in the case of multiple correlated assets is quite cumbersome. We will be content to illustrate the solution of a numerical example with three assets.

Example 5. Let us reconsider the three risky assets of Example 4, but now, instead of assuming independent rates of return, let us assume that there are correlations of .5 between assets 1 and 2, −.3 between assets 1 and 3, and −.2

between assets 2 and 3. Find the optimal portfolio for the same investor (with risk aversion 8).

Solution. In the objective function we can write $x_3 = 1 - x_1 - x_2$ again, which produces the following function of two variables:

$$
\begin{aligned}
g\left(x_1, x_2\right) =\ & .03x_1 + .05x_2 + .07\left(1 - x_1 - x_2\right) \\
& - 8(.001x_1^2 + .002x_2^2 + .003(1 - x_1 - x_2)^2 \\
& + 2x_1x_2(.5)\sqrt{.001}\sqrt{.002} + 2x_1(1 - x_1 - x_2)(-.3)\sqrt{.001}\sqrt{.003} \\
& + 2x_2(1 - x_1 - x_2)(-.2)\sqrt{.002}\sqrt{.003}.
\end{aligned}
$$

To find the points of maximization, we must set both partial derivatives $\frac{\partial g}{\partial x_1}$ and $\frac{\partial g}{\partial x_2}$ equal to zero. We can compute:

$$
\begin{aligned}
\frac{\partial g}{\partial x_1} =\ & .03 - .07 - 8(2(.001)x_1 - 2(.003)\left(1 - x_1 - x_2\right) \\
& + \sqrt{.001}\sqrt{.002}x_2 - .6\sqrt{.001}\sqrt{.003}\left(1 - 2x_1 - x_2\right) \\
& - .4\sqrt{.002}\sqrt{.003}\left(-x_2\right));
\end{aligned}
$$

$$
\begin{aligned}
\frac{\partial g}{\partial x_2} =\ & .05 - .07 - 8(2(.002)x_2 - 2(.003)\left(1 - x_1 - x_2\right) \\
& + \sqrt{.001}\sqrt{.002}x_1 - .6\sqrt{.001}\sqrt{.003}\left(-x_1\right) \\
& - .4\sqrt{.002}\sqrt{.003}\left(1 - x_1 - 2x_2\right)).
\end{aligned}
$$

Despite the complication of these two expressions, they do just produce linear equations in x_1 and x_2 when set to zero. After simplification, they become:

$$
\begin{cases}
.0806277x_1 + .0754659x_2 = .0163138 \\
.0754659x_1 + .0956767x_2 = .0358384
\end{cases}
$$

The solution of this system is roughly $x_1 = -.566$, $x_2 = .821$, and hence $x_3 = 1 - x_1 - x_2 = .745$. Thus, even more short selling is required than in the independent case: An investor with \$100 would sell short \$56.60 worth of asset 1, investing in \$82.10 worth of asset 2 and \$74.50 of asset 3. The investor's expected rate of return would then be:

$$
.03x_1 + .05x_2 + .07\left(1 - x_1 - x_2\right) = .03(-.566) + .05(.821) + .07(.745) = .076.
$$

This is quite fascinating. By allowing the investor to sell short the first, most conservative asset, he can realize a higher rate of return than even the most risky asset, asset 3.   ∎

We end this section and this chapter with an interesting general theorem about the optimal portfolio in the case of one non-risky asset and several independent risky assets. It is rather impressive that elementary tools can achieve such a powerful result. After the theorem we will comment on a surprising implication.

Theorem 1. Let there be n assets, the first of which is non-risky with sure rate of return r, and the rest of which have mutually independent rates of return with means μ_i and variances σ_i^2. Then the optimal portfolio for an investor with risk aversion a is:

$$x_i = \frac{\mu_i - r}{2a \cdot \sigma_i^2}, i = 2, 3, ..., n; x_1 = 1 - \sum_{i=2}^{n} x_i. \qquad (4.27)$$

Proof. In the independent case we have the objective function:

$$f(x_1, x_2, ..., x_n) = \sum_{i=1}^{n} x_i \mu_i - a \cdot \left(\sum_{i=1}^{n} x_i^2 \sigma_i^2 \right),$$

and for the risk-free asset we have $\sigma_1^2 = 0$ and $\mu_1 = r$. Writing $x_1 = 1 - \sum_{i=2}^{n} x_i$, we can restate the objective as an unconstrained function of $n - 1$ variables:

$$\begin{aligned} g(x_2, x_3, ..., x_n) &= r\left(1 - \sum_{i=2}^{n} x_i\right) + \sum_{i=2}^{n} x_i \mu_i - a \cdot \left(\sum_{i=2}^{n} x_i^2 \sigma_i^2 \right) \\ &= r + \sum_{i=2}^{n} x_i (\mu_i - r) - a \cdot \left(\sum_{i=2}^{n} x_i^2 \sigma_i^2 \right). \end{aligned}$$

The partial derivatives of g are:

$$\begin{aligned} \frac{\partial g}{\partial x_2} &= (\mu_2 - r) - 2a \cdot \sigma_2^2 x_2 \\ \frac{\partial g}{\partial x_3} &= (\mu_3 - r) - 2a \cdot \sigma_3^2 x_3 \\ &\vdots \\ \frac{\partial g}{\partial x_n} &= (\mu_n - r) - 2a \cdot \sigma_n^2 x_n. \end{aligned}$$

Setting all of these to zero produces linear equations in the individual variables x_i that all have the same form:

$$2a \cdot \sigma_i^2 x_i = (\mu_i - r) \Longrightarrow x_i = \frac{\mu_i - r}{2a \cdot \sigma_i^2}, i = 2, 3, ..., n.$$

The proportion x_1 in the non-risky asset is therefore $x_1 = 1 - x_2 - \cdots \cdots - x_n$, which completes the proof.

Notice that formula (4.27) makes very clear that the proportion of wealth given to asset i increases with its mean rate of return μ_i, decreases as the risk aversion a grows, and decreases as its variance σ_i^2 increases. All of these results are consistent with intuition.

Example 6. Consider a market with a non-risky asset whose rate of return is $r = .005$, and which has four risky assets with mean rates of return $\mu_2 = .02$, $\mu_3 = .03$, $\mu_4 = .05$, and $\mu_5 = .08$ and variances $\sigma_2^2 = .0006$, $\sigma_3^2 = .0015$, $\sigma_4^2 = .0070$, and $\sigma_5^2 = .01$. Find the optimal portfolio for an investor of risk aversion 15, assuming that the asset rates of return are independent random variables.

Solution. This is a direct application of Theorem 1. We compute:

$$x_2 = \frac{\mu_2 - r}{2a \cdot \sigma_2^2} = \frac{.02 - .005}{30(.0006)} = .833;$$

$$x_3 = \frac{\mu_3 - r}{2a \cdot \sigma_3^2} = \frac{.03 - .005}{30(.0015)} = .556;$$

$$x_4 = \frac{\mu_4 - r}{2a \cdot \sigma_4^2} = \frac{.05 - .005}{30(.0070)} = .214;$$

$$x_5 = \frac{\mu_5 - r}{2a \cdot \sigma_5^2} = \frac{.08 - .005}{30(.01)} = .250.$$

The total of these portfolio weights is about 1.853, which makes $x_1 = 1 - 1.853 = -.853$. So the investor is advised to be short on the risk-free asset, i.e., to borrow cash. An investor wanting to make a net \$1000 investment will borrow an additional \$853, and purchase \$833, \$556, \$214, and \$250 worth of the four risky assets. ∎

Something very strange follows from Theorem 1. Consider the ratio between the proportion of wealth in a particular asset i and the total proportion invested in all risky assets. This is:

$$x_i^* = \frac{\dfrac{\mu_i - r}{2a \cdot \sigma_i^2}}{\displaystyle\sum_{j=2}^{n} \dfrac{\mu_j - r}{2a \cdot \sigma_j^2}} = \frac{\dfrac{\mu_i - r}{\sigma_i^2}}{\displaystyle\sum_{j=2}^{n} \dfrac{\mu_j - r}{\sigma_j^2}}. \tag{4.28}$$

The expression on the right was obtained by simply canceling the common factors of 2 and a. But this means that the share of wealth devoted to risky asset i among all risky assets does not depend on the investor's risk aversion. Said another way, all investors will hold risky assets in the same balance with one another in their portfolios. The only difference is that investors with larger risk aversions will devote a smaller proportion of their wealth to the risky assets in total, reserving more for the non-risky asset. The collection of values x_i^* in equation (4.28) expresses the so-called **market portfolio** of risky assets that every optimizing investor will hold.

In higher level courses, this theorem and its generalization to correlated assets set the stage for the so-called **capital market theory**, in which relationships are derived between the mean and variance of the rate of return on individual assets in the market, and the mean and variance for the whole market, which are influenced as well by the correlations between the individual assets and the whole market. Capital market theory has become one of the cornerstones of the subject of finance, as evidenced by the award of a Nobel Prize in Economics in 1990 to its developers Merton Miller, Harry Markowitz, and William Sharpe.

Important Terms and Concepts

Portfolio objective function - Maximize $E[R_p] - a \cdot \text{Var}(R_p)$, where this objective is fully written in formula (4.22) as a function of the weights x_i given to each asset in the portfolio.

Finding optimal portfolios - After forming the portfolio objective for a specific group of assets, solve for one weight in terms of all the others, e.g., $x_1 = 1 - \sum_{i=2}^{n} x_i$, substitute into the objective function, and set all partial derivatives with respect to the x_is equal to zero.

Independent asset case - The explicit formulas for the optimal portfolio weights in the case where assets are independent and there is a non-risky asset are as in formula (4.27).

Market portfolio - In the case where a risk-free asset with rate of return r exists and the risky assets are independent, the proportion of wealth in risky asset i among total wealth devoted to all risky assets is:

$$x_i^* = \frac{(\mu_i - r)/\sigma_i^2}{\sum_{j=2}^{n} (\mu_j - r)/\sigma_j^2}.$$

Exercises 4.3

1. Consider an investor of risk aversion 10 who is interested in splitting his wealth between a non-risky asset with rate of return .04 and a risky asset with mean rate of return .08 and standard deviation .1. How should he do it, and what are the mean and variance of his rate of return?

2. Using the data in Example 1, what risk aversion corresponds to an optimal portfolio in which exactly half of the wealth is devoted to each asset?

3. Suppose that two risky assets have mean rates of return .03 and .05, variances .004 and .010, and correlation $-.2$. Find the optimal portfolio for an investor whose risk aversion is 4.

4. Assume that there are two risky assets with two possible outcomes as below. Find the optimal portfolio for an investor whose risk aversion is 7. Does your answer depend on the risk aversion?

$$
\begin{array}{ccc}
(R_1, R_2) & (.07, .04) & (.03, .05) \\
\text{probability} & 2/5 & 3/5
\end{array}
$$

5. Derive a general formula for the proportion of wealth to optimally devote to the risky asset in a case where there is a single non-risky asset with rate of return r and a single risky asset with parameters μ and σ^2.

6. Consider the case of two independent assets, the first of which has mean rate of return .04 and standard deviation .08, and the second of which has mean rate of return .06 and standard deviation .1. What risk aversion is necessary for exactly 100% of the wealth to be given to the riskier asset?

7. Suppose that an investor is interested in three assets; one is a risk-free asset with rate of return .02, the second is a risky asset with mean rate of return .04 and variance .002, and the other with mean rate of return .05 and variance .003. The correlation between the two risky assets is $-.2$. Write the portfolio objective function in terms of variables x_2 and x_3, the proportions of wealth held in the risky assets. Use calculus to find the optimal portfolios for an investor with risk aversion: (a) $a = 2$; (b) $a = 10$.

8. Suppose there are two independent risky assets with means .04 and .06 and standard deviations .08 and .12. Use calculus to derive the optimal portfolio for an investor with risk aversion 12.

9. With the same independent risky assets as in the previous exercise and a non-risky asset with rate $r = .02$, use: (a) calculus; and (b) Theorem 1 to find the optimal portfolio for the same investor with risk aversion 12.

10. Show in general that the optimal portfolio weights $x_1, x_2, ...x_n$ for the case in which there are n independent risky assets with means μ_i and variances σ_i^2 satisfy the equations below.

$$x_i = \frac{\mu_i - \mu_1 + 2a \cdot \sigma_1^2 x_1}{2a \cdot \sigma_i^2}, i = 2, 3, ..., n;$$

$$x_1 = \frac{1 - \sum_{j=2}^{n} \frac{\mu_j - \mu_1}{2a \cdot \sigma_j^2}}{1 + \sum_{j=2}^{n} \frac{\sigma_1^2}{\sigma_j^2}}.$$

11. Find the market portfolio if the prevailing market risk-free rate of return is $r = .01$, and there are five independent risky assets with mean rates of return $\mu_1 = .025$, $\mu_2 = .043$, $\mu_3 = .051$, $\mu_4 = .063$, and $\mu_5 = .068$ and standard deviations $\sigma_1 = .042$, $\sigma_2 = .065$, $\sigma_3 = .091$, $\sigma_4 = .122$, and $\sigma_5 = .146$.

12. Let there be three independent risky assets with mean rates of return .03, .04, and .06, together with a non-risky asset with deterministic rate of return .02. Assume that the variances of the rates of return on the three risky assets are .0036, .0049, and .0081, respectively. (a) Characterize the market portfolio. (b) Find the optimal portfolio for an investor of risk aversion $a = 10$. (c) At what risk aversion will an investor save 25% of his wealth for the risk-free asset?

13. In the scenario of the Exercise 12, find the portfolio of the three independent risky assets that has smallest variance. What is that variance?

14. In the scenario of Exercise 12, find the portfolio that has the smallest variance for a given mean return of .05.

15. Suppose that a market has n independent risky assets and a non-risky asset. What happens to the market portfolio if all risky assets have their rates of return multiplied by a constant c and also the risk-free asset rate of return r is multiplied by the same c. Be specific and justify your response.

Chapter 5

Valuation of Derivatives

Much of the field of financial mathematics has been devoted in recent years to the theory and application of *financial derivatives*, that is, assets whose values depend on some other underlying asset. Derivatives provide opportunities for investment speculation that can produce great rewards (in percentage terms), but also incur great losses. Derivatives can also be used in tandem with other assets to *hedge*, that is, to provide protection against risk. In this chapter we will introduce derivatives in a discrete-time environment whose properties are fairly well known. In Section 5.1, the basic language of derivative pricing theory and the key idea of *arbitrage* are introduced. Anti-arbitrage assumptions will be used in Section 5.2 to find derivative values in single time period models. These results are generalized rather easily to multiple time period models for simple derivatives in Section 5.3. Some more complicated kinds of derivative assets are discussed in Section 5.4, as well as the use of simulation techniques to estimate the values of derivatives. There are a number of very good sources that will allow you to continue your study of derivative valuation, including Baxter and Rennie [1], Pliska [9], Roman [10], Ross [11], and Stampfli and Goodman [14]. The last of these does a particularly good job in illustrating the chaining method that we will study for multiple time period problems.

New kinds of derivatives and new applications arise frequently. So we can only hope to gain acquaintance with the main ideas here, and we also limit ourselves to discrete-time. To extend our valuation techniques to continuous-time situations requires more advanced mathematics, including probability theory, analysis, and stochastic processes. The groundbreaking work is this area was done by Fischer Black and Myron Scholes in a 1973 paper, "The Pricing of Options and Corporate Liabilities." Another mathematical economist, Robert Merton, also made seminal contributions to the problem, and the three were awarded a Nobel Prize in Economics in 1997 (posthumously in Black's case) for their work.

5.1   Basic Terminology and Ideas

5.1.1   Derivative Assets

For now we will just consider two times, a beginning (time 0) and an end (time 1). Assume that the effective interest rate per period on risk-free investments is r, so that x dollars in a risk-free asset grows to $(1 + r)x$ from time 0 to 1. There is one underlying asset, whose value S_0 at time 0 is assumed to be known. The market price of this asset at time 1 is denoted S_1, which is a random variable.

Definition 1. A ***contingent claim***, or ***derivative***, is a contract on the underlying asset which is a function of the values that the asset takes on.

Thus, in the case of a single time period, a contingent claim is some function $V(S_0, S_1)$ that we think of as giving the final, or claim value, of the derivative, which is known if the full path of the underlying asset is known. Some claims depend only on S_1, and not on the full path. But, since these derivatives are traded in the market, we would like to value them fairly at times prior to the final time; here, at time 0. We use V_0 to denote the value of the derivative at the initial time, and V_1 to denote its value at time 1.

Two types of derivatives that we will consider are ***futures*** and ***options***.

In a ***futures contract***, two parties, the ***buyer*** and the ***seller***, agree to exchange an asset at an agreed upon price E at a fixed date in the future. The buyer must buy the asset and the seller must sell it, or else a cash settlement is given by one to the other, governed by the relationship between the contract price and the current market price when the contract is to be executed. Futures are available on commodities ranging from soybeans to gold, and on non-physical assets like stock market indices. For example, a buyer in a futures contract on soybeans may agree to buy a unit of this commodity at price $E = \$1000$ 1 year from now, and the seller is requred to supply the soybeans. If, after that year, the market price of a unit of soybeans is \$1100, then the buyer pays \$1000 to the seller, obtains the unit, and can sell it for a profit of \$100. The seller, on the other hand, must purchase the soybeans in the open market for \$1100, receiving only \$1000 when they are delivered to the buyer, for a loss of \$100. The buyer in the futures contract is therefore speculating that the price of soybeans will go up, while the seller believes that it will go down. If the contract allows for cash settlement, then the goods themselves need not be exchanged. The contract is settled by a transfer of \$100 from the seller to the buyer. The question is: given the price of the underlying asset initially, what is the "fair" price F that the buyer and seller should agree upon initially? By the end of this section, we will answer that question.

Example 1. A buyer and a seller agree on a futures contract based on the value of the S&P 500 index. Its current value is \$1500, and the optimistic buyer agrees to "buy" a unit of the index for \$1600 from the seller in 1 year, by means of executing a cash settlement. If the value of the index after a year is \$1550, then the seller's more pessimistic market view has prevailed, and the buyer must give the seller $50 (that is, the difference $1600 - $1550 between the contract price and the market price). If, on the other hand, the value of the index is $1650, then the buyer is entitled to receive $50 from the seller. ■

It should be clear from this example that, if S_1 is the final price of the underlying asset and the futures contract price is F, then

$$V_1 = \text{final value of future to buyer } = S_1 - F. \tag{5.1}$$

If $S_1 < F$, then the buyer loses and the seller wins, while, if $S_1 > F$, the seller loses and the buyer wins. For the seller, the order of subtraction is reversed and his value is $F - S_1 = -(S_1 - F)$. One way to look at the futures contract is to say that the buyer is long and the seller is short on the future. The buyer, who owns $+1$ unit of the future, profits by $S_1 - F$, and the seller, who owns -1 unit of the future, profits by $-(S_1 - F)$. This allows us to unify the claim value of the future to be $S_1 - F$ as in formula (5.1).

Futures contracts have existed for hundreds of years and are negotiated on exchanges such as the Chicago Board of Trade (CBOT). Their original purpose was to protect agricultural concerns from variations in prices; for instance, if an individual or business was invested in hogs, and was worried about a price decline, that party might sell futures at price F so that the hogs could be unloaded for no less than F per unit.

Whereas the participants in a futures contract must execute the deal at the expiration time of the contract, an **option** puts that decision in the hands of the holder of the option.

Definition 2. An **option** on an asset is the right, but not the obligation, to buy or sell the asset for a contracted price E at or before a contracted time T. E is called the **strike price**, or **exercise price**.

In simple **European options**, exercise takes place exactly at the exercise time T at the discretion of the holder of the option. So-called **American options** may be exercised by the option holder at any time up to T (which means that their values are harder to determine). Choice has value for the owner of an option. The premium that is required by the issuer of the option in exchange for this choice is the price that the option purchaser must pay. In Section 2, we show how to find that price.

Options are categorized in another way, as follows:

Definition 3. A **call option** is an option to buy an asset. A **put option** is an option to sell the asset.

It is easy to derive formulas for the final claim values of call and put options. For the call, if the exercise time is $T = 1$, the strike price is E, and the price of the underlying asset at that time is S_1, then, if $S_1 < E$, the call is worthless because, if the call holder really wants to purchase the asset, it is cheaper to buy it in the market than to exercise the option. If $S_1 \geq E$, then the option holder can exercise the option, buying the asset at E and reselling at S_1, for a profit of $S_1 - E$. It follows that the claim value of a call option is:

$$V_1^{\text{call}} = \max\left(S_1 - E, 0\right). \tag{5.2}$$

For the put option, if the asset price $S_1 \geq E$, then it is better for the option holder to sell the asset in the market than to exercise the option and sell at price E (if the option holder even owns the asset, which is often not the case). So the put is worthless in this case. But, if $S_1 < E$, then the option holder can buy the asset in the market at price S_1 and exercise the option to sell it at price E for a net payoff of $E - S_1$. Thus, the final claim value of a put option is:

$$V_1^{\text{put}} = \max\left(E - S_1, 0\right). \tag{5.3}$$

An option is called "in the money" if it is valuable to execute it. For example, in a call option, if $S_1 = \$20$ and $E = \$18$, then the option is in the money because executing it profits $S_1 - E = \$2$. If it would incur a loss to execute the option, it is referred to as "out of the money." If a put option had exercise price $E = \$40$, and the market price was $S_1 = \$45$, then it would be a losing proposition to sell the asset by executing the option instead of selling in the market, so that option is out of the money. When the investor is indifferent between executing the option or not, that is, when $S_1 = E$, the option is said to be "at the money." These terms are also applied at times earlier than the expiration time. For example, an option on an asset currently priced at $S_0 = \$25$ whose strike price is also $E = \$25$ is also said to be at the money.

Graphs of the claim values of the call and the put option are in Figure 5.1. These graphs do not take into account the initial cost of purchasing options that we have yet to determine, which would have the effect of translating the graphs downward by an amount equal to that cost. Notice that the call value is unbounded, while the put is bounded above by the strike price.

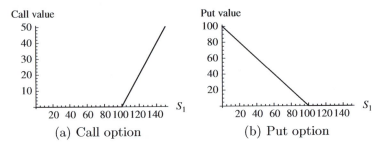

Figure 5.1: Final claim values of European options, strike price $E = \$100$.

Options are traded on markets like the Chicago Board Options Exchange (CBOE) and are also brokered privately. Because of the relationships between option values and the values of the assets on which they are based, companies and individuals purchase options to reduce or hedge the risks they otherwise carry. For instance, like futures, put options may be used by individuals or corporations to guarantee in advance a market price for their goods. Unlike futures, owners of put options can choose to execute the sale only if it is to their benefit. Call options may be used to guarantee that the prices of raw materials needed by the holder of the call are bounded; if the market price rises above the exercise price E, the option allows its holder to purchase the materials for just E. Both puts and calls can also be used for speculation, as shown in the next example.

Example 2. Lorraine believes that the price of Stock A is on its way down. She purchases 500 put options on this stock, which is currently selling for $50 per share. Each option costs her $1, and the options can be exercised 6 months from now at strike price $45. What is Lorraine's effective annual rate of return if the market price at the deadline is: (a) $47; (b) $44; (c) $43?

Solution. We will compute in each case the 6-month rate of return r and convert to an effective annual rate by the expression $(1 + r)^2 - 1$.

(a) If the final price is $47 per share, then the put option is worthless, because she would not purchase the shares in the market at $47 only to sell them immediately for $45 apiece as in her option contract. But she has paid $500 and lost it all, for a rate of return of $-1 = -100\%$ per semi-annum, and therefore $(1 - 1)^2 - 1 = -100\%$ yearly. Observe that this will be the case for any final market price of $45 or above.

(b) If instead the final price is $44 per share, then she can exercise all of her 500 put options, buying 500 shares of Stock A for $500 \cdot \$44$ and reselling them to the issuer of the put option for $500 \cdot \$45$ for a gain of $500 \cdot \$1 = \$500$. Or she can take a cash settlement of $500 from the issuer of the put. She had invested $500 in the options, which gives her a net value of 0 and so a rate of return of 0, both semi-annually and annually.

(c) Lorraine's best outcome among these three occurs if the per-share price is $43. Then, reasoning as in part (b), she earns $500(\$2) = \$1000$ on the transaction at the end, having invested $500. Her semi-annual rate of return is $(1000 - 500)/500 = 1 = 100\%$. This translates to an annual effective rate of $(1 + 1)^2 - 1 = 3 = 300\%$. Clearly, speculating in options has both its risks and its rewards. ∎

One of the purposes of derivatives is to provide a kind of insurance against bad outcomes. The next example shows how desirable this can be.

Example 3. A dealer is contracted to sell 10 futures based on the NYSE Dow Jones index. The futures price is $15,000, with a cash settlement as the culmination of his gamble. Fearing that the index may go up, the dealer purchases

10 call options on the index with strike price \$15,000, due to expire at the same time as the futures. Each option costs \$40. What is the profit or loss for this dealer if the market value of the index at the common time of expiration of the two derivatives is \$16,000? \$14,000? What is the dealer's profit or loss if he had not purchased the call options?

Solution. The dealer has essentially purchased $10 \cdot \$40 = \$400$ of insurance against the rise of the market. If the Dow index ends at \$16,000, then he chooses to execute the call options, purchases 10 "shares" of the index at \$15,000 per share, and sells them to the futures contract buyer for the same \$15,000 apiece. The dealer loses all of his \$400, for a rate of return of -100% for the period, but the magnitude in dollars of the loss is relatively small. If he had not purchased the call options, he would have had to "buy" 10 shares of the index at \$16,000 each and sell them to the futures partner at \$15,000 each, losing a total of $10(\$1000) = \$10,000$, which is a much more serious loss. In the case that the final value of the Dow Jones index is \$14,000, the dealer can afford to let the options expire unexecuted, buy 10 shares of the index at \$14,000 each, sell them to the futures partner at \$15,000 each, profiting $10(\$1000) = \$10,000$. Deducting the cost of the options, his overall profit is \$9,600. In this case, if he had never bought the options, he would have earned the full profit of \$10,000. So we see that using the call options as insurance in this manner limits the downside while achieving nearly the same upside. ∎

5.1.2   Arbitrage

In the examples of the last subsection, in order to illustrate principles, we were picking option and future prices out of the air. It stands to reason that since the claim values of derivatives are well-defined functions of the underlying assets on which they are based, their prices ought to be predictable. A first guess at a fair price for a derivative might be the expected value of its final claim value. But just a little more thought reveals that, since the claim value is earned in the future, we should multiply the expected value by a present value discounting factor.

It turns out that this is not quite enough. We will see the details in the next section, but essentially we must price derivatives to guard against the possibility that investors could earn unlimited riskless profits with no initial net investment. We call this situation **arbitrage**.

Definition 4. A financial strategy or situation in which, with no initial expenditure, there is a positive probability of a gain and no probability of a loss is called an **arbitrage** strategy.

You might think of arbitrage as a free lunch, or something for nothing. Note that the definition only requires that there is a positive probability of something for nothing, and zero probability of a loss. We don't insist that a positive gain must happen, only that it is possible to occur with no risk of loss.

Arbitrage opportunities should be short-lived in a real market. They typically involve assets that are either systematically overpriced or underpriced given their real value. If such an opportunity arises, astute investors called **arbitrageurs** will quickly rush in and their competition will drive up the price of an undervalued asset due to increased demand, or reduce the price of an overvalued asset due to increased desire to sell.

Arbitrage is a violation of an economic principle referred to as the **Law of One Price**. This states the following: Two investments that yield the same payments at the same times must have the same initial price. Why? If not, sell the investment with the higher price, buy the one with the lower price, and pocket the difference (investing in a risk-free asset to improve value if you like). The investment that is sold requires payments to the buyer. But these are the same payments earned by the arbitrageur who has bought the cheaper investment, so this clever arbitrageur can turn over that money, relieving all obligations and still coming away with money in his pocket.

The arguments above may seem somewhat abstract. Let us see what arbitrage looks like in a concrete situation.

Example 4. Show that, if the effective yield rate on zero-coupon bonds is 4.5%, and a local bank is willing to lend money up to $50,000 at an interest rate of 3% nominal per year compounded monthly, then an arbitrage opportunity exists.

Solution. The local bank is a cheap source of money which can fund the purchase of the more desirable bonds. The price of zero-coupon $50,000 bonds satisfies:

$$P = (1 + .045)^{-1} \cdot \$50,000 = \$47,846.90.$$

So the investor could buy bills of face value $50,000 at this initial price, borrowing the funds from the bank to do so.

The investor has no initial expenses. By the end of the year, the principal on the loan from the bank has grown to:

$$\$47,846.90 \left(1 + \frac{.03}{12}\right)^{12} = \$49,302.20.$$

The investor cashes in the bonds for $50,000, repays the loan, and pockets $50,000 - \$49,302.20 = \$697.80$ at the end, with no initial outlay. ∎

The exercise set for this section contains other interesting examples of arbitrage scenarios.

5.1.3 Arbitrage Valuation of Futures

As a first example of using arbitrage considerations to price derivative assets, we look at the case of futures. Recall that, in a futures contract, a buyer and a seller agree on a price at which the buyer will buy a unit of an asset at time 1. No money changes hands at the beginning. We would like to find a single price F, if there is one, upon which they can agree such that neither the buyer nor the seller has an arbitrage opportunity. The next example points the way.

Example 5. Suppose that the current price of an asset is $S_0 = \$50$ and the risk-free effective rate is 5%. Find the futures price for a single period future on the asset.

Solution. Both the buyer and seller know that the asset will be exchanged for F. The buyer will be worried that the market price at time 1, S_1, will be less than F and the seller will be worried that S_1 will be greater than F. The seller can guard against this event by pre-buying a share now at the current market price of \$50, borrowing the money to do so, and negotiating F to be high enough to cover his debt at the end, namely, $\$50(1 + .05)$. The buyer, on the other hand, would want to obtain protection by selling short the asset now for \$50, investing that at the risk-free rate to accrue to $\$50(1+.05)$, having enough to pay the seller for the asset, and transfering the asset to the short sales partner to eliminate that obligation. So it is in the interest of both parties for the futures price to be $\$50(1.05) = \$52.50$. ∎

The argument of the last example suggests that a fair settlement price for a futures contract on an asset whose current value is S_0 is $S_0(1+r)$, where r is the risk-free rate per period, so that the futures price is valued as if the asset grows at the risk-free interest rate. We can also see this using an arbitrage argument. Again suppose that $S_0 = \$50$ and $r = .05$. Suppose that the futures price is more than $\$50(1.05) = \$52.50$; let's say that it is \$53. The seller in the transaction enters into a contract to sell the buyer a share at this price. The seller can buy a share now at \$50, borrowing the money to do so. At time 1, the seller gives the share to the futures contract partner, collecting \$53, and pays back the loan, which has accumulated to $\$50(1.05) = \$52.50$. The net position of the seller at the end, with no outlay at all at the start, is $\$53 - \$52.50 = \$.50$. Modest as that is, if the transaction could be duplicated many times, then large arbitrage profits could occur for the seller.

On the other hand, if the futures price is less than \$52.50, say \$52, then the buyer in this contract can sell short a share now at \$50, invest the proceeds, and enter into a futures contract to buy a share at time 1 for \$52. At the end, the buyer takes the share that is bought from the futures seller at \$52 and gives it to the short-selling partner to remove that obligation. The buyer collects the proceeds from the \$50 investment. The buyer's final position is $\$50(1.05) - \$52 = \$.50$, so an arbitrage situation has arisen for the buyer. The fair price clearly must be \$52.50 exactly in order to avoid either kind of "something-for-nothing" deal.

We end this section by using the preceding reasoning in order to prove the futures valuation formula in general.

Theorem 1. Suppose that an asset is currently priced at S_0, and the risk-free rate of interest per period is r. The unique arbitrage-free price of settlement in a futures contract executed at time 1 on the asset is:

$$F = S_0 \cdot (1 + r). \tag{5.4}$$

Proof. Suppose instead that $F > S_0(1 + r)$. The seller of the futures contract can buy a share now and keep it for delivery at time 1, financing the purchase by borrowing S_0 now. By the time the loan is due at time 1, the seller is ready to deliver the share, pay back the amount $S_0(1 + r)$ of the loan with interest, and pocket the difference of $F - S_0(1+r)$, with no risk. This violates the no-arbitrage assumption. Now suppose that $F < S_0(1+r)$. The buyer of the futures contract can sell short a share at S_0, invest the money received at the risk-free rate, and then at time 1 pay F to the seller of the futures contract, deliver the share of stock to the buyer of the short sale, receiving the difference $S_0(1 + r) - F$ in cash, with no risk, again contradicting the no-arbitrage assumption.

Important Terms and Concepts

Underlying asset - An asset upon which another constructed asset (contingent claim, below) is based.

Derivative or contingent claim - A secondary asset whose value is determined by the behavior of its underlying asset.

Future - A contract in which a buyer and a seller agree in advance upon a price F at which a transaction for an underlying asset will be made later. If the value of the underlying asset initially is S_0, then the arbitrage-free transaction price of the future is $F = S_0 \cdot (1 + r)$.

Option - A contract that has been issued by one party and bought by another called the holder, in which the option holder has the right, but not the obligation, to buy or to sell an underlying asset at a strike price E at a later time T.

Call option - An option to buy the underlying asset.

Put option - An option to sell the underlying asset.

European option - An option that can only be exercised at the expiration time T.

American option - An option that can be exercised by the holder at any time prior to T.

Arbitrage - An opportunity to make riskless profits with no possibility of loss, while investing no money initially.

Exercises 5.1

1. A dealer sells a futures contract on an asset with current price $1000 and futures price $1050. Suppose that the possible asset values at the execution time of the contract are $950, $1000, $1050, or $1100. Find the dealer's profit in each case. What is the dealer's expected profit if the four final asset values are equally likely to occur?

2. Sid buys a share of a stock index futures contract from his dealer Nancy with futures price $20,000. The current market index value for his shares is $19,500. Find Sid's profit in each of the cases where the final value of the shares is $19,000, $21,000, $22,000. If these prices occur, respectively, with probabilities $\frac{1}{2}, \frac{1}{4}$, and $\frac{1}{4}$, what is Sid's expected profit?

3. Kelly purchases 50 European call options on an asset currently priced at $20 per share. The options cost her $1.50 each and the strike price is $25. Find her rate of return in each of the cases where the final price per share is: (a) $22; (b) $25; (c) $28; (d) $30. Compare these rates of return to the rates that she would have earned if she had just bought four shares of the original asset rather than the options.

4. Harry has contracted to be the seller in a futures contract for which an asset is to be exchanged for $F = \$500$. Fearing a rise in the price of the asset, he buys a call option for $5 that allows him to buy the asset for $520. Find his profit in each of the cases where the final asset value is: (a) $480; (b) $520; (c) $550; (d) $600. Compare to the profits or losses that he would have had if he had not purchased the options.

5. Vicki is a buyer in a futures contract in which an asset is to be exchanged for $1000. Fearing a decline in the market, she buys a put option for $4 which allows her to sell the asset for $980. Find Vicki's profit in each of the cases where the final asset value is: (a) $1050; (b) $980; (c) $950; (d) $900.

6. Catherine wants to buy a new refrigerator for $1000. She notices that one bank has a special loan offer of 3% nominal annual yearly interest rate compounded monthly, and another bank offers a risk-free, 1-year certificate of deposit at a 5% effective annual interest rate. Describe how Catherine can get her refrigerator for free in a year.

7. In Exercise 1, construct an arbitrage opportunity if the risk-free rate is: (a) .04; (b) .06.

8. In Exercise 2, construct an arbitrage opportunity if the risk-free rate is: (a) .02; (b) .03.

9. Generalize the argument for the existence of arbitrage in Example 4, in which there is a zero-coupon bond with yield rate j per period and maturity of one period, and riskless borrowing rate per period of $r < j$. How much actual riskless profit occurs if the investor buys a bond of face value F? What happens to the argument if there is some positive probability p that the bond defaults, that is, does not pay the face value?

10. Suppose that two stocks, Shark Industries and PilotFish, Inc., tend to move together in the market. A share of Shark sells now for $52 and a share of PilotFish for $13. There are three possible price pair scenarios at the end of 1 week, as in the table below. Find a portfolio of the two stocks that results in an arbitrage opportunity.

	scenario 1	scenario 2	scenario 3
Shark	$60	$52	$47
PilotFish	$15	$13	$10

11. Suppose that asset X has possible one-period rates of return of .05, .04, or .03, occurring with equal probabilities. Show that there is an arbitrage opportunity if the effective risk-free rate per period is $r = .03$.

12. Stock A has potential values of $a_1, a_2, ..., a_n$ at the end of one period, and Stock B has potential values $b_1 = ca_1, b_2 = ca_2, ..., b_n = ca_n$, for the corresponding outcomes for Stock A. If the initial price of a share of stock A is $50, find, with careful justification, the initial share price of stock B.

13. Suppose that there is a risk-free asset whose price is $10 at time zero and whose rate of return is 4%. A stock is currently priced at $80. Suppose also that, in the market, the futures price was $84. Explain how arbitrageurs can leap in and make a riskless profit.

14. Suppose that there is a portfolio of two risky assets that is long 10 shares of the first asset and short 20 shares of the second, such that there are only two outcomes at the end of one time period: $P_1 = \$20$ and $P_2 = \$5$, or $P_1 = \$22$ and $P_2 = \$6$, where P_1 and P_2 are the prices of the two assets in the portfolio at the end of the period. A risk-free asset with effective rate of return 1% per period exists. What must be the initial value of the risky portfolio, assuming that arbitrage is not possible?

15. A bank is willing to loan cash at an effective interest rate of 5% up to $20,000. Show how to obtain risk-free profits of as much as either $3000 or $4000 with no initial net investment, if a risky asset is available currently priced at $20 per share which will go up in value by $4 or $5 with equal probability by the end of

the year.

16. In a **butterfly spread option**, the investor believes that the current option price will not change very much between now and the expiration time and wants to construct a derivative that profits when that happens. These derivatives are available on the market, but are also constructible from several standard call options. Suppose that an asset is currently priced at $50. Consider an option portfolio containing four call options with the same expiration time (time 1), in which an investor (i) purchases a call option with strike price $45; (ii) purchases a call option with strike price $55; and (iii) issues, or is short, two call options with strike price $50. Calculate the final claim value on this portfolio of options if the final asset price is each of: $58, $52, $50, $48, and $42. Devise a general formula for the claim value as a function of the asset price at time 1, and graph the function.

Exercises 17–21 develop the properties of another derivative that we will call a **forward**. Holding a forward is similar to being the buyer in a future, except there is an asymmetry between the holder and the issuer of the forward. At time 0, the price of a share of a particular asset is S_0. Like the future, at time 1, the holder of the forward contract must pay a fixed amount of money, the **exercise price** E, to the issuer of the forward, who must deliver one share of the asset to the holder. But the issuer charges the holder a premium for engaging in the contract; call it V_0. The problem is: What is the "fair" price V_0 that the issuer can charge the holder at time 0 for the forward contract?

17. If the market price of the asset at time 1 is $S_1 > E$, then what are the consequences for the holder of a forward? The issuer? What if $S_1 \leq E$? Can you write a general expression for the claim value of a forward at time 1?

18. Let r be the effective risk-free interest rate for the period corresponding to the execution date of the forward. Consider a portfolio consisting of one forward contract and the following amount of money in a risk-free asset:

$$E \cdot (1+r)^{-1}.$$

The total value at time 0 is then $V_0 + E \cdot (1+r)^{-1}$.

(a) Find the total value of this portfolio at time 1.

(b) Using the result of part (a) and the Law of One Price, what does the initial value of the portfolio also have to be?

(c) Conclude that the initial price of the forward must be $V_0 + E \cdot (1+r)^{-1}$.

19. Use anti-arbitrage arguments to give an alternative proof of the formula $V_0 = S_0 - E \cdot (1+r)^{-1}$ for the initial price of a forward on an asset with begin-

ning value S_0, and exercise price E, assuming risk-free rate r.

20. Suppose that there is a risk-free asset whose price is \$10 at time zero, and whose rate of return is 4%. A stock is currently priced at \$80. Show that an arbitrage opportunity exists if a forward contract with exercise price \$82 on that stock is currently available for a price of \$1.

21. A forward contract on a stock is currently priced at \$1 and the current price of the stock is \$25. The exercise price on the forward is \$24 and the risk-free rate is 3%. Show how to construct an arbitrage opportunity.

5.2 Single-Period Options

In Chapter 3 we worked with the ***binomial branch process*** model for the motion of a risky asset. Recall that the asset begins at a price S_0 at time 0, then moves "up" by a factor of $(1 + b)$ or "down" by a factor of $(1 + a)$, where $b > a$, with probabilities p and $1 - p$, respectively. This law of motion continues for as many periods as we wish. Figure 5.2 shows the tree of possible paths of the process for two time steps. Note that, if we allow a large number n of very short time steps, then we have $n + 1$ possible final states, and so we can model in relatively good detail the possible values of a real asset.

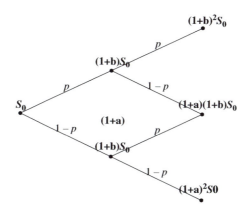

Figure 5.2: Binomial branch process model of price of asset.

In Section 5.3 we will consider multiple-time period options whose underlying assets satisfy this model, in which the special structure of the binomial branch process simplifies the derivations and computations of the value of the options. For now we will start with a single time step, which turns out to contain all of the crucial elements that we will need for the the more general model.

5.2.1   Pricing Strategies

Consider derivatives like the European options whose claim value at expiration is a function only of the final value of the underlying asset. It is very tempting to call on previous financial reasoning to guess that the initial value of the derivative should be the expected present value under the prevailing risk-free rate r of the final claim value of the derivative. For the European call option expiring at time 1, for instance, if the final price S_1 exceeds E, then the owner of the option can exercise it, buy the share at price E, then immediately resell it at the higher price S_1 for a profit of $S_1 - E$. But, if $S_1 < E$, then it does not pay to exercise the option to buy, so the option becomes valueless. We saw in the last section that a function that describes the final claim value is $\max(S_1 - E, 0)$. Therefore, the present value of the call option is the expectation

$$E\left[(1+r)^{-1}\max(S_1 - E, 0)\right]. \tag{5.5}$$

So it appears that we must know something about the probabilistic structure of the price process in order to compute this expectation.

Surprisingly, this is incorrect. We will see that arbitrage can exist if the expectation in formula (5.5) is used to value the option. But the formula is almost correct. Let us probe a bit farther. The diagram in Figure 5.3 will help you to read and understand the following discussion.

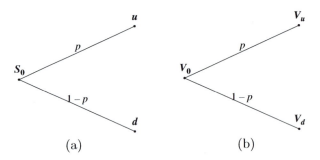

Figure 5.3: Single-step trees for (a) underlying asset and (b) derivative.

Assume the simplest possible model with one time step between the inital time 0 and exercise time $T = 1$. To simplify notation, let the two possible stock prices at time 1 be denoted by u and d and let the two corresponding values of the option at time 1 be V_u and V_d. (Notice that we are allowing for any kind of single-period contingent claim here.) Consider an initial portfolio of Δ units of the stock and b units of a risk-free asset with rate of return r. Its value at time 0 is:

$$\Delta \cdot S_0 + b. \tag{5.6}$$

We can design this portfolio to replicate the value of the option at time 1, whichever of the two possible outcomes occurs, by carefully choosing Δ and b. Since the risk-free asset must grow to a value of $b(1 + r)$ at time 1, the parameters Δ and b must satisfy the linear system:

$$\begin{cases} \Delta \cdot u + b(1 + r) = V_u \\ \Delta \cdot d + b(1 + r) = V_d. \end{cases} \tag{5.7}$$

Solving simultaneously,

$$\Delta = \frac{V_u - V_d}{u - d},$$
$$b = (1 + r)^{-1} (V_d - \Delta \cdot d) \tag{5.8}$$
$$= (1 + r)^{-1} \left(V_d - \frac{V_u - V_d}{u - d} \cdot d \right).$$

Since our option must have the same final value V_1 at time T as this portfolio, the Law of One Price shows that it must have the same initial value V_0. Therefore, the option value at time 0 is:

$$V_0 = \Delta S_0 + b = \frac{V_u - V_d}{u - d} S_0 + (1 + r)^{-1} \left(V_d - \frac{V_u - V_d}{u - d} \cdot d \right).$$

A little rearrangement of the second term gives the formula in the next theorem.

Theorem 1. If an underlying asset satisfies a binomial model in one time period with initial value S_0 and possible next values u and d, if the risk-free interest rate per period is r, and if a derivative is constructed on this asset whose possible claim values are V_u and V_d at time 1 when the asset value is u or d, respectively, then the arbitrage-free initial value of the derivative is:

$$V_0 = \frac{V_u - V_d}{u - d} S_0 + (1 + r)^{-1} \frac{u V_d - d V_u}{u - d}. \tag{5.9}$$

Formula (5.9) is called the **replicating portfolio formula** for the initial value of the derivative, because of the argument that was used to derive it. What is remarkable about the formula is that the probability p of an up move by the risky asset does not figure in at all. The anti-arbitrage argument encapsulated in the Law of One Price made it unnecessary.

Example 1. Let the underlying asset have initial price \$35 and assume that it is a binomial branch process with "up" rate $b = .04$ and "down" rate $a = .02$. The risk-free rate is $r = .03$. Let us find the value of a European call option with exercise time 1 and strike price $E = \$36$.

Solution. The possible values u and d of the underlying asset at time 1 are:

$$u = (1.04)(\$35) = \$36.40; d = (1.02)(\$35) = \$35.70.$$

The option will be in the money only for the "up" state $S_1 = \$36.40$. Thus, $V_u = \$36.40 - \$36 = \$.40$ and $V_d = 0$. Substituting into the replicating portfolio formula yields:

$$V_0 = \frac{\$.40 - 0}{\$36.40 - \$35.70} \cdot \$35$$
$$+ (1.03)^{-1} \frac{(\$36.40)(0) - (\$35.70)(\$.40)}{\$36.40 - \$35.70}$$
$$= \$.194175.$$

In Exercise 1 you will argue that, if the option had been valued by computing the expected present value of the claim value, then there is arbitrage. In an arbitrage-free economic universe, $\$.194175$ is the enforced initial price of the option. ■

We will need the result of the next example shortly.

Example 2. In the previous example of a one-step binomial branch process, the parameters a, b, and r satisfied the inequality $a < r < b$. Show in general for the binomial branch model that, if this was not the case, then an arbitrage opportunity exists.

Solution. First, if $r \leq a$, then both $(1 + b)S_0$ and $(1 + a)S_0$ are at least as large as $(1 + r)S_0$, with the former being strictly larger. The risky asset is a strictly better investment than the risk-free asset. So create arbitrage by borrowing an amount S_0 at time 0 at the risk-free rate and using the money to buy a share of the risky asset. Consequently, there is no initial expenditure by the investor. As long as the up probability $p > 0$, there is a positive probability that the risky asset will be worth strictly more than $(1 + r)S_0$, and even in the down case it is at least as large. Hence, at time 1 the asset can be sold, the loan principal of $(1 + r)S_0$ can be repaid, no loss can occur, and there is a positive probability of a gain that is strictly positive.

You are asked in Exercise 2 to show that, if $r \geq b$, then arbitrage is possible as well. Thus, r must lie strictly between a and b. ■

The portfolio replication technique that led to Theorem 1 seems to solve our problem completely, but a little more investigation will pay benefits. Start with formula (5.9) again, draw out $(1 + r)^{-1}$ as a common factor, and group terms involving V_u and V_d together to produce the following alternative form of the derivative valuation formula.

$$
\begin{aligned}
V_0 &= \frac{V_u - V_d}{u - d} S_0 + (1+r)^{-1} \frac{uV_d - dV_u}{u - d} \\
&= (1+r)^{-1} \left((1+r) \frac{V_u - V_d}{u - d} S_0 + \frac{uV_d - dV_u}{u - d} \right) \\
&= (1+r)^{-1} \left(\frac{S_0(1+r)V_u - S_0(1+r)V_d}{u - d} + \frac{uV_d - dV_u}{u - d} \right) \\
&= (1+r)^{-1} \left(\frac{S_0(1+r)V_u - dV_u}{u - d} + \frac{uV_d - S_0(1+r)V_d}{u - d} \right) \\
&= (1+r)^{-1} \left(V_u \left(\frac{S_0(1+r) - d}{u - d} \right) + V_d \left(\frac{u - S_0(1+r)}{u - d} \right) \right).
\end{aligned}
\tag{5.10}
$$

Notice the interesting fact that, if you add together the coefficients of V_u and V_d in this formula, you get 1. Also, as in Example 2, it makes sense to assume that $S_0(1+r) > d$ and $u > S_0(1+r)$, else arbitrage opportunities would result. So, defining

$$
q = \frac{S_0(1+r) - d}{u - d},
\tag{5.11}
$$

it follows that q is a probability, which we call the **risk-neutral probability**. The expression for V_0 in formula (5.10) then reduces to formula (5.12) in the next theorem.

Theorem 2. With the notation and assumptions of Theorem 1, and the probability q defined by formula (5.11), the arbitrage-free initial value of the derivative can be written as:

$$
V_0 = (1+r)^{-1} \left(V_u \cdot q + V_d(1 - q) \right) = E_q \left[(1+r)^{-1} V_1 \right],
\tag{5.12}
$$

which is an expected present value of the claim value of the option using the risk-neutral probability q in place of the asset "up" probability p.

Theorem 2 says that we can still think of the initial value of the derivative as an expected present value of the claim random variable V_1, but it is q instead of the natural probability p that is used in the expectation. Formula (5.12) is called the **risk-neutral probability formula** for the value of the derivative. As we have shown, it is equivalent to the replicating portfolio formula, but can be easier to compute with. Notice in particular how simple the risk-neutral probability q is in the usual binomial branch notation:

$$
\begin{aligned}
q &= \frac{S_0(1+r) - d}{u - d} \\
&= \frac{S_0(1+r) - S_0(1+a)}{S_0(1+b) - S_0(1+a)} \\
&= \frac{(1+r) - (1+a)}{(1+b) - (1+a)} \\
&= \frac{r - a}{b - a},
\end{aligned}
\tag{5.13}
$$

and

$$1 - q = 1 - \frac{r-a}{b-a} = \frac{b-a}{b-a} - \frac{r-a}{b-a} = \frac{b-r}{b-a}. \tag{5.14}$$

Example 3. Use the risk-neutral probability formula to compute the initial value of a European put option in one period with strike price \$80, assuming that the risk-free rate is .04, and the underlying asset is a binomial branch process with initial value \$80 and up and down rates $b = .10$, $a = -.02$.

Solution. The possible values of the underlying asset at time 1 are:

$$\$80(1.1) = \$88; \$80(.98) = \$78.40.$$

Since the exercise price is \$80, in the "up" case the put is worthless, and in the "down" case the claim value is \$80 − \$78.40 = \$1.60. By formula (5.14), the risk-neutral probability is:

$$q = \frac{r-a}{b-a} = \frac{.04 - (-.02)}{.10 - (-.02)} = \frac{.06}{.12} = .5.$$

Then $1 - q = .5$ as well, and the option value is the expected present value:

$$E_q \left[(1+r)^{-1} V_1 \right] = (1.04)^{-1} ((.5)(0) + (.5)(\$1.60)) = \$.77. \ \blacksquare$$

There is an interesting consequence of the risk-neutral probability form of the valuation equation, which can help you remember the result and also motivate the language "risk-neutral" that is in the common lexicon but still does not quite capture the idea. Notice that the expected value under q of the price S_1 of the underlying asset at time 1:

$$
\begin{aligned}
E_q \left[S_1 \right] &= qu + (1-q)d \\
&= \left(\frac{S_0(1+r) - d}{u-d} \right) u + \left(\frac{u - S_0(1+r)}{u-d} \right) d \\
&= \frac{S_0(1+r)u - ud}{u-d} + \frac{ud - S_0(1+r)d}{u-d} \\
&= \frac{S_0(1+r)u - S_0(1+r)d}{u-d} \\
&= S_0 \cdot \frac{(1+r)u - (1+r)d}{u-d} \\
&= S_0 \cdot \frac{(1+r)(u-d)}{u-d} \\
&= S_0(1+r).
\end{aligned}
$$

Looking at the final right-hand side of the derivation, we see that, for the purpose of valuing the derivative, the underlying asset process is expected to move like a risk-free asset under the probability measure induced by q. Perhaps the language should be "risk-free" probability instead of "risk-neutral," but that is not in common use. It follows directly from this formula that:

$$E_q\left[(1+r)^{-1}S_1\right] = S_0 = (1+r)^0 S_0, \tag{5.15}$$

that is, the present value process $\tilde{S}_n = (1+r)^{-n}S_n$ is constant in expectation, at least for one time period. Processes with this property are called **martingales**, and we will give a more careful definition and exploit the martingale property in the next section on multiple time period models.

Example 4. Suppose that a risky asset with current price $50 is such that the price in the next period is either $54 or $48. The current risk-free rate is $r = .02$ per period. A derivative pays either $2 in the case where the underlying asset moves to $54 at time 1, or $-$1 if the asset moves to $48. Find the probability that makes the risky asset into a martingale (in present value terms), and find the arbitrage-free initial value of the derivative.

Solution. To make this asset into a martingale, q must satisfy:

$$(1.02)^{-1}E_q\left[S_1\right] = S_0$$

$$\Longrightarrow \quad (1.02)^{-1}(q \cdot \$54 + (1-q) \cdot \$48) = \$50$$
$$\Longrightarrow \quad q \cdot \$54 + (1-q) \cdot \$48 = 1.02 \cdot \$50 = \$51.$$

Solving the simplified linear equation $q \cdot \$6 + \$48 = \$51$ gives $q = \frac{1}{2}$. This means that the initial value of the derivative must be:

$$(1.02)^{-1}E_q\left[V_1\right] = (1.02)^{-1}(.5 \cdot \$2 + .5 \cdot (-\$1)) = (1.02)^{-1} \cdot \$.50 = \$.49. \ \blacksquare$$

5.2.2   Put–Call Parity

It turns out that there is a relationship between the initial prices of European call options and put options sharing the same exercise price and time. The next example shows how this comes about.

Example 5. Consider a portfolio in which we sell a call option with exercise price E at its arbitrage-free price denoted by C, we purchase a share of stock at S_0, and we also purchase a put option with the same exercise price as the call, at its arbitrage-free price P. The exercise time on both options is $T = 1$. A risk-free asset with per-period rate r is available. What is the value of this portfolio at time 1 in each of the scenarios where the final stock price is at least E and where it is less than E? What implication does this have regarding the relationship between the call price and the put price?

Solution. The initial value of this portfolio is $S_0 - C + P$. If $S_1 \geq E$, then the share of stock is worth S_1, the call is worth $S_1 - E$, and the put is worthless. The value of the portfolio in this case is:

$$S_1 - (S_1 - E) + 0 = E.$$

In the case that $S_1 < E$, the stock is worth S_1, the call is worthless, and the put is worth $E - S_1$. So the final portfolio value is:

$$S_1 - 0 + (E - S_1) = E.$$

In both cases the final value of this portfolio must be E. To avoid arbitrage, the initial value of the portfolio must be the present value of E, so that:

$$S_0 - C + P = (1 + r)^{-1}E.$$

Rearranging, we obtain:

$$P + S_0 = C + (1 + r)^{-1}E. \quad \blacksquare \tag{5.16}$$

Equation (5.16), called the **put–call parity formula**, relates the initial price of the put to the initial price of the call, allowing us to compute one easily if we know the other. In Example 1, for instance, where the initial price was \$35, the risk-free rate was .03, and the exercise price of the call was \$36, we found that the initial price of the call was about \$.19. Substituting the known data into the put-call parity formula, we obtain, for the value of a put option with the same exercise price,

$$P = C + (1 + r)^{-1}E - S_0 = \$.19 + (1.03)^{-1} \cdot \$36 - \$35 = \$.14.$$

Put-call parity remains true in the multiple time period problem, which we will study in the next section.

5.2.3 Δ–Hedging

Before we leave this section we would like to highlight and quantify the idea of hedging away risk by combining underlying assets with derivatives on them. We can take advantage of the portfolio replication form of the derivative value, which, as above, says that the initial value of a derivative is the same as a portfolio of Δ units of an underlying risky asset and b units of cash $V_0 = \Delta S_0 + b$, where:

$$\Delta = \frac{V_u - V_d}{u - d}, b = (1 + r)^{-1}\left(V_d - \frac{V_u - V_d}{u - d} \cdot d\right). \tag{5.17}$$

The final values of the replicating portfolio and the derivative agree as well, which lets us conclude in rough terms that:

Δ units of underlying asset $+ b$ units of riskless asset $= 1$ unit of derivative.

$$\tag{5.18}$$

Thus,

1 unit of derivative $-\Delta$ units of underlying asset $= b$ units of riskless asset.

$$(5.19)$$

Consequently, a portfolio that is short in the underlying asset and long in the derivative in a ratio of $\Delta/1$ will be the same as a risk-free investment. The risk has been hedged away. Because of the fact that Δ is the notation in common use for the ratio between high and low claim value of the derivative and high and low value of the underlying asset, this technique for eliminating risk is referred to as Δ-*hedging*.

The next example illustrates the complete Δ-hedge that we have described, as well as the consequences of doing a partial hedge.

Example 6. Consider a single-period model in which the underlying asset begins at price $100 and can reach a price of either $110 or $90 at the end of one period. A risk-free asset is available with per period rate of return 2%. Ted has sold short 10 shares of the asset and invested the proceeds in the risk-free asset, but he is worried that the share price will go up and he will incur a loss. Find the number of units of a single-period call option with exercise price $102 that Ted should buy in order to hedge away risk, and check his final position in both the up and down cases. What could he stand to gain or lose if he does not hedge at all? If Ted chooses to buy at most half that many options, what are the possible profits or losses in each of the two cases where the stock goes up or it goes down?

Solution. The parameters given in the problem are $u = \$110$, $d = \$90$, $r = 2\%$, $S_0 = \$100$, $E = \$102$. Since the option is a call option, its possible final values are $V_u = \$110 - \$102 = \$8$ and $V_d = 0$. The critical hedge ratio Δ becomes:

$$\Delta = \frac{V_u - V_d}{u - d} = \frac{\$8 - \$0}{\$110 - \$90} = \frac{8}{20} = .4.$$

This is the ratio between shares of the underlying asset and units x of the option to be bought, so that:

$$\frac{.4}{1} = \frac{10}{x} \implies x = \frac{10}{.4} = 25.$$

If Ted buys 25 options to go along with his short position of 10 shares, he should be protected. Let us see.

The initial price of the 25 options is 25 times the price of one option. Since the risk-neutral probability q is:

$$q = \frac{S_0(1 + r) - d}{u - d} = \frac{\$100(1.02) - \$90}{\$110 - \$90} = \frac{\$102 - \$90}{\$20} = .6,$$

the option price is:

$$E_q\left[(1 + r)^{-1}V_1\right] = (1.02)^{-1}(.6 \cdot \$8 + .4 \cdot 0) = \$4.71.$$

So Ted must pay $25 \cdot (\$4.71) = \$117.75$ for the 25 options. He has also placed the $1000 from the short sale into the riskless asset, giving a value of $\$1000(1.02) = \$1020$ at the end. In the case that the asset goes up to $110, he may execute 10 of the options to buy 10 shares at $102 each, eliminate the short sale, and cash in the remaining 15 options for $15(\$8) = \$120$. His final position is:

$$\$1020 + \$120 - 10(\$102) = \$120.$$

If, on the other hand, the underlying asset goes down to $90 per share, then Ted lets the options expire worthless and buys 10 shares in the market to restore to the short sales partner. His final position in this case is:

$$\$1020 - 10(\$90) = \$120.$$

In either case, he has paid $117.75 for his options and will end at $120 regardless of whether the risky asset goes up or down. It does check that, up to rounding, $(1.02)(\$117.75) = \$120$, so his short position in the underlying asset combined with a purchase of 25 options has resulted in exactly what he would have made by not short selling and investing $117.75 cash in the risk-free asset. By comparison, if Ted had not purchased any options, his final position would be one of:

$$(\text{up case:})\$1020 - 10 \cdot \$110 = -\$80 \text{ or } (\text{down case:})\$1020 - 10 \cdot \$90 = \$120.$$

So he has protected himself against the worst case by Δ-hedging.

Now, if Ted only hedges at most halfway, that is, he buys 12 options for a total of $12 \cdot (\$4.71) = \$56.52$, in the up case he can still execute 10 of them to purchase shares at $102 to restore to the short sales partner and cash in the remaining options for $2 \cdot (\$8) = \$16$. Then his final position would be

$$\$1020 + \$16 - 10(\$102) = \$16.$$

In the down case, where the options are worthless, the final position would again be $\$1020 - 10 \cdot \$90 = \$120$. So, by going halfway, he could either lose $40.52 or gain about $63.48, taking into account the cost $56.52 of the options. Both his worst and best case scenarios are roughly cut in half in comparison to not hedging at all. ∎

Important Terms and Concepts

Binomial branch process - A model for the price behavior of an underlying risky asset in which the asset begins at a price S_0 at time 0, then at time 1 takes on the value $(1 + b)S_0$ with probability p, or $(1 + a)S_0$ with probability $1 - p$. The same construction can be extended to multiple time periods.

Replicating portfolio - A portfolio of assets that has the same final value in all possible scenarios as another asset. The Law of One Price says that the initial value of the portfolio must match the initial value of the asset.

Replicating portfolio formula - The formula $V_0 = \frac{V_u - V_d}{u - d} S_0 + (1 + r)^{-1} \frac{uV_d - dV_u}{u - d}$ for valuation of a derivative with possible claim values V_u and V_d, occurring, respectively, when the underlying asset takes values u and d.

Risk-neutral probability - The probability $q = \frac{S_0(1+r)-d}{u-d}$ under which the discounted asset process is a martingale, which allows the valuation of derivatives to be done by the expectation formula $V_0 = E_q\left[(1+r)^{-1}V_1\right]$. In the binomial branch case, q reduces to $\frac{r-a}{b-a}$.

Put–call parity - The relationship $P + S_0 = C + (1 + r)^{-1}E$ between the initial value P of a European put option and the initial value C of a European call option with the same exercise time and price.

Δ–hedging - Combining a unit of a derivative with a short position of Δ shares of its underlying asset in order to completely hedge away risk.

Exercises 5.2

1. In Example 1, compute the expected present value of the claim value of the option assuming that the up probability is $p = .6$. Argue that, if the option is priced in this way, then there is an arbitrage opportunity.

2. Complete Example 2 by showing that, if the risk-free rate r is at least as large as the up rate b for the risky asset, then arbitrage can be constructed.

3. For a single-period binomial branch asset model with parameters $S_0 = \$10$, up rate $b = .05$, down rate $a = .01$, $p = .6$, and risk-free rate $r = .02$, find the value of a European call option with exercise price $E = \$10.30$.

4. Find the value of a put option with exercise price $E = \$10.20$ on the same asset as in the previous exercise.

5. In a single-period binomial branch call option model with strike price E, $b = .1$, $a = 0$, $r = .05$, and initial underlying price $S_0 = \$20$, if the option is valued at $\$1$, then what must have been the strike price?

6. A call option on a binomial branch asset with parameters $b = .06$, $a = .02$ and initial price $\$80$ is priced initially at $\$.90$. If the exercise price of this option is $E = \$82$, what must have been the risk-free rate?

7. Exercises 17–21 in Section 5.1 introduced a derivative called a forward, which is like a call option with strike price E without the privilege of the option. Its claim value is therefore $S_1 - E$. It was derived in that set of exercises that the initial value of such a forward is $V_0 = S_0 - E \cdot (1 + r)^{-1}$. Use the techniques of this section to rederive this formula for the forward, given a general binomial branch underlying asset.

8. Find the value of a single-period European call option with exercise price $9 if the current value of the underlying asset is $8.50, the risk-free rate is 2%, and the underlying asset follows the binomial branch model with up rate $b = .08$ and down rate $a = 0$. Find the value of a put option with the same exercise price, and check that the put-call parity relation does hold.

9. Suppose that a stock follows the binomial branch model (for a single period) with up rate $b = .08$ and down rate $a = -.02$, that the initial price of the stock is $18, and that the risk-free rate is 5%. Find a replicating portfolio of stock and risk-free asset that replicates the time 1 value of a call option with exercise price $19. What is the value of this portfolio at time 0?

10. An asset begins with price $S_0 = \$75$, and its possible values at time 1 are $70 or $80. As in the section, define the discounted asset process by $\tilde{S}_n = (1+r)^{-n} S_n$, where r is the risk-free interest rate. Find the up probability q that makes the discounted process a martingale, if: (a) $r = 0$; (b) $r = .03$.

11. Find a general formula for the probability q that forces $E_q\left[(1 + r)^{-1} S_1\right] = S_0$, where r is the risk-free rate, and the price S_1 of the underlying asset at time 1 can be either of S_u or S_d with probabilities q and $1 - q$, respectively.

12. A **binary option** is a derivative that pays a fixed amount of money (not dependent on the final price of the underlying asset) in the case that the price exceeds a strike price E, and nothing otherwise. Use the replicating portfolio formula to price a binary option that yields $10 if its underlying asset exceeds $50. Assume the risk-free rate is $r = .03$, the initial price of the underlying asset is $50, and the possible final values of the asset are $55 or $48.

13. A **cash-or-nothing** call option gives to its holder the exercise price E if the price of the underlying asset exceeds E and nothing otherwise. A **stock-or-nothing** call option gives to the holder a share of the underlying asset if its price exceeds E, and nothing otherwise. Find the initial values of each of these two derivatives in a single-period binomial branch model. What final payoff occurs in a portfolio which is long one stock-or-nothing option and short one cash-or-nothing option sharing the same exercise price? What does that say about the initial value of a European call option with exercise price E? For simplicity, consider the case where the up value of the stock exceeds E and the down value does not.

14. Use the result of Example 3 and put–call parity to compute the initial value

of a call option with the same exercise price.

15. Rederive the put–call parity formula (5.16) by arguing that a portfolio that includes a put option with strike price E plus a share of the underlying asset is the same as a portfolio consisting of a call option with the same strike price plus cash in the amount of $(1 + r)^{-1}E$ dollars.

16. Stockbroker Barbara has issued 1000 puts with strike price $28 on an asset whose price is now $S_0 = \$30$. The risk-free rate is currently $r = .04$, and the asset will either go up to $35 or down to $27 next period. Describe how Barbara can completely hedge away the risk she incurs.

17. Steve has bought 90 call options with exercise price $95 on an asset that is currently valued at $S_0 = \$90$ and will either stay the same next period or go up by 10%. The risk-free rate of return is 5%. How many shares of the asset should he sell short so as to be absolutely sure what his final position will be?

18. Tanna has 500 shares of an asset currently priced at $60, and she is concerned that the stock will drop. Assume that the next price will be either 2% greater than the current price or 10% less. The risk-free rate of return per period is 1%. So Tanna considers buying put options in order to hedge away her risk. How many puts with strike price $57 should she buy to eliminate all possibility of loss?

Exercises 19–23 begin the process that will be detailed in Section 5.3 of extending the ideas we have used to the multiple-time period case. Each of these problems will use just two time periods.

19. A stock now priced at $20 either goes up by 10% or down by 10% tomorrow, with probabilities .6 and .4, respectively. Suppose that behavior continues the next day. List the possible prices of the stock after these 2 days and their probabilities of occurrence. If a call option with exercise price $21 that expires in two periods is issued on this asset, what are its possible claim values?

20. Consider a binomial branch asset model in two time periods in which $b = .2$ and $a = -.1$ and the initial price is S_0. Find an up probability q for a single period such that $E_q[S_1] = S_0$, and verify that $E_q[S_2] = S_0$.

21. Assume that a risky asset has the same parameters as in the previous exercise, but now there is given a risk-free interest rate per period $r = .02$. Define the discounted asset process $\tilde{S}_k = (1 + r)^{-k}S_k, k = 0, 1, 2$. Find an up probability q such that $E_q\left[\tilde{S}_1\right] = \tilde{S}_0 = S_0$, and verify that $E_q\left[\tilde{S}_2\right] = \tilde{S}_0 = S_0$.

22. Again assume that a risky asset satisfies a binomial branch model with parameters $u = 1.2$ and $d = .9$, and suppose also that the initial price is $S_0 = \$50$.

Using the probability q from the previous exercise, a risk-free rate of 2%, and a two-period European call option with strike price $55, compute the expected claim value under q, discounted by two periods. We will see later that this is the initial value of the option.

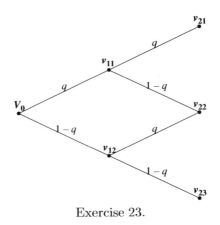

Exercise 23.

23. The answer in the previous exercise can also be found using a backwards stepping algorithm. Assume the same problem data, and fill in the tree in the figure as follows. The nodes are in correspondence to the nodes in the tree for the underlying asset. We use the first subscript to indicate the time, or the level of the tree, and the second to indicate the node within that level. The nodes at level 2 are to be given the final claim value of the two-period call option with strike price $55. Backing up to level 1, each node is to be given the discount factor $(1+r)^{-1}$ times the expected value under the risk-neutral probability q of its child nodes. Similarly, the value at node 0 is computed as $(1+r)^{-1}$ times the expected value of the v_{11} and v_{12} nodes computed in level 1. Check that the V_0 node value is the same as the expected present value in the previous exercise.

5.3   Multiple-Period Options

We would now like to extend the valuation techniques of Section 5.2 to asset processes moving through more than one time period. A typical binomial branch process in three periods for an underlying asset is shown in Figure 5.4, together with a corresponding tree for the derivative values. In this section we use a notational system of double subscripting to name nodes in the tree: the first subscript indicates the level of the tree to which the node belongs, and the second stands for its position within than level. But we just use the initial price S_0 instead of the more complex $s_{0,1}$ for the root node at level 0. Larger second subscripts correspond to smaller node values. At each tree level, the node values form the state space of the random asset price S_i for that level. In the figure, for instance, the level 3 nodes $\{s_{31}, s_{32}, s_{33}, s_{34}\}$ are the state space of the asset value S_3. Our derivative will have values at each time just as the underlying asset does, and the nodes at a level i make up the state space of the derivative value at time i.

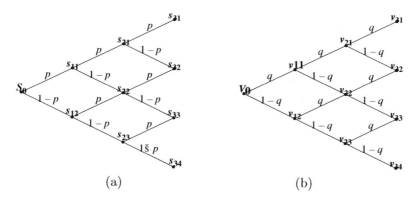

(a)   (b)

Figure 5.4: (a) binomial branch process model of price of asset and (b) derivative tree.

We will show two approaches to computing the initial value V_0 of the derivative which give the same result. One method is a direct computation of an expected present value, and the other is recursive, starting at the right side of the tree and working back. The first is effective when the derivative's value depends only on the final claim value at the end of time, and the second is more adaptable to cases where the derivative depends in some way on the path that the asset process takes to reach its final state at the exercise time.

5.3.1 Martingale Valuation

First, we have the precise definition of martingales.

Definition 1. A sequence of random variables $M_0, M_1, M_2, M_3, \ldots$ is called a *martingale* if. for each $i \geq 1$,

$$E\left[M_i | M_{i-1}\right] = M_{i-1}. \tag{5.20}$$

The defining formula (5.20) is deceptively simple; it has a great deal of content and important implications as well. First, recall that conditional expectation given a random variable is a function of the values of that random variable. The formula is saying that, for each possible value m of the random variable M_{i-1},

$$E\left[M_i | M_{i-1} = m\right] = m. \tag{5.21}$$

So the mean value for the next random variable in the sequence is equal to whatever the current value is. Note as well that a martingale can only satisfy condition (5.20) relative to the probability distribution and associated expectation that is being used. Suppose, as we usually do for asset price models, that the value M_0 at time 0 is known. By the Tower Law for conditional expectation, a martingale satisfies:

$$E\left[M_1\right] = E\left[E\left[M_1|M_0\right]\right] = E\left[M_0\right] = M_0;$$

$$E\left[M_2\right] = E\left[E\left[M_2|M_1\right]\right] = E\left[M_1\right] = M_0;$$

$$E\left[M_3\right] = E\left[E\left[M_3|M_2\right]\right] = E\left[M_2\right] = M_0;$$

etc. so that our definition implies that a martingale is constant in expectation: $E\left[M_i\right] = M_0$ for all times k. This condition is necessary, but not sufficient, for a sequence of random variables to be a martingale. Condition (5.20) is stronger.

In the case of binomial branch processes with price random variables $S_0, S_1, S_2, ...$ the transition structure is a simple one in which the next random asset price is dependent on the current price by:

$$S_{i+1} = \begin{cases} (1+b)S_i & \text{with probability } p; \\ (1+a)S_i & \text{with probability } 1-p. \end{cases} \tag{5.22}$$

For single-time problems, we have already observed the interesting fact that we may ignore the natural probability p that the underlying process goes up, and instead value derivatives as if the risk-neutral probability q was in force. If the riskless interest rate per period is r, then the risk-neutral probability has the simple form:

$$q = \frac{(1+r)S_0 - d}{u - d} = \frac{r - a}{b - a}. \tag{5.23}$$

The key property of q from Section 5.2 that concerns us at the moment is that the discounted price process defined by $\tilde{S}_i = (1+r)^{-i}S_i$ satisfies, in one time period, the equation:

$$E_q\left[\tilde{S}_1\right] = E_q\left[(1+r)^{-1}S_1\right] = S_0. \tag{5.24}$$

Denote the discounted option value process similarly by $\tilde{V}_i = (1+r)^{-i}V_i$, where V_i is the random claim value at time i. Note that $\tilde{V}_0 = V_0$. Then the results of the last section imply that the derivative value process satisfies a similar equation:

$$V_0 = E_q\left[\tilde{V}_1\right] = E_q\left[(1+r)^{-1}V_1\right]. \tag{5.25}$$

Something important is going on here. We might hope that, under q, the discounted asset process $\tilde{S}_i, i = 0, 1, 2, ...$ is a martingale, and so is the discounted option value process $\tilde{V}_i, i = 0, 1, 2, ...$. This would mean that, at the time n of exercise of the derivative:

$$V_0 = E_q\left[\tilde{V}_n\right] = E_q\left[(1+r)^{-n}V_n\right], \tag{5.26}$$

where V_n is the final claim value. Formula (5.26) is the **_martingale valuation formula_** for the initial value of a derivative. All we need to do is to compute the risk-neutral probability q and the distribution of the final claim value V_n at time n under q, and then compute the expected present value of V_n. Here is the formal statement of the result.

Theorem 1. Let the underlying asset follow the binomial branch model (5.22), and let q be the risk-neutral probability in (5.23), where r is the risk-free interest rate. If V_n is the claim value of a derivative with exercise time n, then the initial value of the derivative is:

$$V_0 = E_q\left[(1+r)^{-n}V_n\right].$$

We will give only an informal argument for this theorem. It would follow if we could establish the martingale condition (5.20) for $\tilde{V}_i$ for each time $i \leq n$. To illustrate, consider the case of a derivative expiring at time 3. The trees for the underlying asset and the derivative are shown in Figure 5.4. By the way in which we have numbered the states, for each tree level $i = 1, 2$ in the asset tree prior to the exercise time, we have:

$$s_{i+1,j} = (1+b)s_{i,j}, \quad \text{and} \quad s_{i+1,j+1} = (1+a)s_{i,j}.$$

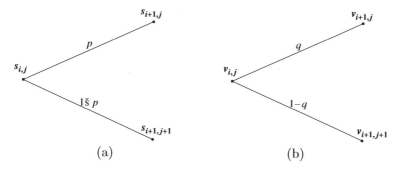

Figure 5.5: (a) single-period asset subtree and (b) associated derivative subtree.

(See Figure 5.5.) This is just a way of denoting the fact that each parent node has two children at the next tree level, with values of either $(1 + b)$ or $(1 + a)$ times the current value. The risk-neutral probability q for node $s_{i,j}$ is:

$$
\begin{aligned}
q &= \frac{(1+r)s_{i,j} - d}{u - d} \\
&= \frac{(1+r)s_{i,j} - s_{i+1,j+1}}{s_{i+1,j} - s_{i+1,j+1}} \\
&= \frac{(1+r)s_{i,j} - (1+a)s_{i,j}}{(1+b)s_{i,j} - (1+a)s_{i,j}} \\
&= \frac{r - a}{b - a},
\end{aligned}
$$

so that each one-period subtree of the whole tree is a copy of the one-period setup of the previous section. The value $v_{i,j}$ of the derivative at this node can be thought of as follows. The derivative can be sold at time $i + 1$, when its values are either $v_{i+1,j}$ or $v_{i+1,j+1}$. Therefore $v_{i,j}$ can be found by the same portfolio replication argument as in Section 5.2, in which there are numbers Δ and b such that:

$$
v_{i,j} = \Delta s_{i,j} + b,
$$

where

$$
\Delta = \frac{V_u - V_d}{u - d} = \frac{v_{i+1,j} - v_{i+1,j+1}}{s_{i+1,j} - s_{i+1,j+1}};
$$

$$
b = (1+r)^{-1}\left(V_d - \Delta \cdot d\right) = (1+r)^{-1}\left(v_{i+1,j+1} - \Delta \cdot s_{i+1,j+1}\right).
$$

That exact argument leads to the valuation of the derivative at node $v_{i,j}$ as:

$$
v_{i,j} = (1+r)^{-1}\left(q \cdot v_{i+1,j} + (1 - q) \cdot v_{i+1,j+1}\right). \tag{5.27}
$$

Notice that this is just another way of writing the conditional expectation equation:

$$
v_{i,j} = E_q\left[(1+r)^{-1} V_{i+1} \mid V_i = v_{i,j}\right], \tag{5.28}
$$

which is the martingale condition (5.20) for the discounted derivative value process $\tilde{V}_i$. From the martingale condition follows the fact that $V_0 = \tilde{V}_0 = E\left[\tilde{V}_3\right] = E\left[(1+r)^{-3} V_3\right]$, since martingales are constant in expectation.

Valuation of derivatives by martingales therefore amounts to simply finding the risk-neutral probability q, using the known claim value of the derivative at its expiration time n as a function of the asset price process to compute the nodes $v_{n,j}$ at the rightmost level of the derivative tree, which are the states of the random variable V_n, and then evaluating the arbitrage-free initial derivative price $V_0 = (1+r)^{-n} E_q\left[V_n\right]$.

Example 1. Suppose that a stock initially priced at \$20 follows a binomial branch process with $b = .03$ and $a = .005$. The risk-free rate per period is $r = .01$. A European call option on this asset is available with strike price \$20.50 and expiration time $n = 2$. Find its current value.

Solution. First, the risk-neutral probability is:

$$q = \frac{.01 - .005}{.03 - .005} = .2.$$

The nodes on the stock price tree are:

$$
\begin{aligned}
S_0 &= \$20; \\
s_{11} &= \$20(1.03) = \$20.60; \\
s_{12} &= \$20(1.005) = \$20.10; \\
s_{21} &= \$20(1.03)^2 = \$21.218; \\
s_{22} &= \$20(1.03)(1.005) = \$20.703; \\
s_{23} &= \$20(1.005)^2 = \$20.2005.
\end{aligned}
$$

Since the derivative is a call option with strike price $20.50, it is in the money for the first two nodes at time 2 and out of the money for the third node. Thus,

$$v_{21} = \$21.218 - \$20.50 = \$.718;$$

$$v_{22} = \$20.703 - \$20.50 = \$.203;$$

$$v_{23} = 0.$$

These are the states of the random claim variable V_2. The probability of two up moves is $q^2 = .2^2 = .04$; the probability of one up and one down is $2q(1 - q) = 2(.2)(.8) = .32$; and the probability of two down moves is $(1 - q)^2 = .8^2 = .64$. Therefore the initial value of the option is the expected present value:

$$
\begin{aligned}
V_0 &= (1 + r)^{-2} E_q [V_2] \\
&= (1.01)^{-2}(.04(\$.718) + .32(\$.203) + .64(0)) \\
&= \$.0918. \quad \blacksquare
\end{aligned}
$$

Example 2. Consider now an underlying asset following a binomial branch model with initial price $50 and up and down rates .04 and $-.02$. There is a risk-free asset with rate of return $r = 1\%$ per period. Find the initial value of a European put option with exercise time 3 and strike price $50.

Solution. The given asset data is $S_0 = \$50$, $b = .04$, and $a = -.02$. Thus, the risk-neutral probability is:

$$q = \frac{r - a}{b - a} = \frac{.01 - (-.02)}{.04 - (-.02)} = \frac{.03}{.06} = .5.$$

To compute the option value, we will need the final asset values at time 3. They are:

$$s_{31} = \$50(1.04)^3 = \$56.24; \ s_{32} = \$50(1.04)^2(.98) = \$53.00;$$

$$s_{33} = \$50(1.04)(.98)^2 = \$49.94; \ s_{34} = \$50(.98)^3 = \$47.06.$$

Hence the possible final claim values are:

$$v_{31} = \max(\$50 - s_{31}, 0) = \$0; \quad v_{32} = \max(\$50 - s_{32}, 0) = \$0;$$
$$v_{33} = \max(\$50 - s_{33}, 0) = \$0.06; \quad v_{34} = \max(\$50 - s_{34}, 0) = \$2.94.$$

As we have seen before, these final states occur with appropriate binomial probabilities; for example, node v_{31} is reached if there are three up moves out of three, node v_{32} is reached if there are exactly two up moves, and nodes v_{33} and v_{34} are reached respectively in the cases of exactly one and exactly zero up moves. The probability mass function of V_3 is therefore:

$$P[V_3 = \$0] = \binom{3}{3}q^3(1-q)^0 + \binom{3}{2}q^2(1-q)^1 = (.5)^3 + 3(.5)^3 = .5.$$

$$P[V_3 = \$0.06] = \binom{3}{1}q^1(1-q)^2 = 3(.5)^3 = .375.$$

$$P[V_3 = \$2.94] = \binom{3}{0}q^0(1-q)^3 = (.5)^3 = .125.$$

The discounted expected value, that is, the initial value of the put option, is:

$$\begin{aligned} V_0 &= (1+r)^{-3}E_q[V_3] \\ &= (1.01)^{-3}(.5(0) + .375(\$0.06) + .125(\$2.94)) \\ &= \$0.379. \ \blacksquare \end{aligned}$$

Consider in general the present value of the European call option with strike price E and exercise time n, which we now know is the expectation:

$$E_q\left[(1+r)^{-n}\max(S_n - E, 0)\right],$$

where q is determined by the equation $q = \frac{r-a}{b-a}$ in the binomial branch model. Since the possible states of the underlying asset are known to be of the form $S_0(1+b)^k(1+a)^{n-k}$ and the probabilities attached to those states are binomial with parameters n and q, we have a concrete formula for the initial value of the call:

$$V_0 = (1+r)^{-n}\sum_{k=0}^{n}\binom{n}{k}q^k(1-q)^{n-k}\max\left(S_0(1+b)^k(1+a)^{n-k} - E, 0\right). \quad (5.29)$$

For instance, suppose that the effective monthly risk-free rate $r = .004$. The value of a 4-month call option with strike price $21 on an asset whose current price is also $21 with weekly up and down rates of $b = .008$ and $a = -.002$ comes out to about $.337:

$$q = \frac{.004 - (-.002)}{.008 - (-.002)} = .6;$$

$$V_0 = (1.004)^{-4} \sum_{k=0}^{4} \binom{4}{k} (.6)^k (.4)^{4-k} \max\left(\$21(1.008)^k(.998)^{4-k} - \$21, 0\right) \approx \$.337.$$

This call value formula is clearly not an easy thing to simplify or even compute with, but with the aid of technology we can look at the dependence of the call value on the problem parameters.

A graph of the call value as a function of the initial price S_0 for the set of parameters in the last paragraph is shown in Figure 5.6. As the initial price of the underlying asset goes up, all other things being equal, the payoff on the option is expected to be higher, and so the option is more valuable.

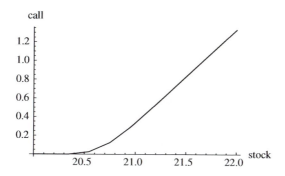

Figure 5.6: Call option value as a function of initial price of underlying asset.

Generalizing the number of periods n until the expiry of the option, and returning to the initial price of $21 for the underlying asset, Figure 5.7 shows that the call value increases as a function of n. Additional time for the underlying asset to increase in value serves to increase the initial value of the option.

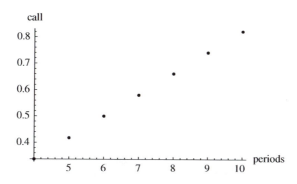

Figure 5.7: Call option value as a function of expiration time of option.

Graphs of the call value as a function of the up rate of return b and the down rate a are shown in Figure 5.8. Although the risk-free probability q changes with the change in these parameters, it is not unexpected that the initial value of the option grows as b grows and falls as a falls, since its value depends on the ability of the underlying asset to exceed the strike price of $21.

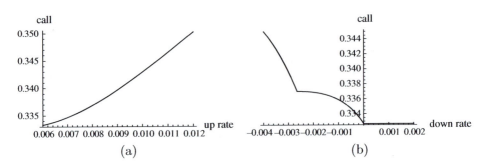

Figure 5.8: (a) call value as a function of b and (b) call value as a function of a.

5.3.2   Valuation by Chaining

An alternative to calculating the expectation $V_0 = (1 + r)^{-n} E_q [V_n]$ directly to value a derivative is to find it recursively by stepping backwards in time through the tree. The technique is suggested by formula (5.27), with a terminal condition added:

$$
\begin{aligned}
v_{i,j} &= (1 + r)^{-1} \left(q \cdot v_{i+1,j} + (1 - q) \cdot v_{i+1,j+1} \right); \\
v_{n,j} &= \text{claim value of derivative for node } j \text{ at time } n.
\end{aligned}
\tag{5.30}
$$

Since the derivative claim value is well-determined as a function of the underlying asset process at the exercise time n, the node values $v_{n,j}$ are known, and they form the initialization step in the procedure. Backing up to level $n - 1$ in the derivative tree, at node j we would have:

$$
v_{n-1,j} = (1 + r)^{-1} \left(q \cdot v_{n,j} + (1 - q) \cdot v_{n,j+1} \right).
$$

It is helpful to refer to part (b) of Figure 5.5. Each node value in the derivative tree is this linear combination of the node values of its child nodes. By the first formula in (5.30), we can continue to move left in the tree to fill in node values for level $n - 2$, $n - 3$, etc., until we reach level 0, at which we will have arrived at the initial value of the derivative.

This technique is referred to as the **chaining method** for valuing derivatives. Although we do not show it here, the single step relationship:

$$
v_{i,j} = E_q \left[(1 + r)^{-1} V_{i+1} | V_i = v_{i,j} \right] = E_q \left[(1 + r)^{-1} V_{i+1} | S_i = s_{i,j} \right]
$$

is extensible by mathematical induction to the following relationship at time n:

$$v_{i,j} = E\left[(1+r)^{-(n-i)}V_n|S_i = s_{i,j}\right].$$

This formula applied at $i = 0$ establishes that the chaining value v_0 at the root of the derivative tree is indeed the expected present value of the claim value at time n.

We will now illustrate the calculations on the data of the first two examples, and of course the answers will come out the same.

Example 3. Consider Example 1, in which $S_0 = \$20$, $b = .03$, $a = .005$, $r = .01$. Use the chaining method to recompute the initial value of a European call option with strike price \$20.50 and expiration time $n = 2$. Make sure that it is the same as the value computed in Example 1.

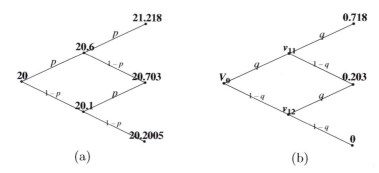

Figure 5.9: Chaining for a two-period call option: (a) stock tree; (b) option tree.

Solution. The two trees are displayed in Figure 5.9. We had found the risk-neutral probability to be $q = .2$, and, from the earlier example, the option nodes at level 2 when the option expires are:

$$
\begin{aligned}
v_{21} &= \text{Max}(\$21.218 - \$20.50, 0) = \$.718; \\
v_{22} &= \text{Max}(\$20.703 - \$20.50, 0) = \$.203; \\
v_{23} &= \text{Max}(\$20.2005 - \$20.50, 0) = 0.
\end{aligned}
$$

The level 1 nodes are found by the chaining relation (5.30):

$$v_{11} = (1+.01)^{-1}(.2 \cdot v_{21} + .8 \cdot v_{22}) = (1.01)^{-1}(.2\cdot(\$.718)+.8\cdot(\$.203)) = \$.30297;$$

$$v_{12} = (1+.01)^{-1}(.2 \cdot v_{22} + .8 \cdot v_{23}) = (1.01)^{-1}(.2 \cdot (\$.203) + .8 \cdot 0) = \$.040198.$$

Similarly, the level 0 node, which is the current value of the call option, is

$$V_0 = (1.01)^{-1}((.2)(\$.30297) + (.8)(\$.040198) = \$.0918.$$

This does match the value computed in Example 1, and we have gained the additional information of the option value at the intermediate nodes at time 1.   ■

Example 4. Recalculate the intial value of the put option in Example 2 by the chaining method, checking that the answer is identical to the one obtained in that example by martingale valuation.

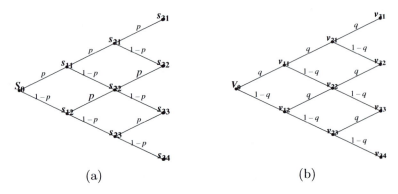

(a)   (b)

Figure 5.10: Chaining for a three-period put option: (a) stock tree; (b) option tree.

Solution. (See Figure 5.10.) In Example 2, we had stock tree parameters $S_0 = \$50$, $b = .04$, and $a = -.02$. We computed that $q = .5$. The backstepping method of chaining needs the final stock values at time 3, which were computed to be:

$$s_{31} = \$56.24;\ s_{32} = \$53.00;\ s_{33} = \$49.94;\ s_{34} = \$47.06.$$

Since the strike price was $50, the final claim values were:

$$
\begin{aligned}
v_{31} &= \ \max\left(\$50 - s_{31}, 0\right) = \$0; \\
v_{32} &= \ \max\left(\$50 - s_{32}, 0\right) = \$0; \\
v_{33} &= \ \max\left(\$50 - s_{33}, 0\right) = \$0.06; \\
v_{34} &= \ \max\left(\$50 - s_{34}, 0\right) = \$2.94.
\end{aligned}
$$

Since the risk-free rate of return was $r = .01$, chaining back to the time 2 nodes produces the values:

$$
\begin{aligned}
v_{21} &= (1+.01)^{-1}(.5 \cdot v_{31} + .5 \cdot v_{32}) \\
&= (1.01)^{-1}(.5 \cdot 0 + .5 \cdot 0) = 0; \\
v_{22} &= (1+.01)^{-1}(.5 \cdot v_{32} + .5 \cdot v_{33}) \\
&= (1.01)^{-1}(.5 \cdot 0 + .5 \cdot (\$.06)) = \$.0297; \\
v_{23} &= (1+.01)^{-1}(.5 \cdot v_{33} + .5 \cdot v_{34}) \\
&= (1.01)^{-1}(.5 \cdot (\$.06) + .5 \cdot (\$2.94)) = \$1.485.
\end{aligned}
$$

At time 1 the two option node values are:

$$
\begin{aligned}
v_{11} &= (1+.01)^{-1}(.5 \cdot v_{21} + .5 \cdot v_{22}) \\
&= (1.01)^{-1}(.5 \cdot 0 + .5 \cdot (\$.0297)) = \$.0147; \\
v_{12} &= (1+.01)^{-1}(.5 \cdot v_{22} + .5 \cdot v_{23}) \\
&= (1.01)^{-1}(.5 \cdot (\$.0297) + .5 \cdot (\$1.485)) = \$.7499.
\end{aligned}
$$

Finally, the time 0 value, which is the initial value of the put, is computed as follows and matches the result of Example 2.

$$
\begin{aligned}
V_0 &= (1+.01)^{-1}(.5 \cdot v_{11} + .5 \cdot v_{12}) \\
&= (1.01)^{-1}(.5 \cdot (\$.0147) + .5 \cdot (\$.7499)) = \$0.379. \ \blacksquare
\end{aligned}
$$

Remarks. (a) It is possible, although tedious, to find a replicating portfolio at each option node. At node j in level i, you find the $\Delta_{i,j}$ and $b_{i,j}$ that make up the replicating portfolio using formulas (5.17) of the previous section. The option value at this node is $\Delta_{i,j} s_{i,j} + b_{i,j}$, where $s_{i,j}$ is the price of the underlying asset at that node. This comes out to be the same value as you would get by chaining. But you do get extra information that way, namely, how to maintain a portfolio of underlying and non-risky assets as time progresses to always have the same value as the option does.

(b) Chaining is particularly well-suited for implementation on spreadsheets. Consider the layout in Figure 5.11 for a simple two-period model. We identify the problem parameters in the block of cells from A1 through B7. As usual, locating parameters in a prominent place facilitates varying them to do "what if" analyses. Two parameters, q in cell B4 and the discount factor $(1 + r)^{-1}$ in cell B7, are computed, while the contents of B1:B3 and B5:B6 would be enetered by the user. The block of cells from A9:C13 contains the stock tree, with the initial price copied down from cell B5 and the other nodes as previous node values times $(1+B3)$ for an up move and $(1+B2)$ for a down move. The cells in the block A16:C20 are the nodes in the option tree for a call option. One starts at the right side, setting the final nodes in column C to the larger of $S_2 - E$ and 0, for the corresponding stock nodes in column C. The formulas in cells B17 and B19 are just the chaining formulas in which the discount factor multiplies the linear combination $q \cdot$ upper child cell $+ (1 - q) \cdot$ lower child cell, and similarly for the root of the option tree in cell A18. You should have no difficulty adapting this template to longer time periods and other options (the latter by simply changing

	A	B	C
1	risk-free	r	
2	down rate	a	
3	up rate	b	
4	risk-neutral	$(\mathrm{B1} - \mathrm{B2})/(\mathrm{B3} - \mathrm{B2})$	
5	init price	S0	
6	exercise price	E	
7	discount	$(1 + \mathrm{B1})^{\wedge}(-1)$	
8			
9			$\mathrm{B10} * (1 + \mathrm{B3})$
10		$\mathrm{A11} * (1 + \mathrm{B3})$	
11	B5		$\mathrm{B12} * (1 + \mathrm{B3})$
12		$\mathrm{A11} * (1 + \mathrm{B2})$	
13			$\mathrm{B12} * (1 + \mathrm{B2})$
14			
15			
16			$\mathrm{MAX}(\mathrm{C9} - \mathrm{B6}, 0)$
17		$\mathrm{B7} * (\mathrm{B4} * \mathrm{C16} + (1 - \mathrm{B4}) * \mathrm{C18})$	
18	$\mathrm{B7} * (\mathrm{B4} * \mathrm{B17} + (1 - \mathrm{B4}) * \mathrm{B19})$		$\mathrm{MAX}(\mathrm{C11} - \mathrm{B6}, 0)$
19		$\mathrm{B7} * (\mathrm{B4} * \mathrm{C18} + (1 - \mathrm{B4}) * \mathrm{C20})$	
20			$\mathrm{MAX}(\mathrm{C13} - \mathrm{B6}, 0)$

Figure 5.11: Template for two-period chaining for a call option on a spreadsheet.

the derivative claim value functions in the rightmost level of the tree). Copying and pasting formulas simplifies the job of filling in larger trees.

The recursive approach is very useful because when we consider "exotic" options more complicated than plain European options, such as the barrier options and American options studied in the next section, the method is quite adaptable to new conditions.

Important Terms and Concepts

Martingale - A sequence of random variables $M_0, M_1, M_2, M_3, \ldots$ satisfying

$$E[M_i | M_{i-1}] = M_{i-1}$$

for each $i \geq 1$.

Martingale valuation - A derivative may be valued at time 0 by $V_0 = E_q\left[\tilde{V}_n\right] = E_q[(1 + r)^{-n} V_n]$, where q is the risk-neutral probability.

Chaining method - A recursive method in which a derivative is given its set of final claim values at the exercise time n, and in prior time periods the value is computed by stepping back using the equation:

$$v_{i,j} = E_q\left[(1+r)^{-1}V_{i+1}|V_i = v_{i,j}\right] = (1+r)^{-1}\left(q \cdot v_{i+1,j} + (1-q) \cdot v_{i+1,j+1}\right).$$

The idea is that the derivative value at any node is the discounted expectation of the values of its child nodes.

Exercises 5.3

1. In an eight-level binomial branch tree with initial price $S_0 = \$60$, $b = .03$, $a = -.02$, compute: (a) s_{22}; (b) s_{41}; (c) s_{63}.

2. For the asset process with parameters as in Exercise 1, if the risk-free rate is .01, compute the conditional expectation $E_q\left[(1.01)^{-2}S_3|S_1\right]$ for each of the two possible values of S_1. What do you find?

3. A risky asset following a binomial branch process is such that it either goes up by a factor of 1.1 with probability $p = .5$ or down by a factor of .9 with probability $1 - p = .5$. Find the distribution, mean, and variance of the price of the asset at time 4, if its initial value is $S_0 = \$35$ at time 0. Do the same assuming that the asset is governed by the risk-neutral probability q instead, where we suppose that the risk-free interest rate is $r = .02$.

4. Use the martingale valuation technique to compute the initial value of a three-period European put option with exercise price $E = \$100$, on an asset following a binomial branch model with parameters $a = .01$, $b = .06$, $S_0 = \$95$. Assume that the risk-free rate is $r = .03$.

5. A European call option with exercise time 2 and strike price $E = \$40$ is available on a risky asset with current price $\$38$, which can either go up by a factor of 1.05 or down by a factor of .95 in each time period. There is a risk-free asset in the market with rate $r = .02$. Find the arbitrage-free initial price of this option by martingale valuation, and by the chaining method.

6. For a general two-step binomial branch process with risk-free rate equal to 0 for simplicity, show by direct computation that $E_q[S_1] = E_q[S_2] = S_0$.

7. Let an asset have initial price $S_0 = \$40$, and suppose that it satisfies a binomial branch model with $b = .10$, $a = -.05$, and "natural" up probability $p = .5$. Let the risk-free interest rate be $r = .04$. Consider a European call option with strike price $\$42$ due to expire in two time periods. Find the expected present value $E_p\left[(1+r)^{-2}V_2\right]$. Show that there is arbitrage if this value is taken as the

initial value of the call option. (Hint: consider a replicating portfolio for the option, which may have to be rebalanced at time 1.)

8. Consider a two-step binomial branch process for an asset of initial price $35 per share, up probability $p = .4$, up rate of return $b = .05$, and down rate of return $a = -.02$. Suppose that the risk-free rate is $r = .02$.   Find the initial value, and the possible values at time 1, of a European call option with exercise time 2 and strike price $36.

9. Consider the call option of Examples 1 and 3. Compute the initial value of the put option with the same strike price using both the martingale and chaining approaches.   Do the put and call prices satisfy the time 2 version of the put-call parity formula $P + S_0 = C + (1 + r)^{-2} E$?

10. Find the initial value of a European put option in two periods, given that $b = .08$, $a = -.04$, $r = .03$, the initial value of the underlying asset is $30, and the exercise price is also $30. Use the time 2 version of the put-call parity formula $P + S_0 = C + (1+r)^{-2} E$ to value the call option with the same parameters.

11. Consider a four-period model in which the underlying asset begins at price $30 and in each period moves up by 5% or down by 3%, independently of the past. A derivative has final values 2, 1, 0, 0, and 0 at the five nodes at time 4. The risk-free rate per period is .01. Find the value of the derivative at each node.

12. Suppose that a risky asset moves according to a two-period binomial branch model with $S_0 = \$50$, $b = .25$, $a = -.05$, and suppose that there is a risk-free asset whose value increases by a factor of $r = .05$ each time period. A European call option has strike price $58 and exercise time 2. Compute the full option tree and the delta-hedging strategy for the sale of one call in the case that the first move was down and the second was up.

13. We suppose in this exercise and the next that there is an underlying asset whose initial price is $100, that the risk-free interest rate is $r = .02$, and that the asset satisfies a three-period binomial branch model with up rate $b = .04$ and down rate $a = -.03$. A derivative exists that pays twice the absolute difference $|S_3 - E|$ between the price of the asset at time 3 and the exercise price $E = \$101$, if that absolute difference exceeds $4, and nothing otherwise. Find the initial price of this derivative.

14. Let the underlying asset be as in the previous exercise, and consider a related derivative that pays as did the previous one, except that it pays when the absolute difference is less than or equal to $4, and does not pay otherwise. Find the initial value of this derivative. What is the claim value at time 3 of a portfolio consisting of one of each of these two derivatives? What must be the initial value of this portfolio in an arbitrage-free market? Check that your answer is consistent with the initial prices you computed in these two exercises.

15. Suppose that, instead of the simple binomial branch model in which an up move always multiplies the current price by a factor $1 + b$ and a down move by a factor $1 + a$, where a and b are constant, we just have a model in which there are two possible next prices for each current price. Assuming, for specificity, that an asset satisfies the three-step model as in the diagram (note that its changes are additive rather than multiplicative), and for simplicity supposing that the risk-free rate is 0, find probabilities on each branch of the tree that make the process into a martingale.

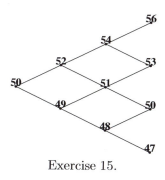

Exercise 15.

16. In the context of Exercise 15, assuming that the martingale valuation technique is still valid, find the initial value of: (a) a two-period put option with strike price $52; (b) a three-period call option with strike price $51.

5.4 Valuation of Exotic Options and Simulation

The methods of Section 5.3 are very effective in the case that both (a) the derivative under consideration is of the European type, in which execution of the derivative can only happen at its expiration time and not before; (b) the derivative value at redemption depends only on the final asset value, and not on the path that the asset took to reach that value. If either of these conditions does not hold, then we have some rethinking to do.

The area of Exotic Options focuses on options that are more complicated than the simple calls and puts that we have looked at. Some of these options are well understood now, and therefore aren't really so "exotic," but we will continue to use the word. The first subsection shows how to value some of the more straightforward exotic options, including American options, barrier options, and Asian options. There is active current research on more sophisticated exotic options as well, which we cannot hope to reach in this introduction.

As derivatives become more intricate and hard to understand, it is useful to be able to find approximate values of the derivatives using the method of simulation. The basic idea is that we suppose that the risk-neutral probability governs the transitions of the underlying asset, and we then replicate the motion of the asset many times, making note of the value of the derivative on each replication. Then we compute an average derivative value among the replications as a estimate of the expectation that is the true value. This process is introduced in the second subsection. For continued study of exotic options, consult, for example, Buchen [3].

5.4.1   Exotic Options

As mentioned above, we will define and value three types of exotic options: *American options*, which can be executed at the discretion of the holder prior to their expiration, *barrier options*, which either become void or come into effect only when the underlying process crosses a boundary, and *Asian options*, whose value depends on the average value of the underlying asset through its motion.

American Options

We begin by reprising an old example.

Example 1. Recall Example 1 of Section 5.3. In it, there was an underlying asset whose initial price was \$20, which followed a binomial branch process with parameters $b = .03$ and $a = .005$. The risk-free rate per period was assumed to be $r = .01$, and a European call option with strike price \$20.50 and expiration time $n = 2$ was valued. An American call option with similar conditions would allow the user to exercise it immediately at time 0 (an unlikely strategy) or to exercise it at time 1, depending on which of the two possible values the underlying asset has at that time. How does the option holder make the decision, and how does that change the value of the option?

Solution. The price tree for this underlying asset is repeated in Figure 5.12(a). In part (b) of the figure, at time 2 the American option has the same claim value as the European option, namely, $\max(S_2 - \$20.50, 0)$, since the option expires at time 2. At earlier times, we reason that, at each internal node of the option tree, we must decide which is better: to execute the option now in light of the current price of the underlying asset, or to wait until the next period, when we will face the same decision. At time 1, we should compare the current value of executing the option, $\max(S_1 - E, 0)$, to the expected present value of the time 2 option value, that is, $(1 + r)^{-1}(q \cdot V_u + (1 - q)V_d)$. But this expression is of course the chaining value of the level 1 node. Whichever is larger determines the course of action; execute immediately if and only if the current execution value exceeds the chaining value.

Let us insert the numbers for the example at hand. Recall that $q = .2$ in this problem. The generalized expressions for the values of the American call option at the level 1 nodes are:

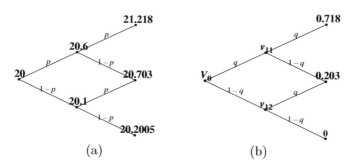

Figure 5.12: A two-period American call option: (a) stock tree; (b) option tree.

$$
\begin{aligned}
v_{11} &= \max\{\max(\$20.60 - \$20.50, 0), \\
&\quad (1.01)^{-1}(.2 \cdot (\$.718) + .8 \cdot (\$.203))\} \\
&= \$.30297; \\
v_{12} &= \max\{\max(\$20.10 - \$20.50, 0), \\
&\quad (1.01)^{-1}(.2 \cdot (\$.203) + .8 \cdot 0)\} \\
&= \$.040198.
\end{aligned}
$$

In both cases the maximum occurs at the chaining value, and so it does not pay to execute the option early. For the level 0 node, the initial value of the option is:

$$
\begin{aligned}
V_0 &= \max\{\max(\$20 - \$20.50, 0), \\
&\quad (1.01)^{-1}((.2)(\$.30297) + (.8)(\$.040198))\} \\
&= \$.0918.
\end{aligned}
$$

Again the maximum occurs at the chaining value, so for this example we find that the American call option is no more valuable than the European call option with the same parameters. It is never optimal to exercise it early. Later we will revisit this idea, which, surprisingly, is a general theorem for calls, but not for puts. ∎

In general, the chaining method for valuing American call options is expressed by the equations:

$$
\begin{aligned}
v_{n,j} &= \max\left(s_{n,j} - E, 0\right); \\
v_{i,j} &= \max\left\{\max\left(s_{i,j} - E, 0\right), \right. \\
&\quad \left. (1+r)^{-1}\left(q \cdot v_{i+1,j} + (1-q) \cdot v_{i+1,j+1}\right)\right\}.
\end{aligned}
\tag{5.31}
$$

Similarly, the algorithm for finding the value of an American put option at each node is:

$$\begin{aligned}
v_{n,j} &= \max\left(E - s_{n,j}, 0\right); \\
v_{i,j} &= \max\left\{\max\left(E - s_{i,j}, 0\right), \right. \\
&\qquad \left. (1+r)^{-1}\left(q \cdot v_{i+1,j} + (1-q) \cdot v_{i+1,j+1}\right)\right\}.
\end{aligned} \tag{5.32}$$

The only difference is in the computation of the immediate claim value at each node; the chaining condition is the same for puts as for calls. American versions of other options are possible, and the algorithm for solving for their values is similar.

Example 2. Suppose that a stock with initial price \$50 obeys the binomial branch model in three periods with $b = .04, a = -.02$, and risk-free rate $r = .01$. Find the value at all nodes of an American put option expiring at time 3 with exercise price \$50. Is there any case in which it is optimal for the holder to exercise the option early?

Solution. We must find the stock price nodes, calculate the time 3 values of the option, then use the chaining relation (5.32). First, the risk-neutral probability is:

$$q = \frac{r-a}{b-a} = \frac{.01 - (-.02)}{.04 - (-.02)} = \frac{.03}{.06} = .5.$$

The stock price tree nodes are:

$$\begin{aligned}
S_0 &= \$50; \\
s_{11} &= \$50(1.04) = \$52; \\
s_{12} &= \$50(.98) = \$49; \\
s_{21} &= \$50(1.04)^2 = \$54.08; \\
s_{22} &= \$50(1.04)(.98) = \$50.96; \\
s_{23} &= \$50(.98)^2 = \$48.02; \\
s_{31} &= \$50(1.04)^3 = \$56.2432; \\
s_{32} &= \$50(1.04)^2(.98) = \$52.9984; \\
s_{33} &= \$50(1.04)(.98)^2 = \$49.9408; \\
s_{34} &= \$50(.98)^3 = \$47.0596.
\end{aligned}$$

Since the put option exercise price is \$50, the values of the option at the level 3 nodes are:

$$\begin{aligned}
v_{31} &= \max\left(\$50 - s_{31}, 0\right) = 0; \\
v_{32} &= \max\left(\$50 - s_{32}, 0\right) = 0; \\
v_{33} &= \max\left(\$50 - s_{33}, 0\right) = \$50 - \$49.9408 = \$.0592; \\
v_{34} &= \max\left(\$50 - s_{34}, 0\right) = \$50 - \$47.0596 = \$2.9404.
\end{aligned}$$

The two trees are shown in Figure 5.13. Chaining backwards, we get at time 2:

$$\begin{aligned}
v_{21} &= \max\left\{\max\left(\$50 - s_{21}, 0\right), (1.01)^{-1}\left(.5v_{31} + .5v_{32}\right)\right\} \\
&= \max\left\{0, (1.01)^{-1}(.5 \cdot 0 + .5 \cdot 0)\right\} \\
&= 0;
\end{aligned}$$

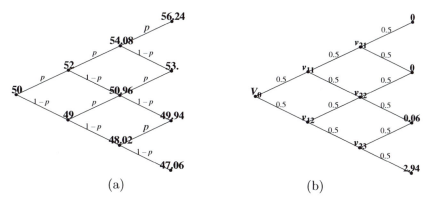

Figure 5.13: (a) Binomial branch process model of price of asset. (b) Put option tree, $E = \$50$.

$$
\begin{aligned}
v_{22} &= \max\left\{\max\left(\$50 - s_{22}, 0\right), (1.01)^{-1}\left(.5v_{32} + .5v_{33}\right)\right\} \\
&= \max\left\{0, (1.01)^{-1}(.5 \cdot 0 + .5 \cdot (\$.0592))\right\} \\
&= \$.02931;
\end{aligned}
$$

$$
\begin{aligned}
v_{23} &= \max\left\{\max\left(\$50 - s_{23}, 0\right), (1.01)^{-1}\left(.5v_{33} + .5v_{34}\right)\right\} \\
&= \max\left\{\$1.98, (1.01)^{-1}(.5 \cdot (\$.0592) + .5 \cdot (\$2.9404))\right\} \\
&= \$1.98.
\end{aligned}
$$

So at node 1 of level 2, neither executing the option nor waiting gives a positive value; at node 2 the maximum is taken on at the chaining value, so it is optimal not to execute the option; but at node 3 the maximum is achieved by immediate exercise. Next, at time 1, we have:

$$
\begin{aligned}
v_{11} &= \max\left\{\max\left(\$50 - s_{11}, 0\right), (1.01)^{-1}\left(.5v_{21} + .5v_{22}\right)\right\} \\
&= \max\left\{0, (1.01)^{-1}(.5 \cdot 0 + .5 \cdot (\$.02931))\right\} \\
&= \$.01451;
\end{aligned}
$$

$$
\begin{aligned}
v_{12} &= \max\left\{\max\left(\$50 - s_{12}, 0\right), (1.01)^{-1}\left(.5v_{22} + .5v_{23}\right)\right\} \\
&= \max\left\{\$1, (1.01)^{-1}(.5 \cdot (\$.02931) + .5 \cdot (\$1.98))\right\} \\
&= \$1.
\end{aligned}
$$

Again the investor should not exercise at the first node, but because for node s_{12} the value of immediate exercise exceeds the expected discounted value of waiting, the option should be exercised if the risky asset moves to this node. Finally, the intial put option value is:

$$
\begin{aligned}
V_0 &= \max\left\{\max\left(\$50 - S_0, 0\right), (1.01)^{-1}\left(.5v_{11} + .5v_{12}\right)\right\} \\
&= \max\left\{0, (1.01)^{-1}(.5 \cdot (\$.01451) + .5 \cdot (\$1))\right\} \\
&= \$.502232.
\end{aligned}
$$

The option should not be exercised at time 0. But we do have additional value for the American option as compared with the European option, because if you check back you will find that the problem parameters here are exactly those of Examples 2 and 4 of the last section in which the initial value of the European option was found to be $V_0 = \$.379$. So the ability to exercise the put early has given it more value, and in particular, this option should be exercised if the asset process moves to either s_{12} or to s_{23}. ■

It stands to reason that the additional choice given to the holder of an American option as opposed to a European option results in a possibly higher, but never lower, value for the American option. The next theorem establishes this, but it also proves the surprising result that, for a call option, the European version of the derivative has the same value as the American. So the computational result in Example 1 above was no coincidence. Example 2 shows that an American put can be more valuable than a corresponding European put, and the theorem assures us that the reverse can never happen.

Theorem 1. Let C_E and C_A denote the initial values of European and American call options on the same underlying asset, with the same expiration times and strike prices. Then (assuming that arbitrage is not possible)

$$
C_E = C_A. \tag{5.33}
$$

Similarly, let P_E and P_A denote the initial values of European and American put options with the same expiration times and strike prices. Then

$$
P_E \leq P_A. \tag{5.34}
$$

Proof. First we will argue that $C_E \leq C_A$. If instead $C_E > C_A$, issue a European call and buy an American call to generate an arbitrage opportunity. The woud-be arbitrageur invests the difference $C_E - C_A$ in the risk-free asset. No expenditure occurs initially. The arbitrageur simply holds the American option until expiration and executes it if and only if the European call partner executes that option. If neither is executed, then the arbitrageur ends with $(C_E - C_A)(1 + r)^n$ dollars without risk, where n is the expiration time. But if both are executed, the arbitrageur can take the E given to him by the European call holder and execute the American option for a price of E, delivering the share of the asset to the European partner. Thus, the final result of the strategy is still $(C_E - C_A)(1 + r)^n > 0$, which is arbitrage.

Now suppose that $C_A > C_E$. If the theorem is true, the American call is overvalued relative to the European call, so that we may try to form an arbitrage opportunity in which we issue an American call, buy a European call, and invest the proceeds $C_A - C_E$ at the risk-free rate. Then there is no net initial expenditure. If the American call is not executed prior to the expiration date n, then its final value is the same as that of the European call, and the arbitrageur has a profit of $(C_A - C_E)(1+r)^n$ at the end. If the American call is executed, say at time i, then the arbitrageur borrows a share of stock to finalize the American call, receives E from the American call partner, and can hold it or even invest it in the risk-free asset for $n - i$ periods to give a total of $E(1+r)^{n-i}$ at time n. This is more than enough to execute the European option to buy the asset at a price E at time n in order to repay the share that was borrowed. This is arbitrage.

The proof of the inequality $P_E \leq P_A$ is left as an exercise.

In the previous section we talked about the implementation of the chaining procedure for European puts and calls on a spreadsheet. It is not hard to adapt the method to the case of American options, and in view of the theorem, we need only be concerned with American puts. Figure 5.14 shows a template in the two-period case. The header information in rows 1–7, in which the user inputs the problem parameters in the appropriate cells in column B, is the same as in the European case. The range of the spreadsheet in which the asset price tree is computed (here rows 9–13) is also the same. In column C, rows 16, 18, and 20 are the usual formulas that compute the final claim value of the put, based on the exercise price in cell B6 and the three final asset prices in column C, rows 9, 11, and 13. In the option tree cells in column B, we see the first change. Instead of just computing the chaining value from the child cells, the spreadsheet will compute those and compare them to the immediate claim value $\max(E - S_1, 0)$, returning the larger of the two. This is also done in cell A18 for the initial value of the American put.

Barrier Options

Consider the case of an investor wanting to purchase a European-variety call option, who, in order to limit the purchase expense, is willing to limit his upside profit by gambling that the underlying asset will never go above a level H. This investor might be matched with a potential issuer of a call option, who wants to protect against the possibility that the underlying asset reaches levels far above the strike price. Both parties may be inclined to agree on a special call option in which, if the underlying asset ever reaches H or greater, the option expires worthless. Such an option should be cheaper for the investor to purchase than an ordinary call, since he is willing to give up on the case where $S_n - E$ is very large, hence reducing the loss potential for the issuer of the option.

This scenario illustrates that there is a potential market for options called **knockouts**, or **barrier options**, which become worthless if their underlying asset crosses a boundary value H. The contract described in the last paragraph is an **"up-and-out" barrier option**. There also exist **"down-and-out" options**, which are voided when the underlying asset falls beneath a boundary value L. Another class of barrier options are called **"up-and-in" options**, which are only activated when the underlying process reaches a value of at least H. The final case is

	A	B	C
1	risk-free	r	
2	down rate	a	
3	up rate	b	
4	risk-neutral	$(B1 - B2)/(B3 - B2)$	
5	init price	S0	
6	exercise price	E	
7	discount	$(1 + B1)^\wedge(-1)$	
8			
9			$B10 * (1 + B3)$
10		$A11 * (1 + B3)$	
11	B5		$B12 * (1 + B3)$
12		$A11 * (1 + B2)$	
13			$B12 * (1 + B2)$
14			
15			
16			$\text{MAX}(B6 - C9, 0)$
17		$\text{MAX}(\text{MAX}(B6 - B10, 0),$ $B7 * (B4 * C16$ $+ (1 - B4) * C18))$	
18	$\text{MAX}(\text{MAX}(B6 - A11, 0),$ $B7 * (B4 * B17$ $+ (1 - B4) * B19))$		$\text{MAX}(B6 - C11, 0)$
19		$\text{MAX}(\text{MAX}(B6 - B12, 0),$ $B7 * (B4 * C18$ $+ (1 - B4) * C20))$	
20			$\text{MAX}(B6 - C13, 0)$

Figure 5.14: Template for chaining for a two-period American put option on a spreadsheet.

the *"down-and-in" barrier option*, where the option takes effect only when the underlying process sinks to a value of L or less. These last two options are referred to generically as *knockin options*. Exercise 10 asks you to make up situations where investors may be inclined to participate in such contracts, as we did with the up-and-out knockout call option.

Notice that these options, unlike plain European options, depend not only on the final value of the underlying asset at the time of expiration of the option, but on the full path of the asset. This is so because, for a knockout option, for example, once the asset crosses the barrier, even if it subsequently crosses back, the option becomes worthless for all times thereafter.

A modified chaining approach still works to value barrier options. The main idea is to zero out nodes that are beyond the boundary as necessary and then execute the chaining algorithm. We illustrate in the next example.

Example 3. A European call option expiring in three time periods has a strike price of \$105. The initial price of the underlying asset is \$100, its binomial branch parameters are $b = .1$, $a = -.1$, and the risk-free rate of interest per period is $r = .05$. But there is an additional stipulation on the option contract that, if the underlying price ever goes below \$95, the option becomes worthless. Find the initial value of this option.

Solution. The problem describes a down-and-out knockout option with lower barrier $L = \$95$. We need to compute the nodes on the asset tree, which is done as usual.

$$
\begin{aligned}
S_0 &= \$100; \\
s_{11} &= \$100(1.1) = \$110; \\
s_{12} &= \$100(.9) = \$90; \\
s_{21} &= \$100(1.1)^2 = \$121; \\
s_{22} &= \$100(1.1)(.9) = \$99; \\
s_{23} &= \$100(.9)^2 = \$81; \\
s_{31} &= \$100(1.1)^3 = \$133.10; \\
s_{32} &= \$100(1.1)^2(.9) = \$108.90; \\
s_{33} &= \$100(1.1)(.9)^2 = \$89.10; \\
s_{34} &= \$100(.9)^3 = \$72.90.
\end{aligned}
$$

By the information given in the problem statement, the risk-neutral probability is:

$$
q = \frac{.05 - (-.1)}{.1 - (-.1)} = \frac{1.5}{.2} = .75.
$$

The underlying asset tree is shown on the left of Figure 5.15. The down-and-out barrier at \$95 is shown as a dashed horizontal line, which is carried over to the option tree on the right of the figure. Below the line, the option is forced to have value 0. Above the line at level 3, the option has the usual call option claim values:

$$
\begin{aligned}
v_{31} &= \max(s_{31} - E, 0) \\
&= \max(\$133.1 - \$105, 0) = \$28.1; \\
v_{32} &= \max(s_{32} - E, 0) \\
&= \max(\$108.9 - \$105, 0) = \$3.9; \\
v_{33} &= 0; \\
v_{34} &= 0.
\end{aligned}
$$

Coincidentally, because of the strike price of \$105, the level 3 nodes v_{33} and v_{34} corresponding to $s_{33} = \$89.1$ and $s_{34} = \$72.9$ have value 0 anyway regardless of the barrier. By chaining, the level 2 node v_{23} where $s_{23} = \$81$ will also have value 0, independently of the barrier condition. But, in contrast to a plain option, this barrier option will also be of value 0 at level 1 node v_{12} where $s_{12} = \$90$, which will affect the initial value V_0.

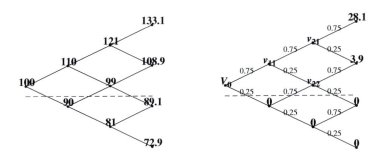

Figure 5.15: Asset and derivative trees for down-and-out knockout call option, $L = \$95$.

We use chaining on the first two nodes at the second level of the option tree; the third node is zero.

$$
\begin{aligned}
v_{21} &= (1+r)^{-1}\left(q \cdot v_{31} + (1-q)v_{32}\right) \\
       &= (1.05)^{-1}((.75) \cdot (\$28.1) + (.25)(\$3.9)) = \$21; \\
v_{22} &= (1+r)^{-1}\left(q \cdot v_{32} + (1-q)v_{33}\right) \\
       &= (1.05)^{-1}((.75) \cdot (\$3.9) + (.25)(0)) = \$2.78571; \\
v_{23} &= 0.
\end{aligned}
$$

Backing up to level 1, chaining gives the option value at the first node, and the second is forced to have value 0:

$$
\begin{aligned}
v_{11} &= (1+r)^{-1}\left(q \cdot v_{21} + (1-q)v_{22}\right) \\
       &= (1.05)^{-1}((.75) \cdot (\$21) + (.25)(\$2.78571)) = \$15.6633; \\
v_{12} &= 0.
\end{aligned}
$$

Last, we chain to find V_0:

$$
\begin{aligned}
V_0 &= (1+r)^{-1}\left(q \cdot v_{11} + (1-q)v_{12}\right) \\
    &= (1.05)^{-1}((.75) \cdot (\$15.6633) + (.25)(0)) = \$11.188.
\end{aligned}
$$

If there had been no barrier, then, as we noted before, levels 2 and 3 of the option tree would have been the same. Node v_{11} is also unchanged, but node v_{12} would have been:

$$
\begin{aligned}
v_{12} &= (1+r)^{-1}\left(q \cdot v_{22} + (1-q)v_{23}\right) \\
       &= (1.05)^{-1}((.75) \cdot (\$2.78571) + (.25)(0)) = \$1.9898,
\end{aligned}
$$

which gives the value of the plain option as:

$$
V_0 = (1.05)^{-1}((.75) \cdot (\$15.6633) + (.25)(\$1.9898)) = \$11.6618.
$$

The fact that an initial downward move makes the option worthless gives an advantage to the issuer over the option holder, leading to a decrease in the initial price of the option. ∎

Example 4. Compute the initial value of a three-period up-and-in knockin put option with barrier $30.50 and strike price $30 on an underlying asset obeying a binomial branch process with initial price $30, up rate $b = .05$, and down rate $a = -.04$. Suppose that the risk-free rate of return is $r = .02$.

Solution. The risk-neutral probability for this situation is:

$$q = \frac{.02 - (-.04)}{.05 - (-.04)} = \frac{.06}{.09} = \frac{2}{3}.$$

To get a sense of the behavior of this barrier option, let us compute the asset nodes and sketch graphs of the trees. We have:

$$
\begin{aligned}
S_0 &= \$30; \\
s_{11} &= \$30(1.05) = \$31.5; \\
s_{12} &= \$30(.96) = \$28.8; \\
s_{21} &= \$30(1.05)^2 = \$33.075; \\
s_{22} &= \$30(1.05)(.96) = \$30.24; \\
s_{23} &= \$30(.96)^2 = \$27.648; \\
s_{31} &= \$30(1.05)^3 = \$34.7288; \\
s_{32} &= \$30(1.05)^2(.96) = \$31.752; \\
s_{33} &= \$30(1.05)(.96)^2 = \$29.0304; \\
s_{34} &= \$30(.96)^3 = \$26.5421.
\end{aligned}
$$

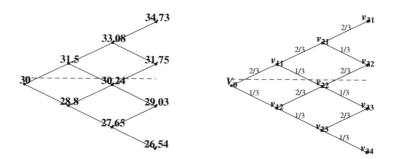

Figure 5.16: Asset and derivative trees for up-and-in knockin put option, $H = \$30.50$.

Remembering that the strike price is $30, the only two nodes at level 3 at which there will be any value for this option are nodes v_{33} and v_{34}. But we must be careful; this option does not even activate until and unless the asset moves above the barrier at \$30.50 shown as a horizontal line on the left in Figure 5.16. To reach node s_{34} requires three downward moves, and in this case the option never becomes active. One can reach the value \$29.03 with two down moves followed by an up move, but then again the option never has the chance to activate. There is in fact only one path for which the option comes into existence and ends with a positive final value, and that is the case where the underlying asset makes one up move and then two down moves to $s_{33} = \$29.03$. If V_3 represents the claim value at time 3, then $V_3 = 0$ except on this one path, whose probability is $(2/3)(1/3)^2$. In this case we can value the option by the martingale technique of finding the discounted expectation of the final claim value, which is:

$$E\left[(1.02)^{-3}V_3\right] = (1.02)^{-3} \cdot \frac{2}{3} \cdot \left(\frac{1}{3}\right)^2 (\$30 - \$29.03) = \$.0677. \ \blacksquare$$

In a spreadsheet, the easy thing to do to value barrier options is to create the sheet in the usual way with formulas appropriate for chaining for the option without a barrier, and just replace the formulas in the option tree by 0 for those nodes whose values are forced to be 0 because of the barrier.

Asian Options

Another kind of path-dependent option gives its holder a reward depending on the average value of the underlying prices during the existence of the option. Such a derivative is called an **Asian option**. Though it is not obvious, an Asian option can also be valued by the martingale technique as a risk-neutral expected present value. But it is necessary to calculate the value of the option on each path, weight that by the (risk-neutral) path probability, and sum and discount to get the expectation. So the valuation of an Asian option is rather computationally intensive. We illustrate with an example.

Example 5. Consider an Asian call option in four periods that pays the larger of $\bar{S} - \$53$ and 0, where $\bar{S}$ is the mean price along the path, and \$53 is the option strike price. In the usual notation, let $r = .03$, $b = .06$, $a = .01$, and let the initial price of the underlying asset be $S_0 = \$50$. Value this Asian option at time 0.

Solution. The stock price values are:

$$
\begin{aligned}
S_0 &= \$50; \\
s_{11} &= \$50(1.06) = \$53; \\
s_{12} &= \$50(1.01) = \$50.5; \\
s_{21} &= \$50(1.06)^2 = \$56.18; \\
s_{22} &= \$50(1.06)(1.01) = \$53.53; \\
s_{23} &= \$50(1.01)^2 = \$51.005; \\
s_{31} &= \$50(1.06)^3 = \$59.5508; \\
s_{32} &= \$50(1.06)^2(1.01) = \$56.7418; \\
s_{33} &= \$50(1.06)(1.01)^2 = \$54.0653; \\
s_{34} &= \$50(1.01)^3 = \$51.5151; \\
s_{41} &= \$50(1.06)^4 = \$63.1238; \\
s_{42} &= \$50(1.06)^3(1.01) = \$60.1463; \\
s_{43} &= \$50(1.06)^2(1.01)^2 = \$57.3092; \\
s_{44} &= \$50(1.06)(1.01)^3 = \$54.606; \\
s_{45} &= \$50(1.01)^4 = \$52.0302.
\end{aligned}
$$

The risk-neutral probability is

$$
q = \frac{.03 - .01}{.06 - .01} = \frac{.02}{.05} = .4.
$$

The martingale approach would say that the value of the Asian option is:

$$
V_0 = E_q \left[(1+r)^{-4} \max \left(\bar{S} - E, 0 \right) \right]. \tag{5.35}
$$

The discount factor $(1+r)^{-4}$ would factor out of the expectation. To compute the rest of the expectation, we need to find all possible values of the average price $\bar{S}$ on all paths and their corresponding path probabilities. These probabilities are used as weights in a weighted average of values of $\max \left(\bar{S} - E, 0 \right)$, which also must be computed.

The following table summarizes the calculations. There are 16 outcomes of the form *udud* specifying up or down at each step that determine the rows of the table. Each path passes through a unique sequence of states S_0, S_1, S_2, S_3, S_4 listed in column 2. The probability of a sequence is $q^{\# \text{ ups}} \cdot (1-q)^{\# \text{ downs}}$, and these are listed in column 3 of the table. For each path, the values of the underlying asset on that particular path are averaged, and the claim value, which is the larger of the average price minus \$53 and 0, is calculated. These are listed in columns 4 and 5. For example, for the path $S_0, s_{11}, s_{21}, s_{31}, s_{41}$ in which all moves are up, the probability of the path is $q^4 = (.4)^4 = .0256$, the average value turns out to be $\bar{S} = (S_0 + s_{11} + s_{21} + s_{31} + s_{41})/5 = \$56.3709$, and the claim value is the larger of $\bar{s}_1 - 53 = \$3.37093$ and 0.

updown	path	prob	average	claim value
uuuu	$S_0, s_{11}, s_{21}, s_{31}, s_{41}$	0.0256	56.371	3.371
uuud	$S_0, s_{11}, s_{21}, s_{31}, s_{42}$	0.0384	55.775	2.775
uudu	$S_0, s_{11}, s_{21}, s_{32}, s_{42}$	0.0384	55.214	2.214
uduu	$S_0, s_{11}, s_{22}, s_{32}, s_{42}$	0.0384	54.68	1.684
duuu	$S_0, s_{12}, s_{22}, s_{32}, s_{42}$	0.0384	54.184	1.184
uudd	$S_0, s_{11}, s_{21}, s_{32}, s_{43}$	0.0576	54.646	1.646
udud	$S_0, s_{11}, s_{22}, s_{32}, s_{43}$	0.0576	54.116	1.116
duud	$S_0, s_{12}, s_{22}, s_{32}, s_{43}$	0.0576	53.616	0.616
dudu	$S_0, s_{12}, s_{22}, s_{33}, s_{43}$	0.0576	53.081	0.081
uddu	$S_0, s_{11}, s_{22}, s_{33}, s_{43}$	0.0576	53.581	0.581
dduu	$S_0, s_{12}, s_{23}, s_{33}, s_{43}$	0.0576	52.576	0
dddu	$S_0, s_{12}, s_{23}, s_{34}, s_{44}$	0.0864	51.525	0
ddud	$S_0, s_{12}, s_{23}, s_{33}, s_{44}$	0.0864	52.535	0
dudd	$S_0, s_{12}, s_{22}, s_{33}, s_{44}$	0.0864	52.540	0
uddd	$S_0, s_{11}, s_{22}, s_{33}, s_{44}$	0.0864	53.040	0.040
dddd	$S_0, s_{12}, s_{23}, s_{34}, s_{45}$	0.1296	51.010	0

The discounted expectation $E_q \left[(1+r)^{-4} \max \left(\bar{S} - E, 0 \right) \right]$ is the discount factor times the sum of products of probabilities and claim values, that is,

$$
\begin{aligned}
V_0 &= (1+r)^{-4}((.0256)(\$3.371) + (.0384)(\$2.775) \\
&\quad + \cdots + (0.1296)(0)) \\
&= \$.554578. \ \blacksquare
\end{aligned}
$$

5.4.2   Approximate Valuation by Simulation

For path-dependent options like barrier and Asian options, if the time horizon is large, the computations are burdensome. We would therefore like to end this section with an introduction to algorithms for valuing options approximately by simulation. These algorithms can be implemented on a number of platforms. All popular languages have randomization facilities and programming structures that allow working programs to be created. We will be content to write detailed step-by-step pseudocode algorithms, and in the digital *Mathematica* version of this text we will show how to implement these algorithms in the *Mathematica* programming language.

What exactly do we mean by approximating the value of a derivative by simulation? The last example in the subsection on Asian options gives a hint. With the martingale probability q in place, if a derivative has time horizon n and claim value V_n, and the risk-free rate per period is r, then the initial derivative value will be:

$$
V_0 = E_q \left[(1+r)^{-n} V_n \right] = (1+r)^{-n} E_q \left[V_n \right]. \tag{5.36}
$$

Expectations are approximated by sample means among many replications of the experiment at hand. This suggests the following simple algorithm.

Simulation Valuation Algorithm

Given an n-period derivative on an underlying asset following the binomial branch model with parameters S_0, a, b, in which the claim value of the derivative is a well-determined function V_n of the path of asset prices, and given a risk-free rate of return r,

1. Compute the risk-neutral probability $q = \frac{r-a}{b-a}$;

2. Repeat a large number m of times:

   (a) Simulate a path of the underlying asset using q as the up probability;

   (b) Compute the value of V_n for that path;

3. Compute the average value $\bar{V}$ of the m simulated V_ns;

4. Return the discounted average value $= (1+r)^{-n} \cdot \bar{V}$.

We would like to add an important fifth step to the basic algorithm which gives information about the closeness of the estimate to the true derivative value. Recall from Chapter 3 that, if $X_1, X_2, ..., X_m$ is a sequence of independent, identically distributed random variables with mean μ and variance σ^2, and $\bar{X} = (X_1 + X_2 + \cdots + X_m)/m$ is the sample mean, then $\bar{X}$ is a random variable itself, with mean and variance:

$$E\left[\bar{X}\right] = \mu; \operatorname{Var}\left(\bar{X}\right) = \frac{\sigma^2}{m}. \tag{5.37}$$

In the context of our simulation algorithm, this means that our estimate $(1+r)^{-n} \cdot \bar{V}$ of the derivative value has mean $(1+r)^{-n} E_q\left[V_n\right]$, so it is pointing at the right target. The mean claim value $\bar{V}$ itself will have variance approximately S^2/m, where S^2 is the sample variance of the simulated V_ns. By properties of variance, the random variable $(1+r)^{-n} \cdot \bar{V}$ has variance approximately $\left((1+r)^{-n}\right)^2 \cdot S^2/m$. In light of Chebyshev's inequality, we may add a step to the algorithm and have it also compute S^2 and return the quantity:

$$2 \cdot \frac{S}{\sqrt{m}} \cdot (1+r)^{-n}, \tag{5.38}$$

which approximates 2 standard deviations of the estimate $(1+r)^{-n} \cdot \bar{V}$. This quantity serves as an approximate upper bound on how far away the true derivative price V_0 can reasonably be from our simulation estimate. Thus, we would have an idea how precise the estimate is. As the number of replications grows, the estimate should be more precise. Here is the revised algorithm.

Simulation Valuation Algorithm with Precision Estimate

Given an n-period derivative on an underlying asset following the binomial branch model with parameters S_0, a, b, in which the claim value of the derivative is a well-determined function V_n of the path of asset prices, and given a risk-free rate of return r,

1. Compute the risk-neutral probability $q = \frac{r-a}{b-a}$;

2. Repeat a large number m of times:

    (a) Simulate a path of the underlying asset using q as the up probability;

    (b) Compute the value of V_n for that path;

3. (a) Compute the average value $\bar{V}$ of the m simulated V_ns;

    (b) Compute the sample variance S^2 of the m simulated V_ns;

4. (a) Return the discounted average value $= (1+r)^{-n} \cdot \bar{V}$.

    (b) Return the precision estimate $= 2 \cdot \frac{S}{\sqrt{m}} \cdot (1+r)^{-n}$.

Example 6. In Example 3 we showed how to value exactly a three-period down-and-out knockout call option with lower barrier $L = \$95$ and strike price $E = \$105$. We had assumed that $S_0 = \$100$, $b = .1$, $a = -.1$, and $r = .05$. The initial value of this option turned out to be $\$11.188$. Let us see how to create a program that will approximate this value.

Solution. The algorithm above is a very useful guide, but we need to think about how the details play out for this particular situation. We had already found the risk-neutral probability to be $q = .75$ in the previous example. The heart of the simulation algorithm is in the details of step 2(a). In Section 3.6 we observed that computer languages provide ways to simulate uniformly distributed random real numbers in the interval $[0,1]$, and therefore we can simulate a single move of the underlying asset by:

1. Simulate U, a uniform$[0,1]$ random variable;

2. If $U \le q$ return an up move, otherwise return a down move.

Then an up move will occur with probability q and a down move with the complementary probability $1 - q$. To return an up move means to multiply the current price by $(1 + b) = 1.1$, otherwise the move is down and the current price is multiplied by $(1 + a) = .9$. We can form a list $\{S_0 = \$100, S_1, S_2, S_3\}$ of simulated asset price values using this strategy.

Computing the value V_n needs some planning. The down-and-out option becomes valueless if any of the simulated prices fall equal to or below the $95 barrier, which occurs if and only if the smallest element of the simulated price list is less than or equal to $95. For outcomes for which the option remains alive, its final value is $\max(S_3 - E, 0)$. A statement that captures this idea is:

$$\text{If } \min(S_0, S_1, S_2, S_3) \le 95, \text{ then } V_n = 0, \text{ else } V_n = \max(S_3 - 105, 0).$$

The rest of the algorithm is straight computation. We are now prepared to give the details of the program to simulate m values of this knockout option and return estimates:

1. Let $q = \frac{r-a}{b-a} = .75$;

2. Initialize an empty list of claim values;

3. Do the following m times:

   (a) Initialize a list of prices as $\{S_0 = 100\}$

   (b) Do the following $n = 3$ times:

      (i) Simulate U, a uniform$[0,1]$ random variable;

      (ii) If $U \le q$, then adjoin $(1 + b) = 1.1$ times the previous price to the list of prices, else adjoin $(1 + a) = .9$ times the previous price;

   (c) If the minimum of the price list is ≤ 95, then let $V = 0$, else let $V = \max(S_3 - 105, 0)$;

   (d) Adjoin V to the list of claim values;

4. Compute summary statistics:

   (a) Compute the average value $\bar{V}$ of the list of claim values;

   (b) Compute the sample variance S^2 of the list of claim values;

5. Output estimates:

   (a) Return the discounted average value $= (1 + r)^{-n} \cdot \bar{V} = (1.05)^{-n} \cdot \bar{V}$;

   (b) Return the precision estimate $= 2 \cdot \frac{S}{\sqrt{m}} \cdot (1.05)^{-n}$.

Good programming practice suggests letting parameters like S_0, r, a, b, the barrier L, the exercise price E, and the number of replications m be input parameters to the program, so that they may be changed as necessary. The number of steps n, that is, the time of expiration of the option, can also be left general. In the *Mathematica* version of this book, one particular run with $m = 2000$ replications produced an estimate of about \$11.16 and a tolerance of about .5, which is a satisfyingly close result in view of the true price of \$11.19. ∎

Example 7. Let us now reconsider the Asian call option of Example 5. The payoff at time 4 was $\max(\bar{S} - 53, 0)$, where $\bar{S}$ was the mean price. We had assumed that $r = .03$, $b = .06$, $a = .01$, and $S_0 = \$50$, and we had calculated $V_0 = \$.554578$. We show how to use simulation to approximate this.

Solution. As in Example 6, we will adapt the generic algorithm to our situation. In fact, most of the work is done already; we just have to change the portion of the algorithm that calculates one claim value. This is in step 3(c). Elsewhere, we adjust the given parameter values. Here is the algorithm.

1. Let $q = \frac{r-a}{b-a} = .4$;

2. Initialize an empty list of claim values;

3. Do the following m times:

   (a) Initialize a list of prices as $\{S_0 = 50\}$;

   (b) Do the following $n = 4$ times:

      (i) Simulate U, a uniform[0,1] random variable;

      (ii) If $U \le q$, then adjoin $(1 + b) = 1.06$ times the previous price to the list of prices, else adjoin $(1 + a) = 1.01$ times the previous price;

   (c) Let $V = \max\left(\frac{S_0+S_1+S_2+S_3+S_4}{5} - 53, 0\right)$;

   (d) Adjoin V to the list of claim values;

4. Compute summary statistics:

   (a) Compute the average value $\bar{V}$ of the list of claim values;

   (b) Compute the sample variance S^2 of the list of claim values;

5. Output estimates:

(a) Return the discounted average value $= (1+r)^{-n} \cdot \bar{V} = (1.03)^{-4} \cdot \bar{V}$;

(b) Return the precision estimate $= 2 \cdot \frac{S}{\sqrt{m}} \cdot (1.03)^{-n} = 2 \cdot \frac{S}{\sqrt{m}} \cdot (1.03)^{-4}$.

A run of $m = 800$ paths in *Mathematica* resulted in an estimated value of about $\$.542$ with a precision estimate of .057, which is in line with the exact answer of $\$.554578$. ∎

The exercise set has several problems in which you are asked to modify the algorithm to suit other purposes. You should try to implement your algorithms in a programming environment with which you are familiar to really see the power of simulation techniques. This is only one of many areas to pursue as you extend your ability to solve problems in Financial Mathematics.

Important Terms and Concepts

American option - A derivative that may be executed prior to its expiration time by the holder. American options may be of either the call or put variety.

Barrier option - An option that either becomes void when, or remains void until, its underlying asset crosses a barrier.

Up-and-out knockout - A type of barrier option which becomes worthless if the underlying asset ever rises above a level H.

Down-and-out knockout - A type of barrier option which becomes worthless if the underlying asset ever falls below a level L.

Up-and-in knockin - A type of barrier option which goes into effect only if the underlying asset ever rises above a level H.

Down-and-in knockin - A type of barrier option which goes into effect only if the underlying asset ever falls below a level L.

Asian option - An option that has final claim value that depends on the average of the underlying asset values that were traversed on its path.

Approximation of derivative values by simulation - The process of simulating many paths of the underlying asset and estimating the derivative value as the discounted average claim value of the derivative over all of the simulated paths.

Exercises 5.4

1. Suppose that an underlying asset follows a binomial branch model with initial price $S_0 = \$40$ and up and down rates $b = .03$, $a = -.01$, and the risk-free rate is $r = .01$. Find the initial value of an American call option that expires at time 3, with strike price $42. Check that this option has the same value as a three-period European call option with the same strike price.

2. Consider an American put option of two periods with strike price $20 on an asset whose current price is $22, with binomial branch parameters $b = .04$, $a = -.05$. Assume that the risk-free rate is $r = .01$. Find the initial value of the option. Is it ever optimal to exercise the option early?

3. Assume that an American put option exists on an asset with initial price $100 and parameters $a = 0, b = .1$. The risk-free rate is 4%, and the option expires at time 4 and has exercise price $115. Find the value of the option at all nodes of the price tree, noting whether it is ever optimal to exercise the option early.

4. Find the initial value of a three-period American put option with strike price $85 on the underlying asset whose motion is displayed in the tree in the figure. Assume that $r = .03$. (Note that this is not a binomial branch tree of the usual form.)

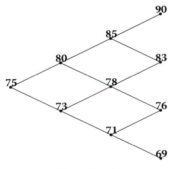

Exercise 4.

5. In Exercise 13 of Section 5.3 we described a European version of a derivative that pays twice the absolute difference between the price of an underlying asset and the exercise price $E = \$101$, if that absolute difference exceeds $4, and nothing otherwise. The asset was assumed to have initial price $S_0 = \$100$, the up and down rates of return were .04 and $-.03$, and the risk-free rate was .02. Compute the initial value of an American version of this derivative with expiration time 3.

6. In Exercise 11 of Section 5.3 we introduced a four-period binomial branch asset whose initial price was $S_0 = \$30$ and whose parameters were $b = .05$, and $a = -.03$. We had assumed that $r = .01$. A derivative had values 2, 1, 0, 0,

and 0 at the five nodes at time 4, so its claim values had the structure of a call option, but in light of the final asset values, there was no single strike price E for a call that would give these claim values. Find the value of the American version of this derivative at each node if the claim values are as below for nodes top to bottom at each of the earlier levels. Is it ever optimal to exercise the derivative early, or, as in the American call option, is it always optimal to wait to the end to execute it?

$$
\begin{aligned}
&\text{time 3:} &&1, .5, 0, 0; \\
&\text{time 2:} &&.3, .1, 0; \\
&\text{time 1:} &&.2, 0; \\
&\text{time 0:} &&0
\end{aligned}
$$

7. Prove that, if there is no arbitrage, the inequality $P_E \le P_A$ holds, where P_E and P_A are, respectively, the intial values of European and American put options with the same strike prices and expiration times.

8. Does a replicating portfolio for an American put option exist? Using the data in Example 1 and a two-period put option with strike price $E = \$20.80$, try to construct a Δ-hedge at the time 0 node and each of the time 1 nodes from the option issuer's point of view. (Hint: if you use the same strategy to compute Δ for each node as in the European case, then the amount of cash b for the issuer to hold must adjust according to whether the holder is expected to execute the option or not at the node.)

9. A put option has an up-and-out barrier condition that says that, if the underlying asset ever exceeds \$100 in value, the put expires worthless. Otherwise the strike price is \$80 at time 3. If the initial value of the underlying asset is \$90, its up and down rates are $b = .1$, $a = -.1$, and the risk-free rate is $r = .05$, use chaining to find the value of this barrier put option.

10. Under what circumstances might a match be made between a potential option holder and a potential option issuer for: (a) a down-and-out put option; (b) an up-and-in call option; (c) a down-and-in call option?

11. In Example 4, try to use chaining to value the option, setting option tree nodes to be 0 as appropriate. Check that you get the same option value as in the example.

12. An asset begins at price \$60 and follows a binomial branch model with up and down rates $b = .04$, $a = -.02$. The risk-free rate of interest is $r = .01$. Find the initial value of a three-period up-and-out call option with barrier $H = \$66$ and strike price $E = \$62$.

13. In an economy with risk-free rate $r = .03$, there is a down-and-in knockin put option which expires in four periods and has strike price $E = \$48$ and barrier $L = \$49$. Its underlying asset follows a binomial branch process with parameters

$S_0 = \$50$, $b = .05$, $a = -.03$. Find the initial value of the option by sketching the asset tree and identifying the paths over which the option will become active.

14. In Exercise 1 we found the common value of a three-period European and American call option with strike price \$42 on an underlying asset with $S_0 = \$40$, $b = .03$, and $a = -.01$, where the risk-free rate was $r = .01$. The initial value of these options turned out to be \$.2116. Suppose that a barrier option of up-and-in type is defined on this underlying asset with barrier $H = \$41$. Compute the initial value of the barrier option, making note of whether it is cheaper than the ordinary European option. Explain the result.

15. In Exercise 4 of Section 5.3 we valued a three-period put option on a binomial branch asset with parameters $a = .01$, $b = .06$, $S_0 = \$95$. The risk-free rate of return was assumed to be .03. Now suppose that there is a down-and-out barrier put option on this asset with strike price \$103 and lower barrier $L = \$96$ that applies only at time 1 and after. Find the value of this option using: (a) chaining; (b) the martingale method.

16. Show that the sum of the values of an up-and-in barrier option and an up-and-out barrier option with common exercise time n, strike price E, and barrier H equals the value of a plain European option with this exercise time and strike price. Do the same for a down-and-in and a down-and-out barrier option with lower barrier L. (Hint: think of martingale valuation and computing expected values by summing on paths.)

17. Use the result of Exercise 16 to value the three-period up-and-in call corresponding to the barrier option in Exercise 12.

18. Use the result of Exercise 16 to value the four-period down-and-out put corresponding to the barrier option in Exercise 13. Explain intuitively why you are getting this result.

19. An asset follows a binomial branch process with initial value $S_0 = \$90$, $a = -.1$, $b = .05$. There is a risk-free asset with rate of return $r = .01$. Find the initial value of a three-period Asian call option on this asset with strike price \$93.

20. A risky asset exists with initial price \$40 and binomial branch parameters $a = -.02$, $b = .04$. The risk-free rate is $r = .01$. Find the initial value of a derivative which pays \$41 minus the average price of the asset over two periods (including the initial value and the next two) if that quantity is non-negative, or zero otherwise.

21. An Asian put option in three periods exists with exercise price \$65 on a risky asset with parameters $S_0 = \$70$, $a = -.08$, $b = .04$. Assuming a risk-free rate of $r = .02$, find the initial value of the option.

22. Find the initial value of a four-period Asian call option with strike price $66 on an underlying binomial branch asset with parameters $S_0 = \$62$, $a = 0$, $b = .04$. Assume that the risk-free rate is .01.

23. Consider the non-binomial branch asset in Exercise 4. Find the initial value of an Asian call option with strike price $80 on this asset, assuming again a non-risky rate of $r = .03$.

24. For a two-period Asian call option with strike price $51 on an underlying asset with the same parameters as Example 5, find the initial value using the strategy of itemizing all paths. Next, use the chaining approach, initializing the time 2 option nodes with the potential final claim values for an exploded tree in which all paths are shown separately (meaning that the ud path is separated from the du path), and check that you get the same value.

25. Consider again a risky asset following the parameters of Exercise 19. The same risk-free rate is in place. A derivative pays the largest positive difference between the asset value and the strike price of $93 during the times 0, 1, 2, and 3 if that difference is positive, else it pays nothing. Find the initial value of this derivative.

26. A **Bermudan option** is a hybrid of an American and a European-style option. This derivative allows the investor to exercise the option early, but only in certain prespecifed periods, not necessarily all periods. Develop an approach that allows you to value a four-period Bermudan call option with strike price $25 which is able to be exercised only at times 2 and 4. Assume that the risk-free interest rate is $r = .03$, and the underlying asset follows the binomial branch model with initial price $22 and up and down rates $b = .05$, $a = .01$.

27. What should happen to the tolerance in a simulation approximation if the number of replications is: (a) multiplied by 4; (b) multiplied by 9; (c) multiplied by 16?

28. Write an algorithm to simulate the four-period down-and-in put from Exercise 13.

29. Write an algorithm to simulate the derivative from Exercise 13 of Section 5.3.

30. Consider the derivative of Exercise 25, which pays the largest positive difference between the asset value and the strike price of $93 during the times 0, 1, 2, and 3 if that difference is positive. The underlying asset has initial value $S_0 = \$90$, $a = -.1$, $b = .05$, and the risk-free rate of return $r = .01$. Write an algorithm to simulate initial values of this derivative.

31. A derivative exists on an asset that moves according to a binomial branch process in eight periods, with initial price $40 and up and down rates $b = .04$,

$a = -.01$. The payoffs of this derivative are \$5, \$4, \$3, 0, 0, 0, \$2, \$3, and \$4, respectively, on the nine possible states $s_{81}, s_{82}, ..., s_{89}$. The risk-free rate is $r = .01$. Write an algorithm to simulate a desired number m of replications of the initial value of this derivative.

Answers to Selected Exercises

Section 1.1

1. The desired effective rates are .05, .1025, .1576.
2. (a) The final value of the investment is \$1180; (b) The effective rate over the full 8 years is 18%.
3. The interest rate must be at least 6%.
4. On $[t_0, t_1]$ the rate of return is .03125.
   On $[t_0, t_2]$ the rate of return is .0625.
   On $[t_0, t_3]$ the rate of return is .09375.
   On $[t_1, t_2]$ the rate of return is .030303.
   On $[t_2, t_3]$ the rate of return is .0294118.
5. The amounts that Tim owes at the end of each year are \$3210, \$3420, \$3630. The effective rates are .07, .0654206, .0614035.
6. The present value of \$5000 four periods from now is \$4274.02.
7. Let y be the yearly simple interest rate. Then $y \geq \frac{2}{11} = .181818$.

Section 1.2

1. (a) is geometric with initial term 6 and common ratio $-\frac{1}{3}$. The general term is $6 \cdot \left(-\frac{1}{3}\right)^n$; (b) is not geometric; (c) is not geometric.
2. $\left(\frac{1}{2}\right)^4 - \left(\frac{1}{2}\right)^{60}$.
3. It takes at least four terms. The sum can never exceed 7.
4. $a \cdot \frac{q - q^{m+1}}{1 - q}$.   **5.** About 92 periods.
6. The number of years is 18.4472. The effective yearly rate is .0613636.
7. (a) For yearly compounding, the future value is \$3581.70. (b) For monthly compounding, the future value is \$3638.79. (c) For daily compounding, the future value is \$3644.06. The limit is about \$3644.24.
8. The necessary amount is \$486,510.
9. The rate of return per quarter is .00971. The yearly effective rate is .0394.
10. About \$80,260.   **12.** .0500842.

14. $r \approx 18.27\%$. His salaries in the second and third years are \$4,730,800 and \$6,459,430.

15. .0241.   **16**. .0526. \$842.33.

17. (a) \$10,992.70; (b) \$10,980.00; (c) \$10,976.80.

18. \$1018.64.

19. $x = \frac{\$50{,}000(1.04)^{15}}{\left(1 + \frac{.07}{12}\right)^{180}} = \$31{,}607.20$.

20. .111.   **21**. \$22.61.   **23**. The second contract.

24. \$951.63.   **25**. \$319,581.

26. (a) 1047.62 items; 4.762%.

Section 1.3

1. \$795.05.   **2**. \$4503.67.   **3**. \$1264.56.   **4**. \$14,795.50.

5. For Tim, the payment is \$547.21. For Tom with a 37-year horizon, the payment is \$551.54. If Tom wants to retire at the same time, then the monthly payment is \$608.12.

7. \$125,848.

8. Sam can carry out his plan for about 441 months. At the last, he will consume \$8975.68.

10. The second bank.

11. The company would save about \$110,000.

12. \$1672.24, \$1462.60.   **13**. \$1,005,000.   **14**. \$1090.90.

15. \$1332.42, \$1329.75, \$1327.10, \$1324.44, \$1321.79, \$1319.15, \$1316.51, \$1313.88, \$1311.25, \$1308.63, \$1306.01, \$1303.40.

17. \$1110.65.   **18**. \$12,793.10.

19. PV $= nP/(1+r)$; PV $= nP$.

23. \$22.61.   **25**. \$8270.87.

27. The final value in this retirement account is \$82,365.60.

28. The present value is \$33,351.60. The future value is \$66,033.90.

29. \$22,497.30.

Section 1.4

1. The nominal rate can be no more than 7.416%.

2. \$397.08.

3. The payments go down by about \$119. The total interest saved is \$35,826.

4. The statement would look like:

     Previous Balance: \$225.62

     Payments: \$100

     Other Credits: \$0.00

     Purchases, Balance Transfers, & Other Charges: \$0.00

     Interest Charged: \$1.53

     New Balance: \$127.15

     Minimum Payment: \$10

5. .0695. **6**. $36,742.80.

8.

month	balance	interest	new balance	payment	final balance
1	$10,000	$41.67	$10,041.67	$299.71	$9741.96
2	$9741.96	$40.59	$9782.55	$299.71	$9482.84
3	$9482.84	$39.52	$9522.36	$299.71	$9222.65
⋮	⋮	⋮	⋮	⋮	⋮

9. 48. **10**. $615.72, 251 months.

11. $718.90. **12**. $455.60; $20,261.20; $9307.20.

13. $4097.22; $229.49.

14. Balance: $10,449; principal: $3379.18; interest: $267.26.

17. For the bank credit card option, MaryAnn's payments are $225.53. Using store credit, she owes $3108.99.

18. $151.89, $171.89, and $191.89.

19. $232.38, $241.68, and $251.34.

20. His total outlay per week is about $3.00, and his total cost is around $156.

21. $489,254. **24**. (b) $226.44.

Section 1.5

1. $B = \$1000 \cdot \frac{(1+r)^4 - 1}{r}$; r. **2**. $i = .02657$. **3**. $i \approx .0543$.

4. $i = .04$. **5**. 1253.10; $i = .00238$.

7. There is no internal rate of return.

8. The internal rate of return is not uniquely determined.

11. .023. **12**. .2808. **14**. .00689.

15. 39.5%. **16**. .0297. **17**. .029.

18. Nominal yearly rate: 14.12%.

19. (a) .0617; (b) .0623. **20**. $\left(1 + \frac{y}{12}\right)^{12} - 1$.

Section 1.6

1. 8.1 years. **3**. (a) 5.13%; (b) 7.7%.

4. Just one compounding suffices.

5. $3729.58. **6**. .0524. **8**. $534.11.

9. $2178.79. **10**. $286.99. **11**. $64,687.10.

12. $t = 3.80$. **13**. 7.44%. **14**. .0217.

16. $68,833,900. **17**. (a) $68,731.30, $25,284.80; (b) $40,000.

19. $12,326.20. **20**. $15,775.60; $23,534.50.

21. $p_0 \cdot \frac{e^{yT} - 1}{y} + a \cdot \frac{e^{yT} - yT - 1}{y^2}$. **22**. $y = .0312$.

23. A bit over $11 per month is saved.

Section 2.1

1. $P = \$948.71$.   **2.** About 2.36%.   **3.** $4179.13.

4. (a) $1,067,890; (b) about 5.8%.   **6.** .0355.   **7.** $F_4 = \$2000$.

8. $\frac{F}{(1+j)^n} + Fr \cdot \frac{1-\left(\frac{1+i}{1+j}\right)^n}{j-i}$.   **9.** $1038.80.

11. The semi-annual internal rate of return is 2.0%.

12. Hilda paid Gladys $92.83. Gladys' yield rate per semi-annum is about 3.44%.

13. $972.46.

14. The semi-annual effective yield rate is $j = .0228245$; Theresa sells the bond to Kate for $988.38; 4.42% expressed annually.

16.

period	coupon	interest	paid	book
0	0.	0.	0.	10817.6
1	250.	216.35	33.65	10783.9
2	250.	215.68	34.32	10749.6
3	250.	214.99	35.01	10714.6
4	250.	214.29	35.71	10678.9
5	250.	213.58	36.42	10642.5
6	250.	212.85	37.15	10605.3
7	250.	212.11	37.89	10567.4
8	250.	211.35	38.65	10528.8
9	250.	210.58	39.42	10489.3
10	250.	209.79	40.21	10449.1
11	250.	208.98	41.02	10408.1
12	250.	208.16	41.84	10366.3
13	250.	207.33	42.67	10323.6
14	250.	206.47	43.53	10280.1
15	250.	205.6	44.4	10235.7
16	250.	204.71	45.29	10190.4
17	250.	203.81	46.19	10144.2
18	250.	202.88	47.12	10097.1
19	250.	201.94	48.06	10049.
20	250.	200.98	49.02	10000.

17. 19 periods.   **18.** $1577.05.   **19.** $3178.64; $48,648.70.

21. (b) $\sum_{i=1}^{m} K_i^c + \frac{1}{j}\left(\sum_{i=1}^{m} g_i C_i - \sum_{i=1}^{m} g_i K_i^c\right)$.

Section 2.2

1. (a) $943.78; (b) $957.54; (c) $984.84.

2. Original bond holder:$j = .0312$. Buyer's yield rate: $j = .0454$.

5. $t = 3.84$.   **7.** (a) $4706.77; (b) $4849.03.   **8.** $4675.96.

10. (a) $j_a = .0251$, $j_t = .0334$; (b) $j_a = j_t = .03$.

12. $j \approx .0354$.   **13.** .038, .037, .036; average yield .0092 per period.
14. 5.29%.   **15.** $j \approx .0119; j \approx .0109$.   **16.** \$95.71.
17. (a) \$1589.01; (b) \$1561.82; \$1578.06.
19. n **20.** 7.20.   **21.** 5.47.   **22.** $-\$38.91; -\$38.71$.

Section 2.3

1. $s_4 = .0011$.   **2.** $s_1 = .042; s_2 = .04502; s_3 = .050127$.
3. $s_1 = .03; s_2 = .019902; s_3 = .0149164; s_4 = .0252616$.
4. (a) $s_2 = .039899$; (b) $s_2 = .0398029$; (c) $s_2 = .0397115$.
7. (a) $j_2 \approx .0199$; (b) $j_2 \approx .0202$.   **8.** \$954.813; $j_4 \approx .0219$.
12. $f_{1,2} = .036, f_{1,3} = .04116$, and $f_{2,3} = .04635$.
13. $f_{1,2} = .0330; f_{1,5} = .0358; f_{1,10} = .0392; f_{2,5} = .0367; f_{2,10} = .040;$
$f_{5,10} = .0420$.
15. $s_3 = .0340; s_4 = .0340; s_7 = .0357$.   **16.** $j_3 \approx .02428$.
19. $P_1 = \$9604.25; s_1 = .0412; P_2 = \$9188.41; s_2 = .0432; P_3 = \$8,756.61;$
$s_3 = .0453; P_4 = \$8,312.95; s_4 = .0473; P_5 = \$7,861.44; s_5 = .0493$.

Section 3.1

1. $\{11.70, 10.91, 10.18, 9.49, 8.85\}$ **3.** (a) 10; (b) 16; (c) 16.
4. (a) 24; (b) 9,979,200.   **5.** 40,320   **6.** 243; 30.
9. (a) $\Omega = \{(U, U, U), (U, U, S), (U, U, D), (U, S, U), (U, S, S), (U, S, D), (U, D, U),$
$(U, D, S), (U, D, D), (S, U, U), (S, U, S), (S, U, D), (S, S, U), (S, S, S), (S, S, D),$
$(S, D, U), (S, D, S), (S, D, D), (D, U, U), (D, U, S), (D, U, D), (D, S, U), (D, S, S),$
$(D, S, D), (D, D, U), (D, D, S), (D, D, D)\};$
(b) (i) 1/3; (ii) 1/9; (iii) 1/27.
10. (a) 5/9; (b) 7/9.   **11.** 120; 36.
12. 125,970; 15,504; .120134; .014786.
14. outcome 800 shekels: probability .05; outcome 500 shekels: probability .10;
outcome 300 shekels: probability .25; outcome 0 shekels: probability .6.
16. (a) 2/9; (b) 4/9.   **18.** .1.
19. $P[S_2 = 41.62] = .16; P[S_2 = 40.39] = .48; P[S_2 = 39.20] = .36$.

Section 3.2

1. $P[X = 0] = \frac{1}{16}, P[X = 1] = \frac{4}{16}, P[X = 2] = \frac{6}{16}, P[X = 3] = \frac{4}{16}, P[X = 4] = \frac{1}{16}$.
2. $P[Y = 0] = \frac{1}{16}, P[Y = 1] = \frac{4}{16}, P[Y = 2] = \frac{6}{16}, P[Y = 3] = \frac{4}{16}, P[Y = 4] = \frac{1}{16}$.
4. (a) The table below summarizes the probability mass function of X the net winnings on one spin.

amt. won	$-\$.25$	$\$999.75$
prob.	.999756	.000244141

Expected winnings $= -.000586$.

(b) For two spins,

amt. won	$-\$.50$	$\$999.50$	1999.50
prob.	.999512	.000488163	5.96048×10^{-8}

5. The probability mass function is:

value of X:	2	3	4	5	6	7	8
probability:	$\frac{1}{16}$	$\frac{2}{16}$	$\frac{3}{16}$	$\frac{4}{16}$	$\frac{3}{16}$	$\frac{2}{16}$	$\frac{1}{16}$

8. (a) $\frac{26}{32}$; (b) $\frac{26}{32}$.

9. $P[X = k] = \binom{4}{k}(.3)^k(.7)^{4-k}, k = 0, 1, 2, 3, 4$. $P[$at most two stocks go up$] = .9163$.

10. $f[47] = 1/8; f[49] = 3/8; f[51] = 3/8; f[53] = 1/8$.

12. The distribution is summarized in the table below. His expected monthly payment is $\$546.84$.

payment	probability
522.58	1/8
534.59	1/4
546.74	1/4
559.02	1/4
571.41	1/8

14.

$$F(x) = \begin{cases} 0 & \text{if } x < 1; \\ 1/15 & \text{if } 1 \le x < 2; \\ 3/15 & \text{if } 2 \le x < 3; \\ 6/15 & \text{if } 3 \le x < 4; \\ 10/15 & \text{if } 4 \le x < 5; \\ 1 & \text{if } x \ge 5. \end{cases}$$

15. $f(0) = .2, f(1) = .1, f(3) = .4, f(5) = .3, P[1 < X \le 5] = .7$.

18. (b) The probability that first die is at least 4 and the total at least 9 is $1/4$.

19. The expected value of the final interest rate is 5.24%.

20. The joint distribution is as in the table below.

state (s_1, s_2)	probability
$(110, 121)$	.25
$(110, 99)$	.25
$(90, 99)$	.25
$(90, 81)$	.25

21. The p.m.f. of the amount lost is:
$f(0) = .922368; f(10000) = .0752954; f(20000) = .00230496;$
$f(30000) = .00003136; f(40000) = .00000016.$
23. (b) The table below gives the p.m.f. of the final portfolio value.

value of portfolio	1475	1510	1520	1540	1545	1560
probability	.1	.2	.3	.2	.1	.1

24. The possible price states are $s(1 + b)^k(1 + a)^{n-k}, k = 0, 1, ..., n$ and their associated probabilities are $\binom{n}{k}p^k(1 - p)^{n-k}$.

Section 3.3

1. The total torque on each side is .135.   **2**. $405.16.
3. $\mu = 2.8; \sigma^2 = 3.36; \sigma = 1.83.$ **4**. $\mu = \$1022.50; \sigma^2 = 431.25; \sigma = 20.77.$
5. $\mu = 4; \sigma^2 = 24/9.$
6. There are two possible solutions: $a = -1$ and $b = 2$; and when $a = 1.4$, $b = -1.6$.
7. $\frac{b+a}{2}$.   **8**. 2.   **12**. 1.7136×10^8; $13,090.50.   **14**. .52381.
15. The mean is $\mu = \frac{3}{2}$ and the variance is $\sigma^2 = \frac{3}{4}$.
16. $\mu = .035208; \sigma^2 = .0002.$ **17**. 0.   **18**. (a) 1; (b) 0.

Section 3.4

1. $\frac{5}{11}$.   **2**. $\frac{1}{5525}; \frac{73}{5525}$.   **3**. (a) .34; (b) $\frac{2}{17}$; (c) .16; (d) $\frac{1}{4}$.
4. $\frac{1}{3}, \frac{2}{3}, \frac{3}{47}, \frac{44}{47} \cdot \frac{1}{4}, \frac{3}{4}, \frac{2}{46}, \frac{44}{46}$.   **6**. (a) $\frac{1}{3}$; (b) $\frac{8}{23}$; (c) $\frac{7}{18}$; (d) $\frac{5}{23}$.
7. $\frac{2}{5}$.   **8**. (a) $\frac{1}{2}$; (b) $\frac{3}{5}$.   **10**. $(1 - p)^2$.
12. The probability as a function of class size n is in the table below.

n	4	5	6	7	8
prob	.273	.419	.564	.695	.802

14. (a) .6561; (b) .0015625; (c) .25515.   **15**. $3p^2(1 - p)$.
18. (a) $P[Q_3 = .01 | Q_2 = .02] = .3; P[Q_3 = .015 | Q_2 = .02] = .7;$
$P[Q_3 = .02 | Q_2 = .02] = 0.$
$P[Q_3 = .01 | Q_2 = .015] = .4; P[Q_3 = .015 | Q_2 = .015] = 0;$
$P[Q_3 = .02 | Q_2 = .015] = .6.$
(b) $P[Q_4 = .01 | Q_3 = .01] = 0; P[Q_4 = .015 | Q_3 = .01] = .5;$
$P[Q_4 = .02 | Q_3 = .01] = .5;$
$P[Q_4 = .01 | Q_3 = .015] = .4; P[Q_4 = .015 | Q_3 = .015] = 0;$
$P[Q_4 = .02 | Q_3 = .015] = .6.$
$P[Q_4 = .01 | Q_3 = .02] = .3; P[Q_4 = .015 | Q_3 = .02] = .7;$
$P[Q_4 = .02 | Q_3 = .02] = 0.$
20. $P[X = \$2382.03] = \frac{9}{16}; P[X = \$2426.98] = \frac{4}{16}; P[X = \$2472.77] = \frac{3}{16}.$
22. (a) 0; (b) 7/8.   **23**. (a) 1/2; (b) 2/9.   **25**. 1/6.
26. (a) .2; (b) .16; (c) .72.   **27**. (a) 52.16; (b) 1.8944.   **28**. $\frac{5}{8}$.

29. $P[X = \$1300] = .16$; $P[X = \$1350] = .096$; $P[X = \$1400] = .384$; $P[X = \$1450] = .36$; $\$1397.20$.

30. (a) .0382; (b) .0000862; (c) .0475; (d) .000156.

31. $c = \frac{1}{36}$; $E[Y|X = x] = \frac{7}{6}$ for each x.

33. $7/2$.   **34**. 10.97; .178.   **35**. 342; 336.

Section 3.5

2. Yes; yes.   **4**. (a) $P[A \cap B]$; (b) $P[A]$; (c) $P[B \cup C]$.   **6**. Yes; no.

7. $P[X = (12, 12)] = .09$; $P[X = (12, 10)] = .18$; $P[X = (12, 8)] = .09$; $P[X = (10, 12)] = .12$; $P[X = (10, 10)] = .24$; $P[X = (10, 8)] = .12$; $P[X = (8, 12)] = .04$; $P[X = (8, 10)] = .08$; $P[X = (8, 8)] = .04$.

8. .0511.

9. In the case of non-payment, the price would be $\$9614.43$, which is $\$110.16$ less than the price under sure coupons.

13. $P[X_5 \geq 51.50] = .087$.

14. mean: 137,015; variance: 5,301,170; standard deviation: 2302.43.

15. (a) .1; (b) .16875; (c) .9.   **16**. mean: 2205; variance: 1,924,000.

17. covariance: -1.40625; correlation: $-.990267$.

18. covariance and correlation are both 0.

19. covariance: .125; correlation: .714286.

21. (a).169282; (b) .18174; (c) .622872.   **22**. .0214; .0192; $-.0052$.

23. no; 0.   **24**. .258161.   **25**. 92.2.

27. mean: 183.75; variance: 548.50.   **28**. $\text{Var}(X)$; 1.

Section 3.6

1. $f(x) = \begin{cases} \frac{1}{4} & \text{if } x = 1, 2, 3, 4 \\ 0 & \text{otherwise.} \end{cases}$ **2**. .0288; .0392.

4. $\frac{1}{4}$.   **5**. $\frac{5}{54}$.   **6**. $\text{Var}(Y) > \text{Var}(\bar{X})$.

7. Underlying distribution: $\mu \approx .00176, \sigma^2 \approx .000151$; distribution of $\bar{X}$: $\mu \approx .00176, \sigma^2 \approx .00000756$.

8. Estimated probability of default: $\frac{5}{33}$; estimated mean and variance of $\hat{p}$: .1515, .00390.

9. $E[\bar{X}] = 2.5$; $\text{Var}(\bar{X}) = .078125$.

10. $E[\bar{X}] = \frac{13}{6} \approx 2.167$; $\text{Var}(\bar{X}) = .04556$.

11. .013; .005.   **12**. 900.   **13**. 22,471.

16. (a) covariance: $-\frac{3}{2}$; correlation: $-.982$; (b) covariance: 0; correlation: 0.

17. covariance: .713377; correlation: .377589.

18. mean ≈ 27.78; variance ≈ 2.30608.

19. The correlation is $\dfrac{\hat{p}_{ij} - \hat{p}_i \cdot \hat{p}_j}{\sqrt{\hat{p}_i(1-\hat{p}_i)} \cdot \sqrt{\hat{p}_j(1-\hat{p}_j)}}$, where $\hat{p}_i = \frac{n_1}{n}$, $\hat{p}_j = \frac{n_2}{n}$, and $\hat{p}_{ij} = \frac{n_{12}}{n}$.

22. $g(1) = 9, g(2) = 8, g(3) = 5, g(4) = 0$.

29. The simulated mean time should come out to around 51.5, and it should take in excess of 8000 replications to estimate to within 1 with high probability.
31. The exact value is $\mu = \frac{104}{3} \approx 34.667$.

Section 4.1

1. $i = .0083$ per month.   **2**. He lost \$90; $i = -.0277$.
3. $i = .0482$.   **4**. \$40.40, \$40.73, \$41.61, \$40.12.
6. .0082, .0216, $-.0358$, overall: $-.0069$.
7. The price must be at least \$40.76.
8. \$220; she will have to sell 10 shares.
10. \$416.67; cash amount: \$250.   **12**. Shirley: .2037; Laverne: .395.
13. \$180.56; \$1780.56 $- 50p$.
14. For Beau, $P_b(37) = -300; P_b(40) = 0; P_b(43) = 300$; for Jeff,
$P_j(37) = 320; P_j(40) = 20; P_j(43) = -280$; identical profit when $x = \$40.10$.
15. (a) The rate of return is independent of the number of shares that are short sold; at least 34 shares for the profit requirement; (b) at most \$23.50.
16. No short selling: .125; short-selling: .3125.

Section 4.2

1. (a) .02 and $-.015$; (b) .02 and $-.0125$.   **2**. $\mu = 20.1; \sigma^2 = .03$.
3. $\mu_x = .01; \mu_y = 0; \sigma_x^2 = .00005; \sigma_y^2 = .00005; \rho = -1$.
4. $\mu_p \approx .00377; \sigma_p^2 \approx .0000556$. **5**. $p = 2/3$.   **7**. $x = 1.2; 1 - x = -.2$.
8. (a) $\mu_p = .021\bar{6}, \sigma_p{}^2 = .0000917$; (b) $\mu_p = .02375, \sigma_p^2 = .0000921875$.
10. (a) $\mu_p = .03, \sigma_p^2 = .000667$; (b) $\mu_p = .03, \sigma_p^2 = .001$.
11. .009167; .0004201.   **12**. $x_1 = 1/3, x_2 = 2/3; \sigma_p = .2667$.
13. (b) $x = -1.1922$.   **14**. $x = -.25$. **15**. $a = 2$.   **16**. $a = 14.5569$.

Section 4.3

1. Optimal weights: non-risky .8; risky .2; mean return: .048; variance: .0004.
2. 18.75.   **3**. $x_1 = .53, x_2 = .47$.   **4**. $x_1 = .2, x_2 = .8$.
5. $x = \frac{\mu - r}{2a \cdot \sigma^2}$.   **6**. 1.
7. (a) $x_1 = -5.27, x_2 = 3.24, x_3 = 3.03$; (b) $x_1 = -.254, x_2 = .648, x_3 = .606$.
8. $x_1 = .652; x_2 = .348$. **9**. .754, .130, .116.
11. .309, .283, .180, .129, .099.
12. (a) .235, .346, .419; (b) .410, .139, .204, .247; (c) 7.865.
13. $x_1 = .459, x_2 = .337, x_3 = .204$. Minimum variance: .00165.
14. $x_1 = .099, x_2 = .3515, x_3 = .5495$.

Section 5.1

1. $100; $50; $0; and $-$50; expected profit $25.
2. $-$1000; $1000; $2000; expected profit $250.
3. (a) with calls: -100%, without calls: 10%; (b) with calls: -100%, without calls: 25%; (c) with calls: 100%, without calls: 40%; (d) with calls: 233%, without calls: 50%.
4. (a) with call: $15, without call: $20; (b) with call: $-$25, without call: $-$20; (c) with call: $-$25, without call: $-$50; (d) with call: $-$25, without call: $-$100.
5. (a) with put: $46, without put: $50; (b) with put: $-$24, without put: $-$20; (c) with put: $-$24, without put: $-$50; (d) with put: $-$24, without put: $-$100.
10. Short 4 shares of PilotFish and long 1 share of Shark.
12. $50c.   **14**. $99.01.
16. $S_1 = \$58$: $0; $S_1 = \$52$: $3; $S_1 = \$50$: $5; $S_1 = \$48$: $3; $S_1 = \$42$: $0.
17. $S_1 - E$.   **18**. (a) S_1.

Section 5.2

1. $.23.   **3**. $.049.   **4**. $.0735.   **5**. $E = 19.95$.
6. $r \approx .0333$.   **8**. Call: $.044; Put: $.368.
9. $\Delta \approx .2444, b \approx -4.1067$; initial value: $.2925.
10. (a) $q = \frac{1}{2}$; (b) $q = .725$.   **12**. $4.85.
13. Cash-or-nothing: $(1 + r)^{-1} \cdot q \cdot E$; stock-or-nothing: $(1 + r)^{-1} \cdot q \cdot (1 + b)S_0$.
14. $3.85.   **16**. Sell short 125 shares now.   **17**. 40.
18. 1200.   **19**. 0 and $3.20.   **20**. $q = \frac{1}{3}$.
21. $q = .4$.   **22**. $2.61438.

Section 5.3

1. (a) $s_{22} = \$60.564$; (b) $s_{41} = \$67.5305$; (c) $s_{63} = \$60(1.03)^4(.98)^2 = \$64.8563$.
2. $61.80; $58.80.
3. Using p, $\mu = \$35, \sigma^2 = 49.74$; using q, $\mu = \$32.28, \sigma^2 = 42.49$.
4. $.419.   **5**. $.8925.
7. The expected present value is $1.4793, which is less than the arbitrage-free value.
8. Time 1: $1.45588 and $.00840; initial value $.8192.
9. $.1879.   **10**. $.384893; $2.11.
11. $v_{41} = 2; v_{42} = 1; v_{43} = 0; v_{44} = 0; v_{45} = 0$;
    $v_{31} = \$1.48515; v_{32} = \$.49505; v_{33} = v_{34} = 0$;
    $v_{21} = \$.980296; v_{22} = \$.245074; v_{23} = 0$;
    $v_{11} = \$.606619; v_{12} = \$.121324$;
    $V_0 = \$.360368$.
12. $v_{21} = \$20.125; v_{22} = \$1.375; v_{23} = 0$;

$v_{11} = \$7.2619; v_{12} = \$.436508.$
$V_0 = \$2.58251.$

13. \$8.317.   **14**. \$1.42133; \$9.73833.

15. $q_{i,j} = \frac{1}{3}$ for each level i and node j.

16. (a) \$2.22; (b) \$.63.

Section 5.4

1. The time 0 value of both options is \$.2116.

2. The initial value of the option is \$.01579, and it is never optimal to exercise the option early.

3. $v_{41} = 0;\ v_{42} = 0;\ v_{43} = 0; v_{44} = \$115 - \$110 = \$5;\ v_{45} = \$115 - \$100 = \$15;$
$\quad v_{31} = 0;\ v_{32} = 0;\ v_{33} = \$5;\ v_{34} = \$15;$
$\quad v_{21} = 0;\ v_{22} = \$5;\ v_{23} = \$15;$
$\quad v_{11} = \$5;\ v_{12} = \$15;\ V_0 = \$15.$

4. \$10.   **5**. \$8.9735.

6. $v_{41} = 2; v_{42} = 1; v_{43} = 0; v_{44} = 0; v_{45} = 0;$
$\quad v_{31} = \$1.69731;\ v_{32} = \$.707214;\ v_{33} = 0;\ v_{34} = 0;$
$\quad v_{21} = \$1.40042;\ v_{22} = \$.500151;\ v_{23} = 0;$
$\quad v_{11} = \$1.13188;\ v_{12} = \$.353714; V_0 = \$.900544.$

8. $\Delta_0 = -1; b_0 = \$20.8;\ \Delta_{11} = -.18835; b_{11} = \$4.08;\ \Delta_{12} = -1; b_{12} = \$20.8.$

9. \$.194228.   **12**. \$.581655.   **13**. \$.0165.   **14**. \$.2116.

15. \$.03637.   **17**. \$.66629.   **18**. 0.   **19**. \$1.55835.

20. \$.739143.   **21**. \$.03873.   **22**. \$.01194.   **23**. \$.6592.

24. \$.65454.   **25**. \$5.43295.   **26**. \$.257153.

27. (a) divided by 2; (b) divided by 3; (c) divided by 4.

Bibliography

[1] Baxter, Martin, and Andrew Rennie. *Financial Calculus.* Cambridge University Press, Cambridge, United Kingdom, 1996.

[2] Broverman, Samuel A. *Mathematics of Investment and Credit, 3rd ed.* Actex Publications, Inc., Winsted, Connecticut, 2004.

[3] Buchen, Peter. *An Introduction to Exotic Option Pricing.* Chapman & Hall/CRC Press, Boca Raton, Florida, 2012.

[4] Çinlar, Erhan. *Introduction to Stochastic Processes.* Prentice-Hall, Inc., Englewood Cliffs, New Jersey, 1975.

[5] Federer Vaaler, Leslie Jane, and James W. Daniel. *Mathematical Interest Theory, 2nd ed.* Mathematical Association of America, Washington, DC, 2007.

[6] Hastings, Kevin J. *Introduction to the Mathematics of Operations Research with Mathematica, 2nd ed.* Chapman & Hall/CRC Press, Boca Raton, Florida, 2006.

[7] Hastings, Kevin J. *Introduction to Probability with Mathematica, 2nd ed.* CRC Press, Boca Raton, Florida, 2010.

[8] Hogg, Robert V., and Elliot A. Tanis. *Probability and Statistical Inference, 8th ed.* Prentice-Hall, Inc., Upper Saddle River, New Jersey, 2010.

[9] Pliska, Stanley R. *Introduction to Mathematical Finance.* Blackwell Publishers, Malden, Massachusetts, 1997.

[10] Roman, Steven. *Introduction to the Mathematics of Finance.* Springer, New York, 2004.

[11] Ross, Sheldon M. *An Elementary Introduction to Mathematical Finance, 2nd ed.* Cambridge University Press, Cambridge, United Kingdom, 2003.

[12] Ross, Sheldon M. *Simulation, 4th ed.* Elsevier/Academic Press, Burlington, Massachusetts, 2006.

[13] Ross, Sheldon M. *Stochastic Processes, 2nd ed.* John Wiley & Sons, Inc., New York, 1996.

[14] Stampfli, Joseph, and Victor Goodman. *The Mathematics of Finance.* Brooks/Cole, Pacific Grove, California, 2001.

[15] Wang, Hui. *Monte Carlo Simulation with Applications to Finance.* Chapman & Hall/CRC Press, Boca Raton, Florida, 2012.

Index